MATHEMATICAL THINKING AND WRITING

RANDALL B. MADDOX

Pepperdine University, Malibu, CA

MATHEMATICAL THINKING AND WRITING

A Transition to Abstract Mathematics

ACADEMIC PRESS

A Harcourt Science and Technology Company

Sponsoring Editor	Barbara Holland
Production Editor	Amy Fleischer
Marketing Coordinator	Stephanie Stevens
Cover/Interior Design	Cat and Mouse
Cover Image	Rosmi Duaso/Timepix
Copyeditor	Editor's Ink
Proofreader	Phyllis Coyne et al.
Composition	Interactive Composition Corporation
Printer	InterCity Press, Inc.

This book is printed on acid-free paper. ∞

Copyright © 2002 by HARCOURT/ACADEMIC PRESS

All rights reserved.
No part of this publication may be reproduced or transmitted in any form or by any means, electronic or mechanical, including photocopy, recording, or any information storage and retrieval system, without permission in writing from the publisher.

Requests for permission to make copies of any part of the work should be mailed to: Permissions Department, Harcourt, Inc., 6277 Sea Harbor Drive, Orlando, Florida 32887-6777.

Academic Press
A Harcourt Science and Technology Company
525 B Street, Suite 1900, San Diego, California 92101-4495, USA
http://www.academicpress.com

Academic Press
Harcourt Place, 32 Jamestown Road, London NW1 7BY, UK
http://www.academicpress.com

Harcourt/Academic Press
A Harcourt Science and Technology Company
200 Wheeler Road, Burlington, Massachusetts 01803, USA
http://www.harcourt-ap.com

Library of Congress Control Number: 2001091290

International Standard Book Number: 0-12-464976-9

PRINTED IN THE UNITED STATES OF AMERICA
01 02 03 04 05 06 IP 9 8 7 6 5 4 3 2 1

For Dean Priest

who unwittingly planted the seed of this book

Contents

Why Read This Book   xiii
Preface   xv

CHAPTER 0   Notation and Assumptions   1

    0.1 Set Terminology and Notation   1
    0.2 Assumptions   5
        0.2.1   Basic algebraic properties of real numbers   5
        0.2.2   Ordering of real numbers   7
        0.2.3   Other assumptions about $\mathbb{R}$ 9

PART I   FOUNDATIONS OF LOGIC AND PROOF WRITING   11

CHAPTER 1   Logic   13

    1.1 Introduction to Logic   13
        1.1.1   Statements   13
        1.1.2   Negation of a statement   15
        1.1.3   Combining statements with AND/OR   15
        1.1.4   Logical equivalence   18
        1.1.5   Tautologies and contradictions   18
    1.2 If-Then Statements   20
        1.2.1   If-then statements   20
        1.2.2   Variations on $p \to q$ 22
        1.2.3   Logical equivalence and tautologies   23

1.3 Universal and Existential Quantifiers 27
1.3.1 The universal quantifier 27
1.3.2 The existential quantifier 29
1.3.3 Unique existence 30
1.4 Negations of Statements 31
1.4.1 Negations of $p \wedge q$ and $p \vee q$ 32
1.4.2 Negations of $p \rightarrow q$ 33
1.4.3 Negations of statements with $\forall$ and $\exists$ 33

CHAPTER 2 Beginner-Level Proofs 38

2.1 Proofs Involving Sets 38
2.1.1 Terms involving sets 38
2.1.2 Direct proofs 41
2.1.3 Proofs by contrapositive 44
2.1.4 Proofs by contradiction 45
2.1.5 Disproving a statement 45
2.2 Indexed Families of Sets 47
2.3 Algebraic and Ordering Properties of $\mathbb{R}$ 53
2.3.1 Basic algebraic properties of real numbers 53
2.3.2 Ordering of the real numbers 56
2.3.3 Absolute value 57
2.4 The Principle of Mathematical Induction 61
2.4.1 The standard PMI 62
2.4.2 Variation of the PMI 64
2.4.3 Strong induction 65
2.5 Equivalence Relations: The Idea of Equality 68
2.5.1 Analyzing equality 68
2.5.2 Equivalence classes 72
2.6 Equality, Addition, and Multiplication in $\mathbb{Q}$ 76
2.6.1 Equality in $\mathbb{Q}$ 77
2.6.2 Well-defined $+$ and $\times$ on $\mathbb{Q}$ 78
2.7 The Division Algorithm and Divisibility 79
2.7.1 Even and odd integers; the division algorithm 79
2.7.2 Divisibility in $\mathbb{Z}$ 81
2.8 Roots and irrational numbers 85
2.8.1 Roots of real numbers 86
2.8.2 Existence of irrational numbers 87
2.9 Relations In General 90

CHAPTER 3 Functions 97

3.1 Definitions and Terminology 97
- 3.1.1 Definition and examples 97
- 3.1.2 Other terminology and notation 101
- 3.1.3 Three important theorems 103

3.2 Composition and Inverse Functions 106
- 3.2.1 Composition of functions 106
- 3.2.2 Inverse functions 108

3.3 Cardinality of Sets 110
- 3.3.1 Finite sets 111
- 3.3.2 Infinite sets 113

3.4 Counting Methods and the Binomial Theorem 118
- 3.4.1 The product rule 118
- 3.4.2 Permutations 122
- 3.4.3 Combinations and partitions 122
- 3.4.4 Counting examples 125
- 3.4.5 The binomial theorem 126

PART II BASIC PRINCIPLES OF ANALYSIS 131

CHAPTER 4 The Real Numbers 133

4.1 The Least Upper Bound Axiom 134
- 4.1.1 Least upper bounds 134
- 4.1.2 The Archimedean property of $\mathbb{R}$ 136
- 4.1.3 Greatest lower bounds 137
- 4.1.4 The LUB and GLB properties applied to finite sets 137

4.2 Sets in $\mathbb{R}$ 140
- 4.2.1 Open and closed sets 140
- 4.2.2 Interior, exterior, and boundary 142

4.3 Limit Points and Closure of Sets 143
- 4.3.1 Closure of sets 144

4.4 Compactness 146

4.5 Sequences in $\mathbb{R}$ 149
- 4.5.1 Monotone sequences 150
- 4.5.2 Bounded sequences 151

4.6 Convergence of Sequences 153
- 4.6.1 Convergence to a real number 154
- 4.6.2 Convergence to $\pm\infty$ 158

4.7 The Nested Interval Property 160
- 4.7.1 From LUB axiom to NIP 161

4.7.2 The NIP applied to subsequences 162
4.7.3 From NIP to LUB axiom 164
4.8 Cauchy Sequences 165
4.8.1 Convergence of Cauchy sequences 166
4.8.2 From completeness to the NIP 168

CHAPTER 5 Functions of a Real Variable 170

5.1 Bounded and Monotone Functions 170
5.1.1 Bounded functions 170
5.1.2 Monotone functions 171
5.2 Limits and Their Basic Properties 173
5.2.1 Definition of limit 173
5.2.2 Basic theorems of limits 175
5.3 More on Limits 180
5.3.1 One-sided limits 180
5.3.2 Sequential limit of f 181
5.4 Limits Involving Infinity 182
5.4.1 Limits at infinity 183
5.4.2 Limits of infinity 185
5.5 Continuity 187
5.5.1 Continuity at a point 188
5.5.2 Continuity on a set 190
5.5.3 One-sided continuity 194
5.6 Implications of Continuity 195
5.6.1 The intermediate value theorem 195
5.6.2 Continuity and open sets 197
5.7 Uniform Continuity 200
5.7.1 Definition and examples 200
5.7.2 Uniform continuity and compact sets 202

PART III BASIC PRINCIPLES OF ALGEBRA 205

CHAPTER 6 Groups 207

6.1 Introduction to Groups 207
6.1.1 Basic characteristics of algebraic structures 208
6.1.2 Groups defined 210
6.1.3 Subgroups 213
6.2 Generated and Cyclic Subgroups 215
6.2.1 Subgroup generated by $A \subseteq G$ 216
6.2.2 Cyclic subgroups 217

- 6.3 **Integers Modulo n and Quotient Groups** 220
  - 6.3.1 Integers modulo n 220
  - 6.3.2 Quotient groups   223
  - 6.3.3 Cosets and Lagrange's theorem   225
- 6.4 **Permutation Groups and Normal Subgroups** 227
  - 6.4.1 Permutation groups   227
  - 6.4.2 The alternating group A_4 229
  - 6.4.3 The dihedral group D_8 230
  - 6.4.4 Normal subgroups   232
  - 6.4.5 Equivalences and implications of normality   233
- 6.5 **Group Morphisms** 236

CHAPTER 7 Rings 243

- 7.1 **Rings and Subrings** 243
  - 7.1.1 Rings defined   243
  - 7.1.2 Examples of rings   245
  - 7.1.3 Subrings   248
- 7.2 **Ring Properties and Fields** 249
  - 7.2.1 Ring properties   249
  - 7.2.2 Fields defined   254
- 7.3 **Ring Extensions** 256
  - 7.3.1 Adjoining roots of ring elements   256
  - 7.3.2 Polynomial rings   258
  - 7.3.3 Degree of a polynomial   259
- 7.4 **Ideals** 260
  - 7.4.1 Definition and examples   260
  - 7.4.2 Generated ideals   262
  - 7.4.3 Prime ideals   264
  - 7.4.4 Maximal ideals   264
- 7.5 **Integral Domains** 267
- 7.6 **UFDs and PIDs** 273
  - 7.6.1 Unique factorization domains   273
  - 7.6.2 Principal ideal domains   274
- 7.7 **Euclidean Domains** 279
  - 7.7.1 Definition and properties   279
  - 7.7.2 Polynomials over a field   282
  - 7.7.3 $\mathbb{Z}[t]$ is a UFD   284
- 7.8 **Ring Morphisms** 287
  - 7.8.1 Properties of ring morphisms   288
- 7.9 **Quotient Rings** 291

Index 299

Why Read This Book?

One of Euclid's geometry students asked a familiar question more than 2000 years ago. After learning the first theorem, he asked, "What shall I get by learning these things?" Euclid didn't have the kind of answer the student was looking for, so he did what anyone would do — he got annoyed and sarcastic. The story goes that he called his slave and said, "Give him threepence since he must make gain out of what he learns."[1]

It is a familiar question: "So how am I ever gonna use this stuff?" I doubt that anyone has ever come up with a good answer because it's really the wrong question. The first question is not what you're going to do with this stuff, but what this stuff is going to do with you.

This book is not a computer users' manual that will make you into a computer industry millionaire. It is not a collection of tax law secrets that will save you thousands of dollars in taxes. It is not even a compilation of important mathematical results for you to stack on top of the other mathematics you have learned. Instead, it's an entrance into a new kingdom, the world of mathematics, where you learn to think and write as the inhabitants do.

Mathematics is a discipline that requires a certain type of thinking and communicating that many appreciate but few develop to a great degree. Developing these skills involves dissecting the components of mathematical language, analyzing their structure, and seeing how they fit together. Once you have become comfortable with these principles, then your own style of mathematical writing can begin to shine through.

Writing mathematics requires a precision that seems a little stifling because at first it might feel like some pedant is forcing you to use prechosen words and phrases to express the things you see clearly with your own mind's eye. Be patient. In time you'll see how adapting to the culture of mathematics and adopting its style of communicating will shape all your thinking and writing. You'll see your skills of critical analysis become more developed and polished. My hope is that these skills will influence the way you organize

---

[1] T.L. Heath, *A History of Greek Mathematics,* Oxford, 1931.

and present your thoughts in everything from English composition papers to late night bull sessions with friends.

Here's an analogy of what the first principles of this book will do for you. Consider a beginning student of the piano. Music is one of the most creative disciplines, and our piano student has been listening to Chopin for some time. She knows she has a true ear and an intuition for music. However, she must begin at the piano by playing scales over and over. These exercises develop her ability to use the piano effectively in order to express the creativity within her. Furthermore, these repetitive tasks familiarize her with the structure of music as an art form, and actually nurture and expand her capacity to express herself in original and creative ways through music. Then, once she has mastered the basic technical skills of hitting the keys, she understands more clearly how really enjoyable music can be. She learns this truth: The aesthetic elements of music cannot be fully realized until the technical skills developed by rote exercises have been mastered and can be relegated to the subconscious.

Your first steps to becoming a mathematician are a lot like those for our pianist. You will first be introduced to the building blocks of mathematical structure, then practice the precision required to communicate mathematics correctly. The drills you perform in this practice will help you see mathematics as a discipline more clearly and equip you to appreciate its beauty.

Let n be a positive integer, and think of this course as a trip through a new country on a bicycle built for n. The purposes of the trip are:

To familiarize you with the territory;

To equip you to explore it on your own;

To give you some panoramic views of the countryside;

To teach you to communicate with the inhabitants; and

To help you begin to carve out your own niche.

If you are willing to do the work, I promise you will enjoy the trip. You'll need help pedaling at first, and occasionally when the hills are steep. But you'll come back a different person, for this material will have done something with you. Then you'll understand that Euclid really got it right after all, and you'll appreciate why his witty response is still fresh and relevant after 2000 years.

Preface

This text is written for a "transition course" in mathematics, where students learn to write proofs and communicate with a level of rigor necessary for success in their upper level mathematics courses. To achieve the primary goals of such a course, this text includes a study of the basic principles of logic, techniques of proof, and fundamental mathematical results and ideas (sets, functions, properties of real numbers), though it goes much further. It is based on two premises: The most important skill students can learn as they approach the cusp between lower- and upper-level courses is how to compose clear and accurate mathematical arguments; and they need more help in developing this skill than they would normally receive by diving into standard upper-level courses. By emphasizing how one writes mathematical prose, it is also designed to prepare students for the task of reading upper-level mathematics texts. Furthermore, it is my hope that transitioning students in this way gives them a view of the mathematical landscape and its beauty, thereby engaging them to take ownership of their pursuit of mathematics.

Why *this* text?

I believe students learn best by doing. In many mathematics courses it is difficult to find enough time for students to discover through their own efforts the mathematics we would lead them to find. However, I believe there is no other effective way for students to learn to write proofs. This text is written for them in a format that allows them to do precisely this.

Two principles of this text are fundamental to its design as a tool whereby students learn by doing. First, it does not do too much for them. Proofs are included in this text for only two reasons. Most of them (especially at the beginning) are sample proofs that students can mimic as they write their own proofs to similar theorems. Students must read them because they will need this technique later. The other proofs are included here because they are too difficult for students of this mathematics skill level to be expected to develop on their own. In most of these instances, however, some climactic detail is omitted and relegated to an exercise.

Second, if students are going to learn by doing, they must be presented with doable tasks. This text is designed to be a sequence of stepping stones placed just the right distance apart. Moving from one stone to the next involves writing a proof. Seeing how to step there comes from reading the exposition and calls on the experience that led the student to the current stone. At first, stones are very close together, and there is much guidance. Progressing through the text, stones become increasingly farther apart, and some of the guidance might be either relegated to a footnote or omitted altogether.

I have written this text with a very deliberate trajectory of style. It is conversational throughout, though the exposition becomes more sophisticated and succinct as students progress through the chapters.

Organization

This text is organized in the following way. Chapter 0 spells out all assumptions to be used in writing proofs. These are not necessarily standard axioms of mathematics, and they are not presented in the context or language of more abstract mathematical structures. They are designed merely to be a starting point for logical development, so that students appreciate quickly that everything we call on is either stated up front as an assumption, or proved from these assumptions. Although Chapter 0 contains much mathematical information, students can probably read it on their own as the course begins, knowing that it is there primarily as a reference.

Part I begins with logic, but does not focus on it. In Chapter 1, truth tables and manipulation of logical symbols are included to give students an understanding of mathematical grammar, of the underlying skeletal structure of mathematical prose, and of equivalent ways of communicating the same mathematical idea. Chapters 2 and 3 put these to use right away in proof writing, and allow the students to cut their teeth on the most basic mathematical ideas. The context of topics in Chapters 2 and 3 is often rather specific, though certainly more broadly applicable. It is designed to ground the students in familiar territory first, then move into generalized structures later. Abstraction is the goal, not the beginning.

Parts II and III are two completely independent paths, the former into analysis, the latter into algebra. Like Antoni Gaudí's *Sagrada Familia,* the unfinished cathedral in Barcelona, Spain, where narrow spires rise from a foundation to give spectacular views, Parts II and III are purposefully designed to rest on the foundation of Part I and climb quickly into analysis or algebra. Many topics and specific results are omitted along the way, but Parts II and III rest securely on the foundation of Part I and allow students to continue to develop their skills at proof writing by climbing to a height where, I hope, they have a nice view of mathematics.

Flexibility

This text can be used in a variety of ways. It is suitable for use in different class settings and there is much flexibility in the material one may choose to cover.

First, because this text speaks directly to the student, it can naturally be used in a setting where students are given responsibility for the momentum of the class. It is written so that students can read the material on their own first, then bring to class the fruits of their work on the exercises, and present these to the instructor and each other for discussion and critique. If class time and size limit the practicality of such a student-driven approach, then certainly other approaches are possible. To illustrate, we may consider three components of a course's activity, and arrange them in several ways. The components are: 1) the students' reading of the material; 2) the instructor's elaboration on the material; and 3) the students' work on the exercises, either to be presented in class or turned in. When I teach from this text, component 1) is first, 3) follows on its heels, and 2) and 3) work in conjunction until a section is finished. Others might want to arrange these components in another order, for example, beginning with 2), then following with 1) and 3).

Which material an instructor would choose to cover will depend on the purpose of the course, personal taste, and how much time there is. Here are two broad options.

1. To proceed quickly into either analysis or algebra, first cover the material from Part I that lays the foundation. Almost all sections and exercises of Part I are necessary for Parts II and III. However, the *Instructor's Guide and Solutions Manual* notes precisely which sections, theorems, and exercises are necessary for each path, and which may be safely omitted without leaving any holes in the logical progression. Of course, even if a particular result is necessary later, one might decide that to omit its proof details does not deprive the students of a valuable learning experience. The instructor might choose simply to elaborate on how one would go about proving a certain theorem, then allow the students to use it as if they had proved it themselves.

2. Cover Part I in its entirety, saving specific analysis and algebra topics for later courses. This option might be most realistic for courses of two or three units where all the Part I topics are required. Even with this approach, there would likely be time to cover the beginnings of Parts II and/or III. This might be the preferred choice for those who do not want to study analysis or algebra with the degree of depth and breadth characteristic of this text.

This book would not have become a reality without the help of many people. First thanks go to Carolyn Vos Strache, Chair of the Natural Science Division at Pepperdine University, for providing me with resources as this project got off the ground. As it has taken shape, this project has come to bear the marks of many people whose meticulous dissection of several drafts inspired many suggestions for improvements. My thanks to colleagues at Pepperdine and from across the country for their hard work in helping shape this volume into its present form:

Bradley Brock, Pepperdine University, Malibu, CA
Julie Glass, California State University, Hayward, CA

Howard Hamilton, California State University, Sacramento, CA
Jayne Ann Harder, University of Texas, Austin, TX
Kevin Iga, Pepperdine University, Malibu, CA
Irene Loomis, University of Tennessee, Chattanooga, TN
Carlton J. Maxson, Texas A&M University, College Station, TX
Bruce Mericle, Minnesota State University, Mankato, MN
Kandasamy Muthuvel, University of Wisconsin, Oshkosh, WI
Kamal Narang, University of Alaska, Anchorage, AK
Travis Thompson, Harding University, Searcy, AR
Steven Williams, Brigham Young University, Provo, UT

I have written this book with the student foremost in mind. Many of my students have shaped this text from the beginning by their hard work in my class. Several students, both my own and from other universities across the country, have also made formal and useful suggestions. Their marks are indelible.

Erik Baumgarten, Texas A&M University, College Station, TX
Justin Greenough, University of Alaska, Anchorage, AK
Reuben Hernandez, Pepperdine University, Malibu, CA
Brian Hostetler, Virginia Tech, Blacksburg, VA
Jennifer Kuske, Pepperdine University, Malibu, CA

Finally, my deepest thanks go to Barbara Holland, senior editor at Harcourt/Academic Press, for making this text a reality. Her ability to read this manuscript through the eyes of the student has been one of my greatest encouragements.

CHAPTER 0

Notation and Assumptions

Suppose you've just opened a new jigsaw puzzle. What are the first things you do? First, you pour all the pieces out of the box. Then you sort through and turn them all face up, taking a quick look at each one to determine whether it's an inside or outside piece, and you arrange them somehow so that you'll have an idea of where certain types of pieces can be found later. You don't study each piece in depth, nor do you start trying to fit any of them together. In short, you just lay all the pieces out on the table and briefly familiarize yourself with each one. This is the point of the game where you merely set the stage, knowing that everything you'll need later has been put in a place where you can find it when you need it.

In this introductory chapter we lay out all the pieces we will use for our work in this course. It's essential that you read it now, in part because you need some preliminary exposure to the ideas, but mostly because you need to have spelled out precisely what you can use without proof in Part I, where this chapter will serve you as a reference. Give this chapter a casual but complete reading for now. You've probably seen most of the ideas before. But don't try to remember it all, and certainly don't expect to understand everything either. That's not the point. Right now, we're just organizing the pieces. The two issues we address in this chapter are: 1) set terminology and notation; and 2) assumptions about the real numbers.

0.1 Set Terminology and Notation

Sets are perhaps the most fundamental mathematical entity. Intuitively we think of a set as a collection of things, where the collection itself is thought of as a single entity. Sets may contain numbers, points in the xy-plane, functions, ice cream cones, steak knives, worms, even other sets. We will denote many of our sets with uppercase letters (A, B, C), or sometimes with scripted letters ($\mathcal{F}, \mathcal{S}, \mathcal{T}$). First we need a way of stating whether a certain thing is or is not in a set.

Definition 0.1.1. If A is a set and x is an entity in A, we write $x \in A$, and say that x is an *element* of A. To write $x \notin A$ is to mean that x is not an element of A.

How can you communicate to someone what the elements of a set are? There are several ways.

1. List them. If there are only a few elements in the set, you can easily list them all. Otherwise, you might start listing the elements and hope that the reader can take the hint and figure out the pattern. For example,

    (a) $\{1, 8, \pi, \text{Monday}\}$

    (b) $\{0, 1, 2, \ldots, 40\}$

    (c) $\{\ldots, -6, -4, -2, 0, 2, 4, 6, \ldots\}$

2. Provide a description of the criteria used to decide whether an entity is to be included. This is how it works:

    (a) $\{x : x \text{ is a real number and } x > -1\}$ This notation should be read "the set of all x such that x is a real number and x is greater than -1." The variable x is just an arbitrary symbol chosen to represent a random element of the set, so that any characteristics it must have can be stated in terms of that symbol.

    (b) $\{p/q : p \text{ and } q \text{ are integers and } q \neq 0\}$ This assumes you know what an *integer* is (p. 5). This is the set of all fractions, integer over integer, where it is expressly stated that the denominator cannot be zero.

    (c) $\{x : P(x)\}$ This is a generic form for this way of describing a set. The expression $P(x)$ represents some specified property that x must have in order to be in the set.

When addressing the elements of a set, or more important, when addressing all things *not* in a particular set, we must have some universal limiting parameters in mind. Although it might not be explicitly stated, we generally consider that there is some *universal set* beyond which we do not concern ourselves. Then, if we're talking about everything not in a set A, we know how far to cast the net. It is limited by our universal set, typically denoted U.

To help visualize sets and how they compare and combine, we sometimes sketch what is called a *Venn diagram*. Given a set A within a universal set U, we may sketch a Venn diagram as in Fig. 0.1.

Given two sets A and B, it just might happen that all elements of A are also elements of B. We write this as $A \subseteq B$ and say that A is a *subset* of B. Equivalently, we may write $B \supseteq A$, and say that B is a *superset* of A. If A is a subset of B, but there are elements of B that are not in A, we say that A is a *proper subset* of B, and write this $A \subset B$. The relationship $A \subseteq B$ can be displayed in the Venn diagram in Fig. 0.2. The region outside A but inside B may or may not have any elements.

Given two arbitrary sets A and B, the standard Venn diagram is like the one in Fig. 0.3. Unless we know some special characteristics of A and B, we'll draw most of our Venn diagrams in this way.

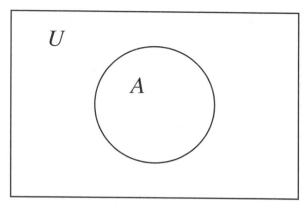

Figure 0.1   A basic Venn diagram.

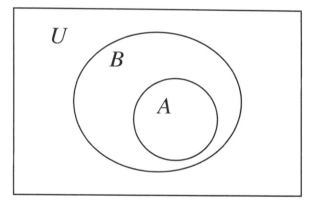

Figure 0.2   Venn diagram with $A \subseteq B$.

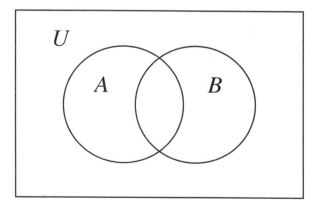

Figure 0.3   A standard Venn diagram with two sets.

Venn diagrams are handy for visualizing new sets formed from old ones. Here are a couple of examples.

Definition 0.1.2. Given a set A, the set A' is called the *complement* of A, and is defined as the set of all elements of U that are not in A (Fig. 0.4). That is,

$$A' = \{x : x \in U \text{ and } x \notin A\} \tag{0.1}$$

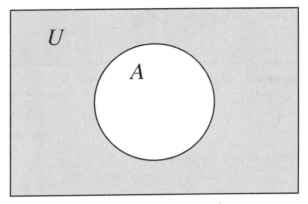

Figure 0.4 Shaded region represents A'.

Definition 0.1.3. Given two sets A and B, we define their *union* $\cup$ and *intersection* $\cap$ in the following way:

$$A \cup B = \{x : x \in A \text{ or } x \in B\} \tag{0.2}$$

$$A \cap B = \{x : x \in A \text{ and } x \in B\} \tag{0.3}$$

An entity is allowed to be in $A \cup B$ precisely when it is in at least one of A, B. An entity is allowed to be in $A \cap B$ precisely when it is in both A and B. See Figs. 0.5 and 0.6.

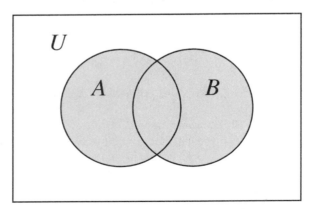

Figure 0.5 Shaded region represents $A \cup B$.

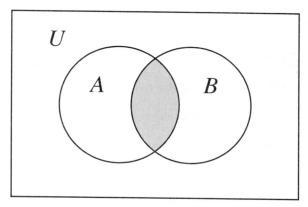

Figure 0.6 Shaded region represents $A \cap B$.

Finally, we provide the notation for commonly used sets. Famous sets we need to know include the following:

Empty set: $\emptyset = \{\}$ (the set with no elements)

Natural numbers: $\mathbb{N} = \{1, 2, 3, \ldots\}$

Whole numbers: $\mathbb{W} = \{0, 1, 2, 3, \ldots\}$

Integers: $\mathbb{Z} = \{\ldots, -3, -2, -1, 0, 1, 2, 3, \ldots\}$

Rational numbers: $\mathbb{Q} = \{p/q : p, q \in \mathbb{Z}, q \neq 0\}$

Real numbers: $\mathbb{R}$ (Explained in what follows)

0.2 Assumptions

One big question we'll face at the outset of this course is what we are allowed to assume and what we must justify with proof. The purpose of this section is to provide a framework for the way you work with real numbers, spelling out the properties you may assume and reminding you of how to visualize them.

0.2.1 Basic algebraic properties of real numbers

The real numbers $\mathbb{R}$, as well its familiar subsets $\mathbb{N}$, $\mathbb{W}$, $\mathbb{Z}$, and $\mathbb{Q}$, are assumed to be endowed with the relation of equality and the operations of addition and multiplication, and to have the following properties. First, equality is assumed to behave in the following way:

(A1) Properties of equality:

(a) For every $a \in \mathbb{R}$, $a = a$ (Reflexive property);

(b) If $a = b$, then $b = a$ (Symmetric property);

(c) If $a = b$ and $b = c$, then $a = c$ (Transitive property). ∎

The first property of addition we assume concerns its predictable behavior, even when the numbers involved can be addressed by more than one name. For example, 3/8 and 6/16 are different names for the same number. We need to know that adding something to 3/8 will always produce the same result as adding it to 6/16. The following property is our way of stating this assumption.

(A2) Addition is well defined: That is, if $a, b, c, d \in \mathbb{R}$, where $a = b$ and $c = d$, then $a + c = b + d$. ∎

A special case of property A2 yields a familiar principle that goes all the way back to your first days of high school algebra: If $a = b$, then, since $c = c$, we have that $a + c = b + c$.

(A3) Closure property of addition: For every $a, b \in \mathbb{R}$, $a + b \in \mathbb{R}$. That is, the sum of two real numbers is still a real number. This closure property also holds for $\mathbb{N}, \mathbb{W}, \mathbb{Z}$, and $\mathbb{Q}$. ∎

(A4) Associative property of addition: For every $a, b, c \in \mathbb{R}$,
$$(a + b) + c = a + (b + c)$$
∎

Addition is what we call a *binary operation*, meaning it combines exactly two numbers to produce a single number result. If we have three numbers a, b, and c to add up, we must split the task into two steps of adding two numbers. Property A4 says it doesn't matter which two, a and b, or b and c, we add first. It motivates us to use the more lax notation $a + b + c$.

(A5) Commutative property of addition: For every $a, b \in \mathbb{R}$, $a + b = b + a$. ∎

If you're not careful, you'll tend to assume order does not matter when two things are combined in a binary operation. There are plenty of situations where order does matter, as we'll see.

(A6) Existence of an additive identity: There exists an element $0 \in \mathbb{R}$ with the property that $a + 0 = a$ for every $a \in \mathbb{R}$. ∎

(A7) Existence of additive inverses: For every $a \in \mathbb{R}$, there exists some $b \in \mathbb{R}$ such that $a + b = 0$. Such an element b is called an *additive inverse* of a, and is typically denoted $-a$ to show its relationship to a. We do *not* assume that only one such b exists. ∎

Properties similar to A2–A7 hold for multiplication.

(A8) Multiplication is well defined: That is, if $a, b, c, d \in \mathbb{R}$, where $a = b$ and $c = d$, then $ac = bd$. ∎

(A9) Closure property of multiplication: For all $a, b \in \mathbb{R}$, $a \cdot b \in \mathbb{R}$. The closure property of multiplication also holds for $\mathbb{N}, \mathbb{W}, \mathbb{Z}$, and $\mathbb{Q}$. ∎

(A10) Associative property of multiplication: For every $a, b, c \in \mathbb{R}$, $(a \cdot b) \cdot c = a \cdot (b \cdot c)$ or $(ab)c = a(bc)$. ∎

(A11) Commutative property of multiplication: For every $a, b \in \mathbb{R}$, $ab = ba$. ∎

(A12) Existence of a multiplicative identity: There exists an element $1 \in \mathbb{R}$ with the property that $a \cdot 1 = a$ for every $a \in \mathbb{R}$. ∎

(A13) Existence of multiplicative inverses: For every $a \in \mathbb{R}$ except $a = 0$, there exists some $b \in \mathbb{R}$ such that $ab = 1$. Such an element b is called a *multiplicative inverse* of a and is typically denoted a^{-1} to show its relationship to a. As with additive inverses, we do not assume that only one such b exists. Furthermore, the assumption that a^{-1} exists for all $a \neq 0$ does not assume that zero does *not* have a multiplicative inverse. It says nothing about zero at all. ∎

The next property describes how addition and multiplication interact.

(A14) Distributive property of multiplication over addition: For every $a, b, c \in \mathbb{R}$, $a(b + c) = (ab) + (ac) = ab + ac$, where the multiplication is assumed to be done before addition in the absence of parentheses. ∎

Property A14 is important because it's the only link between the operations of addition and multiplication. Several important properties of $\mathbb{R}$ owe their existence to this relationship. For example, as we'll see later, the fact that $a \cdot 0 = 0$ for every $a \in \mathbb{R}$ is a direct result of the distributive property, and not something we simply assume.

From addition and multiplication we create the operations of subtraction and division, respectively. Knowing that additive and multiplicative inverses exist (except for 0^{-1}), we write

$$a - b = a + (-b) \tag{0.4}$$

$$a/b = a \cdot b^{-1} \tag{0.5}$$

One very important assumption we need concerns properties A6 and A12. For reasons you'll see later, we need to assume that the additive identity is different from the multiplicative identity. That is, we need the assumption

(A15) $1 \neq 0$. ∎

We'll use these very basic properties to derive some other familiar properties of real numbers in Chapter 2.

0.2.2 Ordering of real numbers

One standard way of comparing two real numbers is with the *greater than* symbol $>$. Intuitively, you think of the statement $a > b$ as meaning that a is to the right of b on the number line. Although this is helpful, the comparison $a > b$ is actually a bit sticky. The nuts and bolts of $>$ are contained in the following. In A16, we make an assumption about how all real numbers compare to zero by $>$, thus giving meaning to the terms *positive* and *negative*. Then in A17 and A18, we make some assumptions about how the positive real numbers behave.

(A16) Trichotomy law: For any $a \in \mathbb{R}$, exactly one of the following is true:

(a) $a > 0$, in which case we say a is *positive*;
(b) $a = 0$;
(c) $0 > a$, in which case we say a is *negative*.

(A17) If $a > 0$ and $b > 0$, then $a + b > 0$. That is, the set of positive real numbers is closed under addition.

(A18) If $a > 0$ and $b > 0$, then $ab > 0$. That is, the set of positive real numbers is closed under multiplication.

Now we can use A16–A18 to give meaning to other statements comparing two arbitrary real numbers a and b.

Definition 0.2.1. If $a, b \in \mathbb{R}$, we say that $a > b$ if $a - b > 0$. The statement $a < b$ means $b > a$. The statement $a \geq b$ means that either $a > b$ or $a = b$. Similarly, $a \leq b$ means either $a < b$ or $a = b$.

The rest of the properties of real numbers are probably not as familiar as the preceding ones, but their roles in the theory of real numbers will be clarified in good time. As with the above properties, we do not try to justify them. We merely accept them and use them as a basis for proofs. A very important property of the whole numbers is the following.

(A19) Well-ordering principle: Any nonempty subset of $\mathbb{W}$ (or $\mathbb{N}$ for that matter) has a smallest element. That is, if $A \subseteq \mathbb{W}$ (or $A \subseteq \mathbb{N}$), and A is nonempty, then there is some number $a \in A$ with the property that $a \leq x$ for all $x \in A$. In particular, 1 is the smallest natural number.

The next property of $\mathbb{R}$ is a bit complicated, but is indispensable in the theory of real numbers. Read it casually the first time, but know that it will be very important in Part II of this text. Suppose $A \subset \mathbb{R}$ is a nonempty set with the property that it is bounded from above. That is, suppose there is some $M \in \mathbb{R}$ with the property that $a \leq M$ for all $a \in A$. For example, let $A = \{x : x^2 < 10\}$. Clearly $M = 4$ is a number such that every $a \in A$ satisfies $a \leq M$. So 4 is an *upper bound* for the set A. There are other upper bounds for A, such as 10, 3.3, and 3.17. The point to be made here is that, among all upper bounds that exist for a set, there is an upper bound that is smallest, and it exists in $\mathbb{R}$. This is stated in the following.

(A20) Least upper bound property of $\mathbb{R}$: If A is a nonempty subset of $\mathbb{R}$ that is bounded from above, then there exists a least upper bound in $\mathbb{R}$. That is, if there

exists some $M \in \mathbb{R}$ with the property that $a \leq M$ for all $a \in A$, then there will also exist some $L \in \mathbb{R}$ with the following properties:

(L1) For every $a \in A$, we have that $a \leq L$, and

(L2) If N is any upper bound for A, it must be that $N \geq L$. ∎

0.2.3 Other assumptions about $\mathbb{R}$

The real numbers are indeed a complicated set. The final two properties of $\mathbb{R}$ we mention are not standard assumptions, and they deserve your attention at some point in your mathematical career. In this text, we assume them.

(A21) The real numbers can be equated with the set of all base 10 decimal representations. That is, every real number can be written in a form like $338.1898\ldots$, where the decimal might or might not terminate, and might or might not fall into a pattern of repetition.

Furthermore, every decimal form you can construct represents a real number. Strangely, though, there might be more than one decimal representation for a certain real number. You might remember that $0.9999\ldots = 1$. (See Section 2.5.1.) The repeating 9 is the only case where more than one decimal representation is possible. We'll assume this. ∎

Our final assumption concerns the existence of roots of real numbers.

(A22) For every positive real number x and any $n \in \mathbb{N}$, there exists a real number solution y to the equation $y^n = x$. Such a solution y is called an nth root of x. The common notation $\sqrt[n]{x}$ will be addressed in Section 2.8. ∎

Notice we make no assumptions about how many such roots of x there are, or what their signs are. Nor do we assume anything about roots of zero or of negative real numbers. We'll derive these from assumption A22.

One final comment about assumptions in mathematics. In a rigorous development of any mathematical theory, some things must be assumed without proof, that is, they must be *axiomatic*, serving as an agreed-on starting place for the mathematician's thinking. In a study of the real numbers, some of the assumptions A1–A22 are standard. Others would be considered standard assumptions only for some subsets of $\mathbb{R}$, perhaps for $\mathbb{W}$. The mathematician would then very painstakingly apply assumptions made to $\mathbb{W}$ in order to expand the same properties to all of $\mathbb{R}$. One assumption in particular, A21, is a most presumptuous one. But let us make no apologies for this. After all, many of the foundational issues in mathematics were addressed very late historically, and this is not a course in the foundations of mathematics. It is a course to teach us how mathematics is done, and to give us some enjoyment of that process. We choose assumptions here that likely coincide with your current idea of a reasonable place to start. In some cases, we'll dig more deeply as we go, though some of the foundational work will come in your later courses.

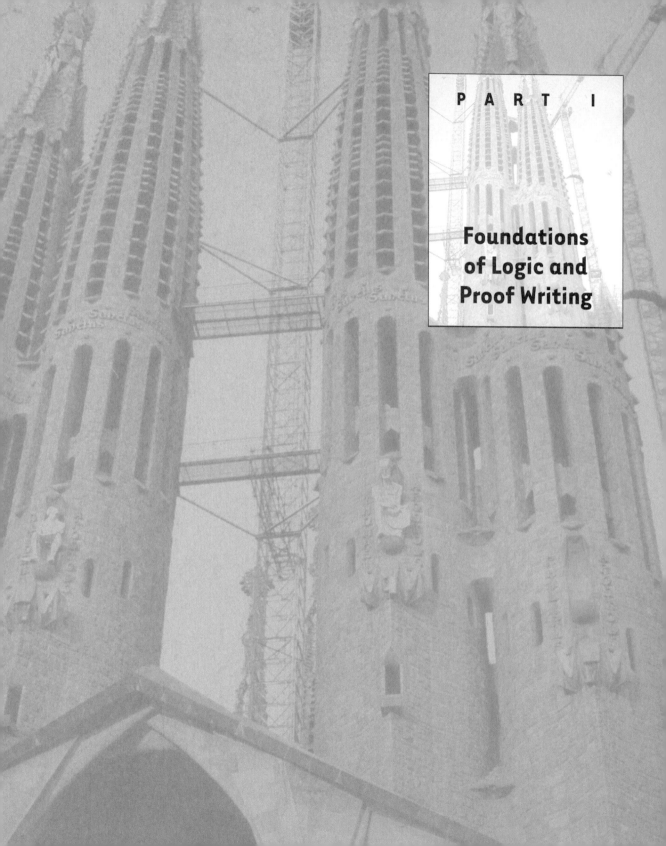

PART I

Foundations of Logic and Proof Writing

Logic

1.1 Introduction to Logic

Mathematicians make as much use of language as anyone else. Not only do they communicate their mathematical work with language, but the use of language is itself part of the mathematical structure. In this chapter, we lay out some of the principles that govern the mathematician's use of language.

1.1.1 Statements

The first issue we address is what kinds of sentences mathematicians use as building blocks for their work. Remember from elementary school grammar that sentences are generally divided into four classes:

Declarative sentences: We also call these *statements*. Here are some examples:

1. Labor Day is the first Monday in September.
2. Earthquakes don't happen in California.
3. Three is greater than seven.
4. The world will end on June 6, 2040.
5. The sky is falling.

One characteristic of statements that jumps out at you is that they generally evoke a reaction such as, "Yeah, that's true," or "No way," or even "That *could* be true, but I don't know for sure." Statement 1 is true, statements 2 and 3 are false, while statements 4 and 5 are uncertain. We cannot know whether statement 4 is true, but we can say that its truth or falsity will be revealed. Statement 5, however, is curious. If you were going to investigate the truth or falsity of this statement, you would immediately be faced with the problem of what the terms mean. How do we define "the sky," and what precisely does it mean to say that it "is falling"?

Imperative sentences: We would call these commands.

1. Don't wash your red bathrobe with your white underwear.
2. Knock three times on the ceiling if you want me.

Interrogative sentences: That is, questions.

1. How much is that doggy in the window?
2. Have you always been that ugly?

Exclamations:

1. What a day!
2. Lions and tigers and bears, Oh My!
3. So, like, whatever.

The mathematician's work centers around the first category, but we have to be careful about exactly which declarative sentences we allow. We will define a *statement* intuitively as a sentence that can be assigned either to the class of things we would call TRUE or to the class of things we would call FALSE. Let's conduct a little thought experiment.

Imagine the set of all conceivable statements, and call it S. Naturally, this set is frighteningly large and complex, but a most important characteristic of its elements is that each one can be placed into exactly one of two subsets: $\mathcal{T}$ (statements called TRUE), and $\mathcal{F}$ (statements called FALSE). Sometimes you might have trouble recognizing whether a sentence even belongs in S. There are, after all, things called paradoxes. For example, "This sentence is false," which cannot be either true or false. If you think the sentence is true, then it is false. However, if it is false, then it is true. We don't want to allow these types of sentences in S.

Now that we have the set of all statements partitioned into $\mathcal{T}$ and $\mathcal{F}$, the true and false ones, respectively, we want to look at relationships between them. Specifically, we want to pick statements from S, change or combine them to make other statements in S, and lay out some understandings of how the truth or falsity of the chosen statements determines the truth or falsity of the alterations and combinations. In the next part of this section, we discuss three ways of doing this:

- the negation of a statement;
- a compound statement formed by joining two given statements with AND; and
- a compound statement formed by joining two given statements with OR.

1.1.2 Negation of a statement

We generally use p, q, r, and so forth, to represent statements symbolically. For example, define a statement p as follows:

p : Megan has rented a car for today.

Now consider the *negation* or *denial* of p, which we can create by a strategic placement of the word NOT somewhere in the statement. We write it this way:

$\neg p$: Megan has not rented a car for today.

Naturally, if p is true, then $\neg p$ is false, and vice versa. We illustrate this in a *truth table* (Table 1.1).

p	$\neg p$
T	F
F	T

Table 1.1

Definition 1.1.1. Given a statement p, we define the statement $\neg p$ (not p) to be false when p is true, and true when p is false, as illustrated in Table 1.1.

1.1.3 Combining statements with AND/OR

When two statements are joined by AND or OR to produce a compound statement, we need a way of deciding whether the compound statement is true or false based on the truth or falsity of its component statements. Let's build these with an example. Define statements p and q as follows:

p : Megan is at least 25 years old.

q : Megan has a valid driver's license.

Now let's create the statement we call "p and q," which we write as

$p \wedge q$: Megan is at least 25 years old, and she has a valid driver's license.

If you know the truth or falsity of p and q individually, how would you be inclined to categorize $p \wedge q$?[1] Naturally, the only way that we would consider $p \wedge q$ to be true is if both p and q are true. In any other instance, we would say $p \wedge q$ is false. Thus, whether

---

[1] Pretend Megan is standing at a car rental counter and must answer yes or no to the question "Are you at least 25 years old and have a valid driver's license?"

p	q	$p \wedge q$
T	T	T
T	F	F
F	T	F
F	F	F

Table 1.2

$p \wedge q$ is in $\mathcal{T}$ or $\mathcal{F}$ depends on whether p and q are in $\mathcal{T}$ or $\mathcal{F}$ individually. We illustrate the results of all different combinations in Truth Table 1.2. Notice how the truth table is constructed with four rows, systematically displaying all possible combinations of T and F for p and q.

Definition 1.1.2. Given two statements p and q, we define the statement $p \wedge q$ (p and q) to be true precisely when both p and q are true, as illustrated in Table 1.2.

Now let's join two statements with OR. Define statements p and q by

p : Megan has insurance that covers her for any car she drives.

q : Megan bought the optional insurance provided by the car rental company.

The compound statement we call "p or q" is written:

$p \vee q$: Megan has insurance that covers her for any car she drives, or she bought the optional insurance provided by the car rental company.

How should we assign T or F to $p \vee q$ based on the truth or falsity of p and q individually?[2] We define $p \vee q$ to be true if *at least one* of p, q is true. See Truth Table 1.3.

p	q	$p \vee q$
T	T	T
T	F	T
F	T	T
F	F	F

Table 1.3

---

[2] Pretend Megan's friend, who is worried about being covered in case of an accident, asks her "Do you have your own insurance, or did you buy the optional coverage provided by the rental company?" Under what circumstances should she say yes?

Definition 1.1.3. Given two statements p and q, we define the statement $p \vee q$ (p or q) to be true precisely when at least one of p and q is true, as illustrated in Table 1.3.

This scenario illustrates why we consider $p \vee q$ to be true even when both p and q are true. We are just as happy if Megan is doubly covered by insurance rather than by one policy only. Because our conversational language is sometimes ambiguous when we use the word OR, we distinguish between two types of OR compound statements. We call $\vee$ the *inclusive* OR. If you want to discuss an OR statement where you mean p or q but not both, use the *exclusive* OR, $p \veebar q$. See Truth Table 1.4. Although we'll not use this very often, it is a nice term to have. Just remember that use of the word OR in mathematical writing always means inclusive OR. For example, when we say $xy = 0$ implies that either $x = 0$ or $y = 0$, we include the possibility that both x and y are zero.

p	q	$p \veebar q$
T	T	F
T	F	T
F	T	T
F	F	F

Table 1.4

Now we can build all kinds of compound statements.

Example 1.1.4. Construct truth tables for the statements

1. $(p \wedge q) \vee (\neg p \wedge \neg q)$,
2. $p \wedge (q \vee r)$,
3. $(p \wedge q) \vee (p \wedge r)$.

Solution: See Table 1.5 for part 1 and Table 1.6 for parts 2 and 3 of Example 1.1.4. Some of the intermediate details are left to you in Exercise 1. Notice how we set up the p, q, and r columns for more than two given statements.

p	q	$\neg p$	$\neg q$	$p \wedge q$	$\neg p \wedge \neg q$	$(p \wedge q) \vee (\neg p \wedge \neg q)$
T	T	F	F	T	F	T
T	F	F	T	F	F	F
F	T	T	F	F	F	F
F	F	T	T	F	T	T

Table 1.5 Solution to Example 1.1.4, part 1

p	q	r	$q \vee r$	$p \wedge (q \vee r)$	$p \wedge q$	$p \wedge r$	$(p \wedge q) \vee (p \wedge r)$
T	T	T	T	T	T	T	T
T	T	F	T	T	T	F	T
T	F	T	T	T			T
T	F	F	F	F			F
F	T	T		F			F
F	T	F		F			F
F	F	T		F			F
F	F	F		F			F

Table 1.6 Solutions to Example 1.1.4, parts 2 and 3

∎

1.1.4 Logical equivalence

There are often several ways to say the same thing. We need to address the situation where two different constructs involving statements should be interpreted as having the same meaning, or as being *logically equivalent*. As a trivial example, consider that $p \wedge q$ should certainly be viewed as having the same meaning as $q \wedge p$. To build a truth table would produce identical columns for $p \wedge q$ and $q \wedge p$. This is the way we define our use of the term *logical equivalence*.

Definition 1.1.5. Two statements are said to be *logically equivalent* if they have precisely the same truth table values. If p and q are logically equivalent, we write $p \Leftrightarrow q$.

Look at parts 2 and 3 of Example 1.1.4. To illustrate the reasonableness of declaring $p \wedge (q \vee r)$ logically equivalent to $(p \wedge q) \vee (p \wedge r)$, consider the following criteria for being allowed to rent a car:

p : Megan has a valid driver's license.

q : Megan has her own insurance policy.

r : Megan bought the rental company's insurance coverage.

Notice that saying "p AND either q or r" has the same meaning to us as "p and q, OR p and r." This is a sort of distributive property; that is, $\wedge$ distributes over $\vee$, exactly like multiplication distributes over addition in real numbers. Exercise 5 asks you to demonstrate that $\vee$ also distributes over $\wedge$.

1.1.5 Tautologies and contradictions

Sometimes a truth table column just happens to have TRUE values all the way down; for example, $p \vee \neg p$. A statement such as "Either Megan has a valid driver's license, or she does not" would make you think, "Of course!" or "Naturally this is a true statement regardless of the circumstances." A statement whose truth table values are all TRUE is called a *tautology*. You'll do the following in Exercise 6.

Example 1.1.6. Show that $\neg(p \wedge q) \vee (p \vee q)$ is a tautology.

A statement such as the one in Example 1.1.6 would be very confusing if expressed in English form. It would read something like,

> Either it is not true that Dave has brown hair and green eyes, or he has either brown hair or green eyes.

One last item. The negation of a tautology is called a *contradiction*. The truth table values of a contradiction are all FALSE. In the same way that a tautology is the kind of statement that makes you think "Of course!," a contradiction makes you think "No way can that ever be true!" A really easy example of a contradiction is $p \wedge \neg p$. Since p and $\neg p$ cannot ever both be true, $p \wedge \neg p$ is always false. Tautologies and contradictions are very useful, as we will begin to see in Section 2.1.

EXERCISES

1. Construct truth tables for the following statements:
   (a) $p \vee (q \vee r)$
   (b) $(p \vee q) \vee r$
   (c) $p \wedge (q \vee r)$
   (d) $(p \wedge q) \vee (p \wedge r)$

2. Which of the statements in Exercise 1 are logically equivalent?

3. Parts (a) and (b) of Exercise 1 show that $\vee$ has the associative property. We can therefore allow ourselves the freedom to write $p \vee q \vee r$ and understand it to mean either $(p \vee q) \vee r$ or $p \vee (q \vee r)$. Does $\wedge$ have the associative property? Verify your answer with a truth table.

4. Below are several logical equivalences that are called *DeMorgan's laws* (a name you'll want to remember). Verify these forms of DeMorgan's laws with truth tables:
   (a) $\neg(p \wedge q); \quad \neg p \vee \neg q$
   (b) $\neg(p \vee q); \quad \neg p \wedge \neg q$
   (c) $\neg(p \wedge q \wedge r); \quad \neg p \vee \neg q \vee \neg r$
   (d) $\neg(p \vee q \vee r); \quad \neg p \wedge \neg q \wedge \neg r$

5. Show that $\vee$ distributes over $\wedge$.

6. Show that $\neg(p \wedge q) \vee (p \vee q)$ from Example 1.1.6 is a tautology.

7. Construct a statement using only $p, q, \wedge, \vee,$ and $\neg$ that is logically equivalent to $p \veebar q$. Demonstrate logical equivalence with a truth table.

8. Use DeMorgan's laws from Exercise 4 as a basis for symbolic substitution and manipulation to show that $\neg[(p \vee q) \wedge r]$ is logically equivalent to $(\neg p \wedge \neg q) \vee \neg r$ by transforming the former statement into the latter. Use a similar technique to construct a statement that is logically equivalent to $\neg[p \vee (q \wedge r)]$.

1.2 If-Then Statements

In this section, we want to do two things: 1) Consider the logical structure of the statement that *p implies* or *necessitates q* and its variations; and 2) return to the idea of logical equivalence and its connection to tautologies. We set the stage with a classic (albeit tired) example.

In my junior high school days, there was a man named Mr. Shephard who would stand in the middle of the street in front of my school every afternoon and flag down cars to try to get a ride to the post office. What made this dangerous stunt notable was the way Mr. Shephard often dressed. If there was even a single cloud in the sky, Mr. S would certainly be wearing his raincoat and carrying his umbrella. However, there were also days without a cloud in the sky on which Mr. S would be wearing his rain attire. When I think back about that period of my life, I am struck with the following fact: On every day that there were any clouds, Mr. S was undoubtedly dressed for rain. Granted, there might have been some clear days that he also dressed for rain, but I am not referring to this possibility in my claim. I'm only making an observation about something that happened on cloudy days.

1.2.1 If-then statements

From this little story, we want first to analyze the correlation between the weather and Mr. Shephard's attire. Specifically, we want to define logically with a truth table what we mean by a statement such as "*If it is a cloudy day, then Mr. S wears his rain gear.*" In general, we consider the statement "If p, then q," which we write as $p \to q$.

Let's isolate one particular day, say day 1. Define statements

$$p_1 : \text{Day 1 is a cloudy day.}$$

$$q_1 : \text{Mr. S wears his rain gear on day 1.}$$

Pretend for the moment that words like IF, THEN, and IMPLIES are not in your vocabulary. How can we piece together a logical statement using only $p_1, q_1, \wedge, \vee,$ and $\neg$ that has the same sense as $p_1 \to q_1$?[3] There are several possible answers. Here are two. Read them as sentences to understand their meaning.

$$\neg p_1 \vee (p_1 \wedge q_1) \tag{1.1}$$

$$\neg p_1 \vee q_1 \tag{1.2}$$

---

[3] One way is to consider the two different weather possibilities and link one of them with Mr. Shephard's expected behavior.

1.2 If-Then Statements

Statements 1.1 and 1.2 are logically equivalent, but 1.2 is simpler, so let's use it as our definition of $p \to q$. We build its truth table:

Definition 1.2.1. The statement $p \to q$ (read "If p, then q" or "p implies q") is defined to be a statement that is logically equivalent to $\neg p \vee q$, as illustrated in Table 1.7. We call p the *hypothesis condition* and q the *conclusion*.

p_1	q_1	$\neg p_1 \vee q_1 \ (\Leftrightarrow p_1 \to q_1)$
T	T	T
T	F	F
F	T	T
F	F	T

Table 1.7

Notice from the definition that constructing a truth table for $\to$ produces FALSE only when the hypothesis condition is true and the conclusion is false.

Example 1.2.2. Construct truth tables for the following statements.

1. $p \to \neg q$
2. $(p \wedge q) \to r$

Solution: See Tables 1.8 and 1.9.

p	q	$\neg q$	$p \to \neg q$
T	T	F	F
T	F	T	T
F	T	F	T
F	F	T	T

Table 1.8 Solution to Example 1.2.2, part 1

p	q	r	$p \wedge q$	$(p \wedge q) \to r$
T	T	T	T	T
T	T	F	T	F
T	F	T	F	T
T	F	F	F	T
F	T	T	F	T
F	T	F	F	T
F	F	T	F	T
F	F	F	F	T

Table 1.9 Solution to Example 1.2.2, part 2

■

1.2.2 Variations on $p \rightarrow q$

Given two statements p and q, we might want to analyze other possible correlations between their truth besides $p \rightarrow q$. Defining:

p : It is a cloudy day,

q : Mr. S dresses for rain,

we might address the following variations of $p \rightarrow q$.

$q \rightarrow p$: If Mr. S dresses for rain, then it is a cloudy day.
(Converse)

$\neg p \rightarrow \neg q$: If it is not a cloudy day, then Mr. S does not dress for rain.
(Inverse)

$\neg q \rightarrow \neg p$: If Mr. S does not dress for rain, then it is not a cloudy day.
(Contrapositive)

Example 1.2.3. Which, if any, of the statements $p \rightarrow q$, $q \rightarrow p$, $\neg p \rightarrow \neg q$, and $\neg q \rightarrow \neg p$ are logically equivalent?

Solution: We construct a truth table (see Table 1.10).

p	q	$\neg p$	$\neg q$	$p \rightarrow q$	$q \rightarrow p$	$\neg p \rightarrow \neg q$	$\neg q \rightarrow \neg p$
T	T	F	F	T	T	T	T
T	F	F	T	F	T	T	F
F	T	T	F	T	F	F	T
F	F	T	T	T	T	T	T

Table 1.10

Notice that the original statement $p \rightarrow q$ is logically equivalent to its contrapositive $\neg q \rightarrow \neg p$, and that the converse $q \rightarrow p$ is logically equivalent to the inverse $\neg p \rightarrow \neg q$. ■

There is one last construct involving if-then. Sometimes we want to consider the statement that we would call true precisely when p and q are either both true or both false. For example, if Mr. S had been firing on all cylinders we could have observed the following: Mr. S dressed for rain if it were cloudy, and if he dressed for rain, then it was a cloudy day. That is, he would have dressed for rain *if and only if* it were cloudy.

Definition 1.2.4. Given statements p and q, the statement $p \leftrightarrow q$ (read "p if and only if q," and often written "p iff q") is defined to be true precisely when p and q are either both true or both false.

How can you use $\to$, $\wedge$, and $\vee$ on p and q to construct a statement that is logically equivalent to $p \leftrightarrow q$?[4] One answer is in the next example, and you'll create others in Exercise 2.

Example 1.2.5. Show that $(p \to q) \wedge (q \to p)$ is logically equivalent to $p \leftrightarrow q$.

Solution: See Table 1.11.

p	q	$p \leftrightarrow q$	$p \to q$	$q \to p$	$(p \to q) \wedge (q \to p)$
T	T	T	T	T	T
T	F	F	F	T	F
F	T	F	T	F	F
F	F	T	T	T	T

Table 1.11

1.2.3 Logical equivalence and tautologies

In Section 1.1, we defined two statements to be logically equivalent if they have exactly the same truth table values. With the definition of $p \leftrightarrow q$, we can now offer an alternate definition of logical equivalence.

Definition 1.2.6. Two statements p and q are logically equivalent if the statement $p \leftrightarrow q$ is a tautology.

As we noted in Section 1.1.4, the significance of statements being logically equivalent is that they are different ways of saying precisely the same thing. If p is logically equivalent to q, then knowing p is true guarantees that q is true, and vice versa.

If $p \leftrightarrow q$ is a tautology, what does that say about $p \to q$ and $q \to p$ separately?[5] If $p \leftrightarrow q$ is a tautology, then $p \to q$ and $q \to p$ must both be tautologies, too. Loosely speaking, the truth of p is sufficiently strong to imply the truth of q and vice versa. That is, in any case where p is true, it can also be noted that q will, without exception, be true, and vice versa.

Example 1.2.7. Show that $p \to q$ and $\neg q \to \neg p$ are logically equivalent using Definition 1.2.6.

Solution: We already know that the truth table columns for $p \to q$ and $\neg q \to \neg p$ are identical, but we are asked to use Definition 1.2.6. Therefore, we construct

---

[4] Take a hint from the sentence, "For example, if Mr. S had been firing on all cylinders ...."
[5] Remember $p \leftrightarrow q$ is defined by $(p \to q) \wedge (q \to p)$.

p	q	$\neg p$	$\neg q$	U	V	$U \to V$	$V \to U$	$(U \to V) \wedge (V \to U)$
T	T	F	F	T	T	T	T	T
T	F	F	T	F	F	T	T	T
F	T	T	F	T	T	T	T	T
F	F	T	T	T	T	T	T	T

Table 1.12

$(p \to q) \leftrightarrow (\neg q \to \neg p)$ and show that it is a tautology. Writing $p \to q$ as U and $\neg q \to \neg p$ as V, the details are displayed in Table 1.12. Since the last column of Table 1.12 is a tautology, U and V are logically equivalent. Notice that the columns $U \to V$ and $V \to U$ are tautologies to make this last column a tautology. ∎

Now let's consider the situation where $p \leftrightarrow q$ is not a tautology, but one of $p \to q$ or $q \to p$ is. If $p \to q$ is a tautology while $q \to p$ is not, then we say that p is a *stronger statement* than q. This means that the truth of p necessitates the truth of q, but the truth of q is not necessarily accompanied by the truth of p.

Example 1.2.8. Which statement is stronger, p or $p \wedge q$? Verify with a truth table (see Table 1.13).

p	q	$p \wedge q$	$(p \wedge q) \to p$	$p \to (p \wedge q)$
T	T	T	T	T
T	F	F	T	F
F	T	F	T	T
F	F	F	T	T

Table 1.13

Solution: Since $(p \wedge q) \to p$ is a tautology while $p \to (p \wedge q)$ is not, $p \wedge q$ is stronger than p. Knowing $p \wedge q$ is true guarantees that p is true, but knowing p is true does not guarantee that $p \wedge q$ is true. ∎

Example 1.2.9. Which statement do you think is stronger, $(p \wedge q) \to r$ or $p \to r$? Determine for sure with a truth table.

Solution: Writing $(p \wedge q) \to r$ as U and $p \to r$ as V, we need truth table values for $U \to V$ and $V \to U$. (See Table 1.14.) Since $V \to U$ is a tautology and $U \to V$ is not, V is stronger than U.

p	q	r	$p \wedge q$	$U : (p \wedge q) \to r$	$V : p \to r$	$U \to V$	$V \to U$
T	T	T	T	T	T	T	T
T	T	F	T	F	F	T	T
T	F	T	F	T	T	T	T
T	F	F	F	T	F	F	T
F	T	T	F	T	T	T	T
F	T	F	F	T	T	T	T
F	F	T	F	T	T	T	T
F	F	F	F	T	T	T	T

Table 1.14

Notice this important fact. Statements U and V from Example 1.2.9 have the same conclusion, but the hypothesis condition for U ($p \wedge q$) is stronger than the hypothesis condition for V (p). However, U is a weaker statement than V. Exercise 4 asks you to investigate and explain from the truth table why an if-then statement is weakened when the hypothesis condition is strengthened.

What is the significance of one statement being stronger than another? Here is an example. Define the following statements:

p : Megan is at least 25 years old.

q : Megan has a valid driver's license.

r : Megan is allowed to rent a car.

What does it mean to say that V ($p \to r$) is stronger than U (($p \wedge q) \to r$)?[6] Statement U says that age and a license will guarantee your eligibility to rent a car. Statement V says that age alone is sufficient to be eligible. Thus, if everyone at least 25 years old can rent a car, then certainly everyone at least 25 with a license can, too. Thus if V is true, so is U. On the other hand, just because licensed people at least 25 years old can rent a car, it does not follow that all people over 25 can do the same. That is, U does not imply V.

Example 1.2.10. Without justifying by proof, state whether the following statements are logically equivalent, whether one is stronger than the other, or neither.

1.
   p : x is an integer that is divisible by 6.

   q : x is an integer that is divisible by 3.

2.
   p : $x^3 - 4x^2 + 4x = 0$

   q : $x \in \{0, 2, 4\}$

---

[6] If age is sufficient for being allowed to rent a car, what does that say about age and a driver's license?

3.
$$p: -2 < x < 2$$
$$q: x^2 < 4$$

Solution:

1. If x is divisible by 6, then it is divisible by both 2 and 3. That is, divisibility by 6 necessitates divisibility by 3. However, 9 is a multiple of 3 that is not divisible by 6, so that divisibility by 3 does not necessitate divisibility by 6. Therefore, p is stronger than q.

2. Factoring and solving the equation in p yields $x = 0$ or $x = 2$. Thus, if $x = 0$ or $x = 2$, then $x \in \{0, 2, 4\}$. But just because $x \in \{0, 2, 4\}$, it does not follow that $x^3 - 4x^2 + 4x = 0$. Thus p is stronger than q.

3. A particular value of x will either make p and q both true or both false. Thus, they are logically equivalent. ∎

EXERCISES

1. The following sentences are alternative ways of expressing an idea of the form $p \to q$ or $p \leftrightarrow q$. For each, define statements p and q and note whether $p \to q$ or $p \leftrightarrow q$ conveys the same sense as the sentence.

   (a) On a clear day, you can see forever.
   (b) I get nervous whenever you look at me that way.
   (c) Every time I turn my back, you sneak away.
   (d) Only fools fall in love.
   (e) There are no refunds on sale items.
   (f) Every time a bell rings, an angel gets his wings.
   (g) I'll wait for you at home, unless you call, in which case I'll leave.
   (h) The only solutions to the equation $x^2 - x = 0$ are nonnegative.
   (i) It only hurts when I laugh.[7]
   (j) Unless you follow the instructions, the cake won't turn out right. (Careful: Just because you follow the instructions doesn't guarantee success. A sudden earthquake could cause it to fall.)

2. Use the information from Table 1.10 to create three statements that are equivalent to $p \leftrightarrow q$ other than $(p \to q) \land (q \to p)$.

3. For each of the following pairs of statements, use a truth table to determine whether they are logically equivalent, one is stronger than the other, or neither.

---

[7] Is this saying that all laughter is painful?

(a) p; $p \vee q$
(b) $p \rightarrow r$; $(p \rightarrow q) \wedge (q \rightarrow r)$
(c) $(p \vee q) \rightarrow r$; $(p \rightarrow r) \vee (q \rightarrow r)$
(d) $(p \vee q) \rightarrow r$; $(p \rightarrow r) \wedge (q \rightarrow r)$
(e) $p \rightarrow (q \wedge r)$; $(p \rightarrow q) \wedge (p \rightarrow r)$
(f) $p \veebar q$; $(p \vee q) \wedge \neg q$
(g) $p \rightarrow (q \vee r)$; $(p \wedge \neg q) \rightarrow r$
(h) $(p \wedge q) \rightarrow r$; $(p \wedge \neg r) \rightarrow \neg q$
(i) $(p \leftrightarrow q) \wedge (r \leftrightarrow s)$; $(p \vee r) \leftrightarrow (q \vee s)$
(j) $p \leftrightarrow q$; $\neg p \leftrightarrow \neg q$

4. In Example 1.2.9, we noted that $p \rightarrow r$ is stronger than $(p \wedge q) \rightarrow r$, while p is weaker than $p \wedge q$. This exercise investigates the reason why an implication statement is strengthened when its hypothesis conditions are weakened.

    (a) Suppose p and q are two statements, and p is stronger than q. What must be true about the truth table entries for p as compared to those for q?[8]
    (b) If p is stronger than q, why does that make $p \rightarrow r$ weaker than $q \rightarrow r$?[9]
    (c) Which of the following pairs of statements is stronger? Explain without using a truth table.
        i. $p \rightarrow r$; $(p \vee q) \rightarrow r$
        ii. $[p \wedge (q \rightarrow s)] \rightarrow t$; $\{p \wedge [(q \wedge r) \rightarrow s]\} \rightarrow t$

1.3 Universal and Existential Quantifiers

The if-then language of Section 1.2 is only one way to address whether the truth of one statement necessitates the truth of another. In this section, we analyze a language construct using words like ALL and SOME. These words are called *quantifiers*. It goes something like this. Consider the two statements:

$$\text{If } x \text{ is a square, then } x \text{ is a rectangle.}$$

$$\text{All squares are rectangles.}$$

They say the same thing, but the latter does not seem to be a compound statement until we rephrase it in the form $p \rightarrow q$.

1.3.1 The universal quantifier

Let's take the story of Mr. S and construct a whole slew of statements, one for each day he went to the post office, and see if cloudy days were always associated with his wearing of

---
[8] The set of F entries for one must be a proper subset of the F entries for the other.
[9] Use your answer to part (a), and the definition of $p \rightarrow q$.

rain gear. We're not investigating whether he dressed this way on clear days; sometimes he did and sometimes he did not. Let's put it together in the following way.

Suppose there were n days that Mr. S hitched a ride to the post office. For each of these days, construct the following statements:

$$p_1 : \text{Day 1 was cloudy.}$$
$$q_1 : \text{Mr. S wore rain gear on day 1.}$$
$$p_2 : \text{Day 2 was cloudy.}$$
$$q_2 : \text{Mr. S wore rain gear on day 2.} \qquad (1.3)$$
$$\vdots$$
$$p_n : \text{Day } n \text{ was cloudy.}$$
$$q_n : \text{Mr. S wore rain gear on day } n.$$

If we consider the statements $\neg p_k \vee q_k$ ($\Leftrightarrow p \to q$) for all $1 \leq k \leq n$, we can link them together to form a huge compound statement that expresses the idea that for every cloudy day, Mr. S wore his rain gear. In the language of $\wedge$ and $\vee$, to say that Mr. S dressed for rain on every cloudy day could be expressed as

$$(\neg p_1 \vee q_1) \wedge (\neg p_2 \vee q_2) \wedge (\neg p_3 \vee q_3) \wedge \cdots \wedge (\neg p_n \vee q_n) \qquad (1.4)$$

or more succinctly as

$$\bigwedge_{k=1}^{n} (\neg p_k \vee q_k) = \bigwedge_{k=1}^{n} (p_k \to q_k). \qquad (1.5)$$

The only way statement 1.5 can be true is if every $p_k \to q_k$ component is true.

We introduce new language and symbols to facilitate such a statement. Consider the following:

For every day that it was cloudy, Mr. S wore his rain gear; and

for all k ($1 \leq k \leq n$) such that day k was cloudy, Mr. S wore his rain gear on day k.

The expressions "for every" and "for all" are called *universal quantifiers*. The shorthand mathematical notation for this is $\forall$.

Let's formalize this further. Let $\mathcal{C}$ be the set of all k ($1 \leq k \leq n$) such that day k was cloudy, that is $\mathcal{C} = \{k : 1 \leq k \leq n \text{ and day } k \text{ was cloudy}\}$. Similarly, let $\mathcal{R} = \{k : 1 \leq k \leq n \text{ and Mr. S dressed for rain on day } k\}$. Then we can reword our statement as

For every $k \in \mathcal{C}, k \in \mathcal{R}$

which we can write mathematically in either of the following ways:

$$(\forall k \in \mathcal{C})(k \in \mathcal{R});$$
$$(\forall k)(k \in \mathcal{C} \to k \in \mathcal{R}).$$

The most general form for a statement involving the universal quantifer would look something like this. If $P(x)$ is some property stated in terms of x, such as "$x \in \mathcal{C} \to x \in \mathcal{R}$," then a general statement involving the universal quantifier would be written

$$(\forall x)(P(x)) \tag{1.6}$$

and would be read, "For all x, $P(x)$." As we said at the beginning of this section, using the universal quantifier is a fancy way of saying "If it was a cloudy day, then Mr. S dressed for rain." Yes, but we'll see in Section 1.4 why we need to understand a statement with the universal quantifier as an $\wedge$-linking of individual $p_k \to q_k$ statements.

1.3.2 The existential quantifier

Now let's go back to our story and consider the clear days. One thing that made Mr. Shephard's antics so amusing to me was that I seem to remember seeing him dressed for rain on a clear day, at least once, that is. We want to build a logical statement that expresses symbolically the thought that there existed at least one clear day when he dressed for rain. How can we consider each day, note whether it was clear and whether we found Mr. Shephard dressed for rain and then address the issue of whether this odd combination of circumstances happened at least once?[10]

The answer is

$$(\neg p_1 \wedge q_1) \vee (\neg p_2 \wedge q_2) \vee \cdots \vee (\neg p_n \wedge q_n) = \bigvee_{k=1}^{n}(\neg p_k \wedge q_k). \tag{1.7}$$

If even for one day both $\neg p_k$ and q_k are true, then Eq. (1.7) will be true. Using our previous notation, and letting $\mathcal{S}$ be the set of all clear (sunny) days, we can express this in any of the following ways:

There exists a day such that it was clear and Mr. S dressed for rain.

There exists $k \in \mathcal{S}$ such that $k \in \mathcal{R}$.

The expression "there exists," written mathematically as $\exists$, is called the *existential quantifier*. Thus the existential statement could be written:

$$(\exists k \in \mathcal{S})(k \in \mathcal{R}) \tag{1.8}$$

$$(\exists k)(k \in \mathcal{S} \wedge k \in \mathcal{R}). \tag{1.9}$$

The most general form of an existential statement involving a property $P(x)$ would be written

$$(\exists x)(P(x)) \tag{1.10}$$

and would be read, "There exists x such that $P(x)$." Note how a statement with the existential quantifier is a $\vee$-linking of many individual statements.

---

[10]The hint is in the sentence itself. Use $\vee$ to join a bunch of pieces that use $\wedge$.

Example 1.3.1. Formalize the following into logical statements involving $\forall, \exists, \wedge,$ and $\vee$:

1. For all $x \in A, x \geq \pi$.
2. There exists $x \in A$ such that either $x < \pi$ or $x > \pi^2$.
3. There exists $x \in \mathbb{N}$ such that, for all $y \in \mathbb{N}, x \leq y$.

Solution:

1. $(\forall x \in A)(x \geq \pi)$.
2. $(\exists x \in A)[(x < \pi) \vee (x > \pi^2)]$.
3. $(\exists x \in \mathbb{N})(\forall y \in \mathbb{N})(x \leq y)$.

∎

1.3.3 Unique existence

Sometimes it's important to know not only that something exists, but that exactly one such thing exists. To communicate that exactly one thing with a certain property exists, we say that it exists *uniquely*. The mathematical statement

$$(\exists! x)(P(x)) \tag{1.11}$$

is read "There exists a unique x such that $P(x)$."

How do we alter the statement $(\exists x)(P(x))$ to include the additional stipulation that there is no more than one such x?[11] The standard way of defining unique existence is the following.

Definition 1.3.2. The statement $(\exists! x)(P(x))$ is defined by

$$[(\exists x)(P(x))] \wedge [(P(x_1) \wedge P(x_2)) \rightarrow (x_1 = x_2)]. \tag{1.12}$$

Thus, unique existence is really a compound AND statement. First of all, it means that some x exists with the property $P(x)$, but also, if we assume that x_1 and x_2 both have the property, then in reality x_1 and x_2 are the same thing.

Example 1.3.3. Reword the following statements of unique existence in a form like that in Definition 1.3.2.

1. There exists a unique real number x such that $x^3 = 8$.
2. For all $a \in \mathbb{R}$, there exists a unique $b \in \mathbb{R}$ such that $a + b = 0$.

Solution:

1. There exists a real number x such that $x^3 = 8$, and if x_1 and x_2 are real numbers such that $x_1^3 = 8$ and $x_2^3 = 8$, then $x_1 = x_2$.

---

[11] Think about what would have to be true of x_1 and x_2 if both $P(x_1)$ and $P(x_2)$ are true.

2. For all $a \in \mathbb{R}$, there exists $b \in \mathbb{R}$ such that $a + b = 0$, and if $b_1, b_2 \in \mathbb{R}$ satisfy $a + b_1 = 0$ and $a + b_2 = 0$, then $b_1 = b_2$. ∎

EXERCISES

1. Express the following statements in the language of universal and existential quantifiers.

   (a) If $x \in A$, then $x \notin B$.
   (b) Everyone in the class is present.
   (c) No element of the set A exceeds m.
   (d) If $n \geq 5$, then $n^2 < 2^n$.
   (e) $A \cap B = \emptyset$.
   (f) The faculty resolution passed unanimously.
   (g) Only fools fall in love.
   (h) Dan has never made an A in a history class.
   (i) Every time a bell rings, an angel gets his wings.
   (j) The only solutions to the equation $x^2 - x = 0$ are nonnegative.
   (k) The equation $x^2 + 1 = 0$ has no solution in $\mathbb{R}$.
   (l) If n is an odd integer, then n^2 is an odd integer.[12]
   (m) If $n \geq 3$, then the equation $x^n + y^n = z^n$ has no integer solutions $x, y, z \in \mathbb{Z}$.

2. Reword each unique existence statement that follows in a form like that of Definition 1.3.2.

   (a) There exists a unique $x \in A \cap B$.
   (b) There exists a unique $x \in \mathbb{R}$ such that $x > 1$ and x is a factor of p.
   (c) The curves $x^2 + y^2 = 1$ and $15y = (x - 11/5)^2$ intersect at exactly one point in the xy-plane.

1.4 Negations of Statements

In Section 1.1 we made our first mention of the negation, or denial, of a statement. Given a statement p, the defining characteristic of its negation $\neg p$ is that the truth table values are exactly the opposite. In this section, we see how to construct negations of compound statements. Then, if someone makes an ugly statement such as

$$(p \wedge q) \rightarrow (r \vee s)$$

---

[12] The terms *even* and *odd* have not been defined yet, though you undoubtedly are familiar with the terms. We'll define them precisely in Section 2.7.

and we want to add NOT to it, we will then have a way of writing

$$\neg[(p \wedge q) \rightarrow (r \vee s)]$$

in a more useful form.

1.4.1 Negations of $p \wedge q$ and $p \vee q$

In Exercise 4 in Section 1.1, you showed that $\neg(p \wedge q)$ is logically equivalent to $\neg p \vee \neg q$, and that $\neg(p \vee q)$ is logically equivalent to $\neg p \wedge \neg q$. These facts are called *DeMorgan's laws,* and they provide us a way of expressing the negations of $\wedge$ and $\vee$ compound statements.

Example 1.4.1. Construct a negation of the following statement:

1. $p \wedge (q \wedge r)$
2. $(p \vee q) \wedge (r \vee s)$

  Solution: Applying DeMorgan's laws, we have

  1. $\neg[p \wedge (q \wedge r)] \Leftrightarrow \neg p \vee \neg(q \wedge r) \Leftrightarrow \neg p \vee (\neg q \vee \neg r)$
  2. $\neg[(p \vee q) \wedge (r \vee s)] \Leftrightarrow \neg(p \vee q) \vee \neg(r \vee s) \Leftrightarrow (\neg p \wedge \neg q) \vee (\neg r \wedge \neg s)$. ∎

Notice that part 1 of Example 1.4.1 is the same as Exercise 4c from Section 1.1. In that exercise you used a truth table, but in Example 1.4.1, we used $\neg(p \wedge q) \Leftrightarrow \neg p \vee \neg q$ as a basis for manipulation of logical symbols.

Example 1.4.2. Use DeMorgan's laws to express in words a negation of the following statements:

1. Jacob has brown hair and blue eyes.
2. Christina will either fax or e-mail the information to you.
3. Megan is at least 25 years old, she has a valid driver's license, and she either has her own insurance or has purchased coverage from the car rental company.

  Solution: Don't hesitate to word the statements in a way that makes them easy to understand:

  1. Either Jacob does not have brown hair, or he does not have blue eyes.
  2. Christina will not fax you the information, and she will not e-mail it to you either.
  3. Either Megan is under 25, or she doesn't have a valid driver's license, or she has no insurance of her own and has not purchased coverage from the car rental company. ∎

1.4.2 Negations of $p \to q$

In Section 1.2 we defined $p \to q$ to be logically equivalent to $\neg p \vee q$. To construct a negation of $p \to q$, we can use this fact with DeMorgan's law:

$$\neg(p \to q) \Leftrightarrow \neg(\neg p \vee q) \Leftrightarrow (\neg\neg p) \wedge \neg q \Leftrightarrow p \wedge \neg q.$$

This might be a little confusing at first, but later you'll want to think of it in the following way. If someone makes a claim that p implies q, then they are, in effect, claiming that the truth of p is always accompanied by the truth of q. If you want to deny that claim, then your task will be to exhibit a situation where p is true and q is false. The techniques that follow for negating $\forall$ and $\exists$ statements will help clear that up.

Example 1.4.3. Construct a negation of the following statements:

1. $p \to (q \wedge r)$
2. $(p \to q) \vee (r \to s)$

   Solution:

   1. $\neg[p \to (q \wedge r)] \Leftrightarrow p \wedge \neg(q \wedge r) \Leftrightarrow p \wedge (\neg q \vee \neg r)$
   2. $\neg[(p \to q) \vee (r \to s)] \Leftrightarrow \neg(p \to q) \wedge \neg(r \to s) \Leftrightarrow (p \wedge \neg q) \wedge (r \wedge \neg s)$ ∎

Example 1.4.4. State in words a negation of the following statements:

1. If it was a cloudy day, then Mr. S dressed for rain.
2. If Megan rents a car, then she either has her own insurance or buys coverage from the car rental company.

   Solution:

   1. It was a cloudy day, and Mr. S did not dress for rain.
   2. Megan rented a car, but she neither has her own insurance, nor has she purchased coverage from the car rental company. ∎

1.4.3 Negations of statements with $\forall$ and $\exists$

Suppose someone makes the following claim:

> Every person in the class passed the first exam.

For you to deny this claim, what sort of statement would you make?[13] You would probably say something like

> At least one person in the class failed the first exam.

---

[13] One slacker is all it takes.

We can formalize the language further. If we let C be the set of students in the class and P be the set of students in the class who passed the first exam, we can write the original statement as

$$(\forall x \in C)(x \in P) \tag{1.13}$$

and its negation would be

$$(\exists x \in C)(x \notin P). \tag{1.14}$$

Using the property notation $P(x)$, we can say the following:

$$\neg[(\forall x)(P(x))] \Leftrightarrow (\exists x)(\neg P(x)). \tag{1.15}$$

Thus, the trick to negating a $\forall$ statement is that the $\neg$ symbol crawls over the $\forall$, and converts it to $\exists$ as it goes.

Example 1.4.5. State in words a negation of the following statements:

1. For all $x \in \mathbb{Z}$, $x^2 \geq 0$.
2. For every $x \in \mathbb{R}$, either x is rational or x is irrational.
3. For all $x \in \mathbb{Z}$, if x is divisible by 6, then x is divisible by 3.

Solution:

1. There exists $x \in \mathbb{Z}$ such that $x^2 < 0$.
2. There exists $x \in \mathbb{R}$ such that x is not rational and x is not irrational.
3. There exists $x \in \mathbb{Z}$ such that x is divisible by 6 but not divisible by 3. ∎

Here's another way to understand the denial of a statement involving $\forall$. In our example of Mr. Shephard's dressing habits in Section 1.2, we defined statements p_k and q_k for $1 \leq k \leq n$ in the following way:

$$p_k : \text{Day } k \text{ was cloudy.}$$

$$q_k : \text{Mr. S wore rain gear on day } k.$$

Remember that the statement "For all k ($1 \leq k \leq n$), $p_k \rightarrow q_k$" can be written as

$$\bigwedge_{k=1}^{n}(p_k \rightarrow q_k) \quad \text{or} \quad \bigwedge_{k=1}^{n}(\neg p_k \vee q_k) \tag{1.16}$$

as in Eq. (1.5). Applying an extended form of DeMorgan's law to this, we would have

$$\neg\left[\bigwedge_{k=1}^{n}(p_k \rightarrow q_k)\right] \Leftrightarrow \neg\left[\bigwedge_{k=1}^{n}(\neg p_k \vee q_k)\right] \Leftrightarrow \bigvee_{k=1}^{n}\neg(\neg p_k \vee q_k) \Leftrightarrow \bigvee_{k=1}^{n}(p_k \wedge \neg q_k). \tag{1.17}$$

Now if statement 1.16 is true, then every $\neg p_k \vee q_k$ component is true. Thus every $p_k \wedge \neg q_k$ in statement 1.17 is false, so that $\bigvee_{k=1}^{n} p_k \wedge \neg q_k$ is false. Conversely, if statement 1.16 is false, it's because at least one $\neg p_k \vee q_k$ is false. Thus at least one $p_k \wedge \neg q_k$ in statement 1.17

is true, which is sufficient for $\bigvee_{k=1}^{n} p_k \wedge \neg q_k$ to be true. So statement 1.16 is true if and only if statement 1.17 is false, and they are indeed negations of each other.

Now we negate statements involving $\exists$. Consider the following:

> Someone traveling in my car didn't chip in enough money for gas.

If we let C be the set of people riding in my car and U be the set of underpayers, the statement becomes

$$(\exists x \in C)(x \in U).$$

How would the negation of this statement read?[14] You could say

> Everyone traveling in my car chipped in enough money for gas,

or

$$(\forall x \in C)(x \notin U).$$

Notice that the existential quantifier $\exists$ became the universal quantifier $\forall$, and the defining characteristic $x \in U$ was negated. Thus we arrive at the following:

$$\neg[(\exists x)(P(x))] \Leftrightarrow (\forall x)(\neg P(x)). \tag{1.18}$$

This looks similar to statement 1.15 in that negating a $\exists$ statement can be done by letting the $\neg$ symbol crawl over the $\exists$, changing it to $\forall$ as it goes.

If someone says to you that a certain something exists, and it is a claim you want to deny, then avoid using the expression "There does not exist." Instead, express your negation in positive language by saying that, for every case, the claimed characteristic is not present.

Example 1.4.6. Construct a negation of the following statements:

1. $(\exists x \in \mathbb{N})(x \leq 0)$.
2. $(\forall \epsilon > 0)(\exists n \in \mathbb{N})(1/n < \epsilon)$.

Solution:

1. $(\forall x \in \mathbb{N})(x > 0)$.
2. $(\exists \epsilon > 0)(\forall n \in \mathbb{N})(1/n \geq \epsilon)$. ∎

Example 1.4.7. State in words a negation of the following statements:

1. Someone in this class cheated on the final exam.
2. For every $x \in A$, it is true that $x \in B$.
3. There exists $x \in \mathbb{N}$ such that, for all $y \in \mathbb{N}$, $x \leq y$.
4. For every integer x, either x is even or x is odd.

---

[14] Avoid saying "no one" if possible. It's better to say what everyone did instead of what no one did.

Solution:

1. Everyone in this class did not cheat on the final exam.
2. There exists $x \in A$ such that $x \notin B$.
3. For every $x \in \mathbb{N}$, there exists $y \in \mathbb{N}$ such that $x > y$.
4. There exists $x \in \mathbb{Z}$ such that x is not even and x is not odd. ∎

One final note. Statements with the universal quantifier $\forall$ can often be written in if-then form, and vice versa. Which form is preferred will depend on the context we're working in. The point to be made here is that negating an if-then statement will produce an existence statement, so you might want to construct the negation of an if-then statement by first wording it in terms that use the universal quantifier.

Example 1.4.8. The statement

If it is a cloudy day, then Mr. S dresses for rain,

is the same as

For every cloudy day, Mr. S dresses for rain.

Similarly the statement

For every x in the class, x passed the first exam,

is the same as

If x is in the class, then x passed the first exam.

Example 1.4.9. State in words a negation of the following statements:

1. If $-1 \leq x \leq 10$, then $x^2 \leq 100$.
2. There exists $M \in \mathbb{R}$ such that, for all $x \in S$, we have that $x \leq M$.
3. For all $a \in A$ and $\epsilon > 0$, there exists $\delta > 0$ such that, if $0 < |x - a| < \delta$, then $|f(x) - L| < \epsilon$.

Solution:

1. There exists x such that $-1 \leq x \leq 10$ and $x^2 > 100$.
2. For all $M \in \mathbb{R}$, there exists $x \in S$ such that $x > M$.
3. There exists $a \in A$ and $\epsilon > 0$ such that, for all $\delta > 0$, there exists an x satisfying $0 < |x - a| < \delta$ and $|f(x) - L| \geq \epsilon$. ∎

EXERCISES

1. Construct a negation of each of the following statements:

    (a) $p \vee (q \wedge r)$

(b) $(p \vee q) \to r$
(c) $p \to (q \vee r)$
(d) $p \leftrightarrow q$

2. Construct a negation of each of the following statements:
   (a) If $x \in A$, then $x \notin B$.
   (b) Everyone in the class is present.
   (c) No element of the set A exceeds m.
   (d) If $n \geq 5$, then $n^2 < 2^n$.
   (e) $A \cap B = \emptyset$.
   (f) The faculty resolution passed unanimously.
   (g) Only fools fall in love.
   (h) Dan has never made an A in a history class.
   (i) Every time a bell rings, an angel gets his wings.
   (j) The only solutions to the equation $x^2 - x = 0$ are nonnegative.
   (k) It only hurts when I laugh.
   (l) If n is an odd integer, then n^2 is an odd integer.
   (m) The equation $x^2 + 1 = 0$ has no solution in $\mathbb{R}$.
   (n) Every integer is either even or odd, but not both.
   (o) If $n \geq 3$, then the equation $x^n + y^n = z^n$ has no integer solutions $x, y, z \in \mathbb{Z}$.
   (p) The equation $x^3 = 10$ has a real number solution.
   (q) There exist $M_1, M_2 \in \mathbb{R}$ such that, for all $x \in S$, $M_1 \leq x \leq M_2$.
   (r) There exists a unique $x \in A \cap B$.

CHAPTER 2

Beginner-Level Proofs

2.1 Proofs Involving Sets

It's time to begin applying the language and logic of Chapter 1 to proof writing. A good place to do this is with sets. In this section, we address two things. First, we return to the set terminology from Chapter 0, and we use it to practice some of the concepts we've learned so far. Second, we get our feet wet by beginning to write proofs. Right off the bat, we'll see three useful techniques for writing proofs: direct proofs; proofs by contrapositive; and proofs by contradiction.

2.1.1 Terms involving sets

Define the word *definition*. The self-referencing nature of this request might make is seem like a hard thing to do, but surely we would agree that a basic feature of a definition is that it gives us a way of substituting a single word for a whole phrase. The single word being defined is created for the express purpose of being *equivalent* to that phrase. You could say:

> Jessica went to Kenya and contracted one of a group of diseases, usually intermittent or remittent, characterized by attacks of chills, fever, and sweating, caused by a parasitic protozoan, which is transferred to the human bloodstream by a mosquito of the genus *Anopheles* and which occupies and destroys red blood cells.

Or you could say simply that Jessica went to Kenya and contracted malaria. Malaria is defined to be that disease. You can get by without the word malaria if you want, but then you have to use a whole lot of words every time you want to address this disease (which we hope is not very often), and you won't get practice in logic. For, you see, a definition is an axiomatic statement of logical equivalence. In this case, we're declaring that the statement

x is malaria $\leftrightarrow x$ is a disease characterized by ...

is a tautology. So, if x is malaria, then x is a disease with those scary symptoms, and if x is a disease with those scary symptoms, then x is malaria.

In Section 0.1 we mentioned some terminology relating to sets. Let's return to these definitions to bring a bit more sophistication to them, and present some new ones.

Definition 2.1.1. Suppose A and B are sets. We say that A is a *subset* of B, written $A \subseteq B$, provided the statement "if $x \in A$, then $x \in B$" is true. That is,

$$(A \subseteq B) \Leftrightarrow (x \in A \rightarrow x \in B) \Leftrightarrow (\forall x \in A)(x \in B). \tag{2.1}$$

We will not always be so wordy in our definitions, but it's helpful to be very thorough at first.

Definition 2.1.2. If A and B are sets, we say that $A = B$ provided $A \subseteq B$ and $B \subseteq A$. That is,

$$A = B \Leftrightarrow (A \subseteq B) \wedge (B \subseteq A)$$
$$\Leftrightarrow (x \in A \rightarrow x \in B) \wedge (x \in B \rightarrow x \in A). \tag{2.2}$$

Definition 2.1.3. If A and B are sets, we define the sets

$$A \cup B = \{x : x \in A \vee x \in B\} \tag{2.3}$$
$$A \cap B = \{x : x \in A \wedge x \in B\}. \tag{2.4}$$

That is,
$$x \in A \cup B \Leftrightarrow (x \in A \vee x \in B) \tag{2.5}$$
$$x \in A \cap B \Leftrightarrow (x \in A \wedge x \in B). \tag{2.6}$$

Definition 2.1.4. If $A \cap B = \emptyset$, then A and B are said to be *disjoint*.

To say $A \cap B = \emptyset$ is to say that there does not exist any $x \in A \cap B$, or, if you prefer, there does not exist $x \in A$ such that $x \in B$ also. That is,

$$\begin{aligned} A \cap B = \emptyset &\Leftrightarrow \neg(\exists x \in A \cap B) \\ &\Leftrightarrow \neg(\exists x \in A)(x \in B) \\ &\Leftrightarrow (\forall x \in A)(x \notin B) \\ &\Leftrightarrow x \in A \rightarrow x \notin B. \end{aligned} \tag{2.7}$$

Notice that the disjointness of A and B is completely summed up by the last line of statement (2.7). It isn't necessary also to include $x \in B \rightarrow x \notin A$, because this is merely the contrapositive of $x \in A \rightarrow x \notin B$.

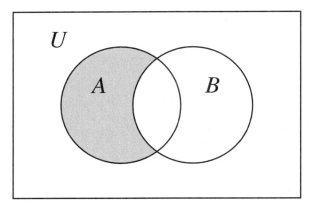

Figure 2.1 Shaded region represents $A \setminus B$.

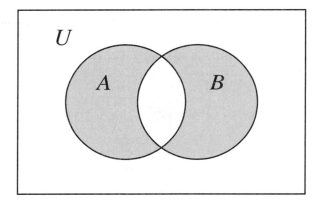

Figure 2.2 Shaded region represents $A \triangle B$.

We mention two other useful sets to construct from two sets A and B. The *difference* $A \setminus B$ is the set of all elements in A that are not in B, and the *symmetric difference* $A \triangle B$ is the set of elements that are either in A or B, but not both. Venn diagrams of these are sketched in Figs. 2.1 and 2.2, respectively. Before you read the definitions of $A \setminus B$ and $A \triangle B$ that follow, see if you can construct them yourself using $\cap$ and $\cup$ and $'$ (complement).

Definition 2.1.5.   If A and B are sets, we define the *difference* of A and B as

$$A \setminus B = \{x : x \in A \text{ and } x \notin B\} = A \cap B'.$$

Definition 2.1.6.   If A and B are sets, we define the *symmetric difference* of A and B as

$$A \triangle B = \{x : x \in A \cap B' \text{ or } x \in A' \cap B\} = (A \cap B') \cup (A' \cap B).$$

Let's practice some statement denials. See Exercise 1 for more similar examples.

Example 2.1.7. Use the preceding definitions to construct a negation of the following statements:

1. $A \subseteq B$
2. $x \in A \cap B$

Solution:

1. The statement $A \subseteq B$ means that, for all $x \in A$, $x \in B$. Thus $A \not\subseteq B$ means that there exists $x \in A$ such that $x \notin B$.

2. The statement $x \in A \cap B$ means that $x \in A$ and $x \in B$. Thus by DeMorgan's Law, $x \notin A \cap B$ means either $x \notin A$ or $x \notin B$. ∎

2.1.2 Direct proofs

Now let's begin to write some proofs of theorems. In general, a theorem is a statement of the form $p \to q$. The hypothesis p will likely be a compound statement, and some of its components might not even be explicitly stated. At the foundational level, our task in writing the proof of a theorem is to show that $p \to q$ is a tautology. However, the final product that we call a proof will be prose, and not look anything like the manipulation of strings of symbols as we did in Chapter 1. In our first examples, we employ the method of *direct proof*. The statement and proof of a theorem proved directly will always look something like this.

Theorem 2.1.8 (Sample). *If p, then q.*

Proof: Suppose p. Then, ... Thus q. ∎

Writing a proof of the theorem is merely the construction of a logical bridge from the hypothesis to the conclusion. In the example theorems that follow, we will dissect every statement and present every detail as completely as possible before we present a proof. This will help us make sure we understand all the logical internals. We'll then write the proof as a cleaned-up presentation of the argument for $p \to q$. As we get more practice, we'll concentrate more on the actual writing of the proof, leaving the logical skeleton concealed. For example, in Theorems 2.1.9 to 2.1.11, we use the following statement definitions:

$$p : x \in A \tag{2.8}$$

$$q : x \in B. \tag{2.9}$$

Theorem 2.1.9. *If $A \subseteq B$, then $A \cap B = A$.*

Here are some preliminary thoughts before we present a proof:

1. In some set proofs, you might want to draw a Venn diagram to convince yourself that the theorem is true. If the Venn diagram is drawn so that A is sketched to lie completely inside B, then the result is apparent. However, the Venn diagram does

not constitute a proof. We have a long way to go before we earn the right to say "Proof by picture."

2. How do we state the hypothesis condition of Theorem 2.1.9 in terms of p and q in the preceding?[1] Also, how do we state the conclusion $A \cap B = A$ in the same terms?[2] The answers:

$$\overbrace{(p \to q)}^{\text{hypothesis}} \to \overbrace{[(p \wedge q) \to p] \wedge [p \to (p \wedge q)]}^{\text{conclusion}}. \tag{2.10}$$

Take a moment to construct a truth table for statement (2.10) (see Exercise 2) and you'll see that it is a tautology. Arguably, this exercise constitutes a proof, but we want more than a proof in the formal language of logic. We want well-written prose.

3. The statement that we're trying to prove in this theorem is an equation involving sets: $A \cap B = A$. This set equality is itself a compound statement about subset inclusion: $A \cap B \subseteq A$ and $A \subseteq A \cap B$. Thus writing our proof will be a two-step task. Whenever we show that one set is a subset of another, say $A \cap B \subseteq A$, we're proving an if-then statement that says if x is an element of one set ($A \cap B$), then x is an element of another set (A). Writing a *subset inclusion* proof of this sort would therefore read something like "Suppose $x \in A \cap B$. Then .... Thus $x \in A$." It's common to see proofs of this sort written using an expression such as "Pick $x \in A \cap B$," which has the imagery of reaching into $A \cap B$ and grabbing an arbitrary element, which we call x. The only thing we know about x is that it's an element of the set where we picked it, but somehow we must be able to conclude that x is an element of A, the set on the other side of the $\subseteq$ symbol. Proofs of this sort, which Theorems 2.1.9 and 2.1.10 will illustrate, are often called *element-chasing* proofs. They show that two sets are equal by chasing an arbitrarily chosen element from one side to the other, then back.

4. A proof is like a road map. The person sketching the map will provide more or less detail depending on the traveler's familiarity with the territory. Because we are just getting started with proofs and want to write clearly and completely, our proofs (especially at first) will often contain the inserted statement "We need to show that ... ," so that the reader can always see where we're going. Even for the most sophisticated mathematical audience, it's not uncommon to have such a phrase included in proofs that are lengthy or complicated. You might want to make occasional use of this phrase at first.

Proof: Suppose $A \subseteq B$. To show $A \cap B = A$, we must apply Definition 2.1.2 and show that $A \cap B \subseteq A$ and $A \subseteq A \cap B$.

($A \cap B \subseteq A$): Pick $x \in A \cap B$. Then $x \in A$ and $x \in B$. Since $x \in A$, we have that $A \cap B \subseteq A$.

($A \subseteq A \cap B$): Pick $x \in A$. Then, because $A \subseteq B$, we have that $x \in B$. Therefore, since $x \in A$ and $x \in B$, it follows that $x \in A \cap B$. Thus $A \subseteq A \cap B$.

---

[1] See Definition 2.1.1.
[2] See Definitions 2.1.2 and 2.1.3.

Since we have shown that $A \cap B \subseteq A$ and $A \subseteq A \cap B$, we have demonstrated that $A \cap B = A$. ∎

The proof of Theorem 2.1.9 is painfully but intentionally wordy so that you can see how its structure derives from the definitions of the terms it uses. We'll quickly dispense with the frequent use of such phrases as "We need to show," provided the previous definitions and theorems should make it obvious what we need to say next. Assuming the reader knows what is to be shown, we have the freedom to be more succinct. Thus a proof of Theorem 2.1.9 would read much better if written in the following way.

Proof (Succinct): Suppose $A \subseteq B$, and pick $x \in A \cap B$. Then $x \in A$, so that $A \cap B \subseteq A$. Now pick $x \in A$. Then since $A \subseteq B$, it is also true that $x \in B$, so that $x \in A \cap B$. Thus $A \subseteq A \cap B$, so that $A \cap B = A$. ∎

Theorem 2.1.10 (DeMorgan's Law). *If A and B are sets, then*

$$(A \cap B)' = A' \cup B'. \tag{2.11}$$

Here are some preliminary thoughts about the logical structure of the theorem.

1. The apparent if-then structure of the theorem includes the hypothesis condition "If A and B are sets." You might argue that the theorem could have been more simply stated as $(A \cap B)' = A' \cup B'$, if the reader only knew that the context of the theorem is sets. Fair enough. In logically analyzing this theorem, let's strip away the phrase "If A and B are sets," and let the theorem simply say $(A \cap B)' = A' \cup B'$. By peeling the phrase away, we have revealed an if-and-only-if statement: $x \in (A \cap B)'$ iff $x \in A' \cup B'$.

2. Using p and q as defined in statement (2.8), this theorem looks a lot like DeMorgan's law for statements:

$$\neg(p \wedge q) \leftrightarrow (\neg p \vee \neg q). \tag{2.12}$$

In fact, restating Theorem 2.1.10 in terms of the p and q in statement (2.8), Theorem 2.1.10 becomes statement (2.12). In Exercise 4 from Section 1.1, you showed that statement (2.12) is a tautology. Thus, in a sense, Theorem 2.1.10 is proved. Still, we write it out in prose.

3. The proof of this theorem will reveal a fork in the road. There will come a point where we are given two possible cases to consider. We must follow each case to the end to show that they both will allow us to arrive at the conclusion.

Proof: We show that $(A \cap B)' \subseteq A' \cup B'$ and $(A \cap B)' \supseteq A' \cup B'$.

($\subseteq$): Pick $x \in (A \cap B)'$. Then $x \notin A \cap B$, so that either $x \notin A$ or $x \notin B$. We consider each case.

(Case $x \notin A$): If $x \notin A$, then $x \in A'$. Thus $x \in A' \cup B'$.
(Case $x \notin B$): If $x \notin B$, then $x \in B'$. Thus $x \in A' \cup B'$.

In either case, we have that $x \in A' \cup B'$, so that $(A \cap B)' \subseteq A' \cup B'$.

($\supseteq$): Pick $x \in A' \cup B'$. Then either $x \in A'$ or $x \in B'$. We consider each case.

(Case $x \in A'$): If $x \in A'$, then $x \notin A$. But if $x \notin A$, then certainly $x \notin A \cap B$. Thus $x \in (A \cap B)'$.

(Case $x \in B'$): If $x \in B'$, then $x \notin B$. But if $x \notin B$, then certainly $x \notin A \cap B$. Thus $x \in (A \cap B)'$.

In either case, we have that $x \in (A \cap B)'$, so that $A' \cup B' \subseteq (A \cap B)'$.

Since we have shown $(A \cap B)' \subseteq A' \cup B'$ and $(A \cap B)' \supseteq A' \cup B'$, we have that $(A \cap B)' = A' \cup B'$. ∎

The fact that the empty set contains no elements can make for some interesting twists in proofs.

Theorem 2.1.11. *For any set A, $A \cup \emptyset = A$.*

Proof: We show $A \cup \emptyset \subseteq A$ and $A \cup \emptyset \supseteq A$.

($\subseteq$): Pick $x \in A \cup \emptyset$. Then either $x \in A$, or $x \in \emptyset$. But since $\emptyset$ contains no elements, it must be that $x \in A$. Thus $A \cup \emptyset \subseteq A$.

($\supseteq$): Clearly, if $x \in A$, then $x \in A \cup \emptyset$. Thus $A \subseteq A \cup \emptyset$.

Therefore, $A \cup \emptyset = A$. ∎

2.1.3 Proofs by contrapositive

In Section 1.2.2, we showed that $p \rightarrow q$ is logically equivalent to $\neg q \rightarrow \neg p$. This suggests that a proof of a theorem of the form $p \rightarrow q$ might be approached contrapositively by showing $\neg q \rightarrow \neg p$ instead. To prove the latter is equivalent to proving the former. In general, a proof by contraposition will go something like this.

Theorem 2.1.12 (Sample). *If p, then q.*

Proof: Suppose $\neg q$. Then, ... Thus $\neg p$. ∎

Whether it's better to attack a theorem directly or contrapositively comes with experience. In proving the following theorem contrapositively, we illustrate how natural it can be to show as a conclusion that something is nonempty, rather than supposing it is empty as a hypothesis condition. To see the reasonableness of Theorem 2.1.13, draw a Venn diagram first.

Theorem 2.1.13. *If $A \subseteq B$, then $A \setminus B = \emptyset$.*

Proof: Suppose $A \setminus B \neq \emptyset$. Then there exists $x \in (A \setminus B)$. Thus $x \in A \cap B'$, so that $x \in A$ and $x \in B'$. But the existence of such an x is precisely the denial of Definition 2.1.1, so that $A \not\subseteq B$. ∎

2.1.4 Proofs by contradiction

Both direct proofs and contrapositive proofs are effectively an effort to show that $p \to q$ is a tautology. Sometimes, however, it's easier to prove $p \to q$ is a tautology by showing $\neg(p \to q)$ is a contradiction. In order to construct a proof by contradiction, we first have to recall the negation of $p \to q$. It's also helpful to remember that an if-then statement might sometimes be better stated using the universal quantifier $\forall$. So, what is another way to write $\neg(p \to q)$? Answer: $p \land \neg q$. Writing a proof by contradiction consists of showing that $p \land \neg q$ is impossible. The following example theorems will all take this form:

Theorem 2.1.14 (Sample). *If p, then q.*

Proof: Suppose p and $\neg q$. Then .... This is a contradiction. Thus $p \to q$. ∎

Theorem 2.1.15. *If A is any set, $\emptyset \subseteq A$.*

Proof: Suppose there exists a set A such that $\emptyset \not\subseteq A$. Then there exists $x \in \emptyset$ such that $x \notin A$. But $\emptyset$ contains no elements. This contradicts the definition of $\emptyset$. Thus $\emptyset \subseteq A$. ∎

2.1.5 Disproving a statement

Not only is it important to be able to prove the truth of statements, we sometimes find ourselves needing to demonstrate that a certain statement is not true. To disprove a statement is simply to prove its negation. Here's an example of a statement that at first glance you might think is true but is not. Think about it before you check the solution, and see if you can demonstrate on your own that it is false.

Example 2.1.16. If A and B are sets such that $A \subseteq B$, then A and B are not disjoint.

Solution: Logically this false statements says

$$(\forall A)(\forall B)[(A \subseteq B) \to (A \cap B \neq \emptyset)]. \qquad (2.13)$$

What is the negation of this statement? It is

$$(\exists A)(\exists B)[(A \subseteq B) \land (A \cap B = \emptyset)]. \qquad (2.14)$$

Disproving statement (2.13) is the same as proving statement (2.14), which means our task is to demonstrate the existence of disjoint sets A and B, where $A \subseteq B$. We must create them. So how about letting $A = \emptyset$ and $B = \{1\}$. Clearly, $A \subseteq B$ and $A \cap B = \emptyset$. ∎

Example 2.1.16 illustrates a very common occurrence in mathematics. Sometimes we might be tempted to believe that a certain statement is true, when in actuality it is not. If such a statement is a universal one as in Example 2.1.16, then to disprove the statement involves demonstrating the existence of what is called a *counterexample*.

EXERCISES

1. Construct the negation of each statement below, as was done in Example 2.1.7.

    (a) $x \in A \cup B$
    (b) A and B are disjoint.
    (c) $A = B$
    (d) $x \in A \backslash B$
    (e) $x \in A \triangle B$
    (f) $A \subseteq B \cup C$
    (g) $A \cup B \subseteq C \cap D$
    (h) If $A \cup C \subseteq B \cup C$, then $A \subseteq B$.

2. Construct a truth table for statement (2.10) on page 42 to verify it is a tautology.

3. Show that the converse of statement (2.10) is also a tautology. State the converse of Theorem 2.1.9 and prove it.

4. Prove the following.

    (a) If $A \subseteq B$, then $A \cup B = B$.
    (b) If $A \subseteq B$ and $B \subseteq C$, then $A \subseteq C$.
    (c) If $A \subseteq B$, then $A' \supseteq B'$.
    (d) (DeMorgan's Law) $(A \cup B)' = A' \cap B'$.
    (e) If $A \cap B = \emptyset$, then $A \triangle B = A \cup B$.

5. Suppose A, B, and C are sets. Consider the following statements:

$$A = B \Leftrightarrow A \cup C = B \cup C \qquad (2.15)$$

$$A = B \Leftrightarrow A \cap C = B \cap C \qquad (2.16)$$

    For each of Eqs. (2.15) and (2.16), one direction of the implication $\Leftrightarrow$ is true and one is false. Prove the direction that is true, and provide a counterexample for the direction that is false.

6. Prove that $\cap$ distributes over $\cup$ and vice versa.

7. If X and Y are disjoint sets, we sometimes write $X \cup Y$ as $X \dot{\cup} Y$. This is a way of talking about the set $X \cup Y$ by tagging it with a little symbol (the dot) that tells the reader the additional information that X and Y are disjoint. So if someone makes a statement like

$$A \cup B = A \dot{\cup} (B \backslash A) \qquad (2.17)$$

    what he is really saying is the compound statement

$$A \cup B = A \cup (B \backslash A) \quad \text{and} \quad A \cap (B \backslash A) = \emptyset. \qquad (2.18)$$

    Prove Eq. (2.17) by showing both parts of Eq. (2.18).

2.2 Indexed Families of Sets

If you're working with no more than three sets at a time as we did in Section 2.1, it's probably sufficient to use A, B, and C to represent them. If you have a set of, say, 10 sets (generally called a *family* or *collection* of sets instead of a set of sets), it might be more sensible to put them into a family and address them as $A_1, A_2, \ldots, A_{10}$. In a case like this, we would say that the set $\{1, 2, 3, \ldots, 10\}$ *indexes* the family of sets. If we write

$$\mathbb{N}_n = \{1, 2, 3, \ldots, n\},$$

then we could write a family of n sets as

$$\{A_1, A_2, A_3, \ldots, A_n\} = \{A_k : k \in \mathbb{N}_n\} = \{A_k\}_{k=1}^n,$$

and we would say that $\mathbb{N}_n$ is an *index set* for the family of sets. This notation has advantages, for then we could write unions and intersections more succinctly:

$$A_1 \cup A_2 \cup A_3 \cup \cdots \cup A_n = \bigcup_{k=1}^n A_k \tag{2.19}$$

$$A_1 \cap A_2 \cap A_3 \cap \cdots \cap A_n = \bigcap_{k=1}^n A_k. \tag{2.20}$$

In Definitions 2.2.3 and 2.2.4, we'll define precisely what we mean by this sort of union and intersection.

We can go even further. It is conceivable we might need to work with infinitely many sets $\{A_1, A_2, A_3, \ldots\}$ that we might want to index with $\mathbb{N}$. For example, if we use the familiar interval notation

$$[a, b] = \{x : x \in \mathbb{R} \text{ and } a \leq x \leq b\},$$

we might talk about the family of intervals $\mathcal{F} = \{A_n\}_{n \in \mathbb{N}}$, where $A_n = [0, 1/n]$. To form the union or intersection of a family of sets indexed by $\mathbb{N}$, we could use notation like that in the preceding:

$$\bigcup_{n=1}^\infty A_n \quad \text{and} \quad \bigcap_{n=1}^\infty A_n, \tag{2.21}$$

or we could write something like

$$\bigcup_{n \in \mathbb{N}} A_n \quad \text{and} \quad \bigcap_{n \in \mathbb{N}} A_n, \tag{2.22}$$

where by Eq. (2.22) we understand that n is allowed to take on all values of the indexing set $\mathbb{N}$.

The notation in Eq. (2.22) is handy when the indexing set is more complicated than $\mathbb{N}$ and does not allow us to think of some index variable n starting at 1 and progressing sequentially off to infinity. For it's conceivable that any set $\mathcal{A}$ can index a family of sets. We can then address individual sets in the family as A_α, where $\alpha \in \mathcal{A}$, and denote the family $\mathcal{F} = \{A_\alpha\}_{\alpha \in \mathcal{A}}$. Union and intersection could then be written as

$$\bigcup_{\alpha \in \mathcal{A}} A_\alpha \quad \text{and} \quad \bigcap_{\alpha \in \mathcal{A}} A_\alpha.$$

Example 2.2.1. One important contribution of the German mathematician Richard Dedekind (1831–1916) is a rigorous foundation of the set of real numbers. He employed what is now called a *Dedekind cut*, whereby the set of rational numbers is "cut" into two pieces. Specifically, for some $r \in \mathbb{R}$, Dedekind worked with sets of the form

$$A_r = \{x : x \in \mathbb{Q} \text{ and } x < r\}, \tag{2.23}$$

$$B_r = \{x : x \in \mathbb{Q} \text{ and } x > r\}. \tag{2.24}$$

For example, although $\sqrt{2}$ is irrational (as you'll see in Theorem 2.8.3), it certainly makes sense to talk about $A_{\sqrt{2}}$, the set of all rational numbers less than $\sqrt{2}$. The set $\mathbb{R}$ is the index set for the families $\{A_r\}_{r \in \mathbb{R}}$ and $\{B_r\}_{r \in \mathbb{R}}$.

Example 2.2.2. Consider the family of intervals

$$\mathcal{F} = \{[-r, r] : r \in \mathbb{R}^+\}, \tag{2.25}$$

where $\mathbb{R}^+$ denotes the positive real numbers. The index set is $\mathbb{R}^+$, and we might want to use the notation $I_r = [-r, r]$ to represent one interval in the family. We might also write $\mathcal{F} = \{I_r\}_{r \in \mathbb{R}^+}$.

Of course, we don't have to worry about having our family of sets so well organized as to be indexed by any set at all. In the same way that we talk about a set A and address an arbitrary element $x \in A$, we can call our family of sets $\mathcal{F}$ and address an arbitrary set in the family as $A \in \mathcal{F}$. Then the union and intersection of the sets in $\mathcal{F}$ could be written as

$$\bigcup_{A \in \mathcal{F}} A \quad \text{and} \quad \bigcap_{A \in \mathcal{F}} A, \tag{2.26}$$

or more simply as

$$\bigcup_{\mathcal{F}} A \quad \text{and} \quad \bigcap_{\mathcal{F}} A. \tag{2.27}$$

The statements in Eqs. (2.26) and (2.27) assume the least structure on the family of sets, so they are a good general notation.

Before we really start digging into families of sets, we're going to set up a particular example, a sort of scenario really, that will help carry us through the complexities of the ideas and notation. Families of sets tend to confuse students at first, and for several reasons. First, when we talk about a family, we're actually talking about a set whose elements are sets. Thus it might be that $x \in A$ and $A \in \mathcal{F}$, but writing $x \in \mathcal{F}$ is wrong, for x is not an element of $\mathcal{F}$, but an element of an element of $\mathcal{F}$. Second, when a family is indexed, the indexing set $\mathcal{A}$ is yet another set to contend with, a set whose elements α serve as something like name tags by which sets in the family are addressed. Therefore, to try to make this as clear as we can, we're going to stage a little production whose cast of characters all represent the set family terms. Then, as we state theorems, the scenario will help us understand what's being said and how to approach the proof.

Let's suppose we're studying the mathematics section of the catalog of Prestigious University. Every mathematics class has a number, so let

$$\mathcal{A} = \{104, 210, 211, 212, 330, 360, 430, 431, 510, 531\}, \qquad (2.28)$$

and think of $\mathcal{A}$ as the set of all the course numbers of mathematics courses offered at PU. This set of course numbers $\mathcal{A}$ will be our index set. We'll address an arbitrarily chosen course number as $\alpha \in \mathcal{A}$. Let A_α be the roster of all PU graduates who passed course number α while they were students at PU. We'll address an arbitrarily chosen student on roster A_α as x. With these role assignments, $\mathcal{F} = \{A_\alpha\}_{\alpha \in \mathcal{A}}$ is the set of all these rosters of students who passed the individual courses. That is,

$$\mathcal{F} = \{A_{104}, A_{210}, A_{211}, \ldots, A_{531}\}. \qquad (2.29)$$

Writing the family of rosters as $\mathcal{F} = \{A_\alpha\}_{\alpha \in \mathcal{A}}$ references each roster in the family in terms of the course number that tags it. Writing $\mathcal{F}$ simply as $\{A\}$ dispenses with the number tags and addresses a particular roster in $\mathcal{F}$ simply as A.

Certainly the family of sets $\mathcal{F}$ and its indexing set $\mathcal{A}$ need not look anything like the ones we've created here. But thinking of $\mathcal{A}$ as in Eq. (2.28) and $\mathcal{F}$ as in Eq. (2.29) should not limit us or make our proofs less than generally applicable if we use the analogy as a tool to help clarify our thinking. Just remember that we'll address sets in the family in one of two ways. To illustrate the first way, sometimes we'll pick some arbitrary $\alpha \in \mathcal{A}$ in order to talk about A_α, which is like choosing an arbitrary course number and talking about the roster for the course with that number. Or perhaps we'll claim the existence of some specific $\alpha_0 \in \mathcal{A}$ in order to talk about A_{α_0}. In our analogy, this is the claim that there is a certain course number whose roster has some property. To illustrate the second way of addressing sets in the family, sometimes we might pick an arbitrary $A \in \mathcal{F}$, which is like choosing an arbitrary mathematics roster without any specific reference to a course number. Or perhaps we'll claim the existence of some specific $A_0 \in \mathcal{F}$. This is the claim that there is a particular mathematics course roster that has a certain property, without making any reference to that course's number.

Having created this little scenario as an aid to our understanding, let's define more terms and derive some results, using the scenario to motivate and get us over some humps. First, how should we define the sets $\cup_{A \in \mathcal{F}} A$ and $\cap_{A \in \mathcal{F}} A$? Or, equivalently, how should we define the following statements?[3]

$$x \in \bigcup_\mathcal{F} A \quad \text{and} \quad x \in \bigcap_\mathcal{F} A. \qquad (2.30)$$

In effect, we want $\cup_{A \in \mathcal{F}} A$ to be the set you get when you take each A in $\mathcal{F}$ and dump all its elements into a single set. For our scenario, $\cup_{A \in \mathcal{F}} A$, or $\cup_{\alpha \in \mathcal{A}} A_\alpha$, however you choose to write it, is the single roster of graduates created by unioning all the individual mathematics class rosters. Maybe you can see that $\cup_\mathcal{F} A$ consists of all graduates who have ever passed a mathematics class at PU. So if x represents a graduate, saying $x \in \cup_\mathcal{F} A$ means that there is some mathematics course at PU that x passed. We can write this in two ways:

$$x \in \bigcup_{A \in \mathcal{F}} A \Leftrightarrow x \in \bigcup_{\alpha \in \mathcal{A}} A_\alpha \Leftrightarrow (\exists A \in \mathcal{F})(x \in A) \qquad (2.31)$$

$$\Leftrightarrow (\exists \alpha \in \mathcal{A})(x \in A_\alpha). \qquad (2.32)$$

---

[3] Think in terms of $\exists$ for the union and $\forall$ for the intersection.

The form of Eq. (2.31) says simply that there is a mathematics course roster on which the name of student x appears, while Eq. (2.32) says there exists a course number $\alpha \in \mathcal{A}$ such that x is on the roster of mathematics course number α. With this, we arrive at the following definition.

Definition 2.2.3. Let $\mathcal{F}$ be a family of sets. Then the *union over $\mathcal{F}$* is defined by

$$\bigcup_{A \in \mathcal{F}} A = \{x : (\exists A \in \mathcal{F})(x \in A)\}. \tag{2.33}$$

If $\mathcal{F}$ is indexed by $\mathcal{A}$, this becomes

$$\bigcup_{A \in \mathcal{F}} A = \bigcup_{\alpha \in \mathcal{A}} A_\alpha = \{x : (\exists \alpha \in \mathcal{A})(x \in A_\alpha)\}. \tag{2.34}$$

Now what does $\bigcap_{A \in \mathcal{F}} A$ (or $\bigcap_{\alpha \in \mathcal{A}} A_\alpha$) correspond to in our scenario? This is the intersection of all rosters, so it's the list of all graduates who passed all mathematics courses at PU. So what does it mean mathematically to say $x \in \bigcap_{\mathcal{F}} A$?

$$x \in \bigcap_{A \in \mathcal{F}} A \Leftrightarrow x \in \bigcap_{\alpha \in \mathcal{A}} A_\alpha \Leftrightarrow (\forall A \in \mathcal{F})(x \in A) \tag{2.35}$$

$$\Leftrightarrow (\forall \alpha \in \mathcal{A})(x \in A_\alpha). \tag{2.36}$$

With this we arrive at the following definition.

Definition 2.2.4. Let $\mathcal{F}$ be a family of sets. Then the *intersection over $\mathcal{F}$* is defined by

$$\bigcap_{A \in \mathcal{F}} A = \{x : (\forall A \in \mathcal{F})(x \in A)\}. \tag{2.37}$$

If $\mathcal{F}$ is indexed by $\mathcal{A}$, this becomes

$$\bigcap_{A \in \mathcal{F}} A = \bigcap_{\alpha \in \mathcal{A}} A_\alpha = \{x : (\forall \alpha \in \mathcal{A})(x \in A_\alpha)\}. \tag{2.38}$$

A lot of the results we proved in Section 2.1 for a family of two or three sets carry over to analogous results for a family of sets of any size. Here are some examples and theorems. Do yourself a favor and try to work them on your own first.

Example 2.2.5. Construct the negation of the statement: $x \in \bigcup_{\mathcal{F}} A$.

Solution: Looking at Eq. (2.32), the statement $x \notin \bigcup_{\mathcal{F}} A$ means

$$(\forall A \in \mathcal{F})(x \notin A).$$

If $\mathcal{F}$ is indexed by $\mathcal{A}$, we use Eq. (2.32) to have

$$(\forall \alpha \in \mathcal{A})(x \notin A_\alpha). \qquad \blacksquare$$

In our mathematics class scenario, what does $x \notin \bigcup_{\mathcal{F}} A$ mean? Graduate x is not on the universal roster of mathematics classes, so x never passed a mathematics class at PU.

That is, for every roster $A \in \mathcal{F}$, x is not on roster A, or, for every course number $\alpha \in \mathcal{A}$, x did not pass the mathematics course numbered α.

Theorem 2.2.6 (DeMorgan's Law). *Suppose $\mathcal{F}$ is a family of sets. Then*

$$\left[\bigcup_{\mathcal{F}} A\right]' = \bigcap_{\mathcal{F}} A'. \tag{2.39}$$

Proof: We prove by mutual subset inclusion.

($\subseteq$): Pick $x \in [\bigcup_{\mathcal{F}} A]'$. Then $x \notin \bigcup_{\mathcal{F}} A$. Therefore, for all $A \in \mathcal{F}$, $x \notin A$. But then $x \in A'$ for every $A \in \mathcal{F}$. Thus, $x \in \bigcap_{\mathcal{F}} A'$.

($\supseteq$): Pick $x \in \bigcap_{\mathcal{F}} A'$. Then $x \in A'$ for every $A \in \mathcal{F}$. Thus, $x \notin A$ for every $A \in \mathcal{F}$, so that $x \notin \bigcup_{\mathcal{F}} A$. Therefore, $x \in [\bigcup_{\mathcal{F}} A]'$. ∎

To see what Theorem 2.2.6 says in our scenario, we need a universal set U in which A' is meaningful. Let U be the roster of all PU graduates. With that, what set is being talked about in Theorem 2.2.6, and what are the two ways of constructing it in Eq. (2.39)? To construct the left-hand side of Eq. (2.39), we first combine all the mathematics rosters into a single roster, then take all the PU graduates *except* these. This is the list of all PU graduates who avoided mathematics altogether while they were at PU. How do we construct the right-hand side of Eq. (2.39)? First, we take each mathematics class roster and consider the complement. For example, A'_{211} is the list of all PU graduates who did not pass Mathematics 211. By taking the intersection of all these complements, we arrive at the list of PU graduates who avoided Math 104 *and* Math 210 *and* ... *and* Math 531; that is, the list of graduates who avoided mathematics altogether.

You'll prove the other form of DeMorgan's law in the following in Exercise 2.

Theorem 2.2.7. *Suppose $\mathcal{F}$ is a family of sets. Then*

$$\left[\bigcap_{\mathcal{F}} A\right]' = \bigcup_{\mathcal{F}} A'. \tag{2.40}$$

For the next theorem, we'll prove part 2 here and leave part 1 to you in Exercise 3.

Theorem 2.2.8. *Let $\mathcal{F}$ be a family of sets indexed by $\mathcal{A}$, and suppose $\mathcal{B} \subseteq \mathcal{A}$. Then,*

1. $\bigcup_{\beta \in \mathcal{B}} A_\beta \subseteq \bigcup_{\alpha \in \mathcal{A}} A_\alpha$,
2. $\bigcap_{\beta \in \mathcal{B}} A_\beta \supseteq \bigcap_{\alpha \in \mathcal{A}} A_\alpha$.

Before we prove part 2 of Theorem 2.2.8, let us see how it relates to our scenario. Since $\mathcal{A}$ is the set of all mathematics course numbers at PU, it might work to think of $\mathcal{B}$ as the set of all course numbers of lower-level mathematics courses at PU. Thus

$$\mathcal{B} = \{104, 210, 211, 212\}. \tag{2.41}$$

With that, what does part 2 of Theorem 2.2.8 say? The set construction on the left-hand side is the intersection of the class rosters across all the lower-level courses, while the right-hand side is the intersection of the rosters of all the mathematics classes.

Therefore, if we pick some $x \in \bigcap_{\alpha \in \mathcal{A}} A_\alpha$, then x is a PU graduate who passed every mathematics course offered. Thus, certainly x passed all the lower-level courses. Here's the proof.

> *Proof:* Pick $x \in \bigcap_{\alpha \in \mathcal{A}} A_\alpha$. We must show that $x \in A_\beta$ for all $\beta \in \mathcal{B}$, so pick $\beta \in \mathcal{B}$. Since $\mathcal{B} \subseteq \mathcal{A}$, it follows that $\beta \in \mathcal{A}$. Therefore, $x \in A_\beta$, because $x \in A_\alpha$ for any $\alpha \in \mathcal{A}$. Since β was chosen arbitrarily, we have shown that $x \in A_\beta$ for all $\beta \in \mathcal{B}$, so that $x \in \bigcap_{\beta \in \mathcal{B}} A_\beta$. ∎

Notice how we showed $x \in \bigcap_{\beta \in \mathcal{B}} A_\beta$ by picking an arbitrary $\beta \in \mathcal{B}$ and showing that $x \in A_\beta$. This shows that $x \in A_\beta$ for all $\beta \in \mathcal{B}$, so that $x \in \bigcap_{\beta \in \mathcal{B}} A_\beta$.

We can write Theorem 2.2.8 in a slightly different form if the family of sets is not indexed. If $\mathcal{F}_1$ is a family of sets and $\mathcal{F}_2 \subseteq \mathcal{F}_1$, we call $\mathcal{F}_2$ a *subfamily* of $\mathcal{F}_1$. Since $\mathcal{B} \subseteq \mathcal{A}$ in Theorem 2.2.8, $\{A_\beta\}_{\beta \in \mathcal{B}}$ is a subfamily of $\{A_\alpha\}_{\alpha \in \mathcal{A}}$. Swapping the notation in Theorem 2.2.8 for an arbitrary family $\mathcal{F}_1$ and a subfamily $\mathcal{F}_2$, we have the following.

Theorem 2.2.9. *Suppose $\mathcal{F}_1$ is a family of sets, and $\mathcal{F}_2$ is a subfamily of $\mathcal{F}_1$. Then,*

1. $\bigcup_{\mathcal{F}_2} A \subseteq \bigcup_{\mathcal{F}_1} A$,
2. $\bigcap_{\mathcal{F}_2} A \supseteq \bigcap_{\mathcal{F}_1} A$.

The next theorem involves two families of sets, both indexed by $\mathcal{A}$, where corresponding sets in the two families are related by subset inclusion. To understand what it's saying, think of B_α as the set of all male PU graduates who passed mathematics course number α. You'll prove both parts in Exercise 4.

Theorem 2.2.10. *Suppose $\mathcal{F}_1 = \{A_\alpha\}_{\alpha \in \mathcal{A}}$ and $\mathcal{F}_2 = \{B_\alpha\}_{\alpha \in \mathcal{A}}$ are two families of sets with the property that $B_\alpha \subseteq A_\alpha$ for every $\alpha \in \mathcal{A}$. Then*

1. $\bigcup_{\alpha \in \mathcal{A}} B_\alpha \subseteq \bigcup_{\alpha \in \mathcal{A}} A_\alpha$
2. $\bigcap_{\alpha \in \mathcal{A}} B_\alpha \subseteq \bigcap_{\alpha \in \mathcal{A}} A_\alpha$

EXERCISES

1. Construct the negation of the statement $x \in \bigcap_{\mathcal{F}} A$.

2. Prove Theorem 2.2.7: Suppose $\mathcal{F}$ is a family of sets. Then

$$\left[\bigcap_{\mathcal{F}} A\right]' = \bigcup_{\mathcal{F}} A'. \tag{2.42}$$

3. Prove part 1 of Theorem 2.2.8: Let $\mathcal{F}$ be a family of sets indexed by $\mathcal{A}$, and suppose $\mathcal{B} \subseteq \mathcal{A}$. Then, $\bigcup_{\beta \in \mathcal{B}} A_\beta \subseteq \bigcup_{\alpha \in \mathcal{A}} A_\alpha$.

4. Prove Theorem 2.2.10: Suppose $\mathcal{F}_1 = \{A_\alpha\}_{\alpha \in \mathcal{A}}$ and $\mathcal{F}_2 = \{B_\alpha\}_{\alpha \in \mathcal{A}}$ are two families of sets with the property that $B_\alpha \subseteq A_\alpha$ for every $\alpha \in \mathcal{A}$. Then

(a) $\bigcup_{\alpha \in A} B_\alpha \subseteq \bigcup_{\alpha \in A} A_\alpha$
(b) $\bigcap_{\alpha \in A} B_\alpha \subseteq \bigcap_{\alpha \in A} A_\alpha$

5. Suppose $\mathcal{F} = \{A\}$ is a family of sets, and suppose C is a set for which $A \subseteq C$ for every $A \in \mathcal{F}$. Show that $\cup_\mathcal{F} A \subseteq C$.[4]

6. Suppose $\mathcal{F} = \{A\}$ is a family of sets, and suppose D is a set for which $D \subseteq A$ for every $A \in \mathcal{F}$. Show that $D \subseteq \cap_\mathcal{F} A$.[5]

2.3 Algebraic and Ordering Properties of $\mathbb{R}$

In this section we turn our attention to proofs of basic algebraic and ordering properties of the real numbers. You might want to take a look at assumptions A1–A22 in Chapter 0 before you continue reading. It is *only* these properties of $\mathbb{R}$ that we can use for free. Anything else, we will have to justify from these assumptions. In this section, we'll make use of assumptions A1–A18 to derive basic and familiar algebraic and ordering properties of the real numbers. The theorems and examples that follow are designed to do two things. First, they will give you a feel for how to write proofs of this sort. The proofs that follow may be a little wordy, but the excessive explanation will probably help you as you begin to write these kinds of proofs. Second, they will serve as assumptions for the exercises and theorems from later sections. Once a theorem is proved, then you are certainly free to use it later, with a proper reference, of course.

2.3.1 Basic algebraic properties of real numbers

Theorem 2.3.1 (Cancellation of addition). *For all $a, b, c \in \mathbb{R}$, if $a + c = b + c$, then $a = b$.*

One of the most basic of the algebraic properties of $\mathbb{R}$, we would like to know that we are free to take a given equation $a + c = b + c$, and, as it were, cancel out the $+c$ from both sides. Notice the theorem is a statement of the form $p \to q$, where p is the statement $a + c = b + c$, and q is the statement $a = b$.

We're going to present two proofs of Theorem 2.3.1, written in two somewhat different styles. The first proof reveals the thought process involved in constructing the proof, but uses several disjoint equations. The second shows how you might clean up these equations to create one extended equation that is the result we want.

Proof 1: Suppose $a + c = b + c$. By property A7, there exists $-c \in \mathbb{R}$ such that $c + (-c) = 0$. Since addition is well-defined (A2), we have that $a + c + (-c) = b + c + (-c)$, which yields $a + 0 = b + 0$, or $a = b$. ∎

---

[4] Think of C as the set of all PU graduates who ever enrolled in a mathematics class.
[5] Think of D as the set of all PU mathematics majors (whom we'll assume would have taken every mathematics course) who graduated with a 4.00 GPA.

Proof 2: Suppose $a + c = b + c$. By property A7, there exists $-c \in \mathbb{R}$ such that $c + (-c) = 0$. We use this together with other properties from A1–A22 to have that

$$a \stackrel{(A6)}{=} a + 0 \stackrel{(A2)}{=} a + [c + (-c)] \stackrel{(A4)}{=} (a + c) + (-c) \\ \stackrel{(A2)}{=} (b + c) + (-c) \stackrel{(A4)}{=} b + [c + (-c)] \stackrel{(A2)}{=} b + 0 \stackrel{(A6)}{=} b. \tag{2.43}$$

By assumption A1 (the transitive property of equality), $a = b$. ∎

Theorem 2.3.1 becomes useful immediately in the following.

Theorem 2.3.2. *For every* $a \in \mathbb{R}$, $a \cdot 0 = 0$.

Proof: Pick $a \in \mathbb{R}$. By properties A6 and A14, we have that

$$0 + a \cdot 0 = a \cdot 0 = a \cdot (0 + 0) = a \cdot 0 + a \cdot 0. \tag{2.44}$$

By Theorem 2.3.1, we may cancel $a \cdot 0$ from both sides of Eq. (2.44) to have $a \cdot 0 = 0$. ∎

We'll eventually get a little lazy about referencing the properties of $\mathbb{R}$ that we need to use. At first, however, we need to make sure for our own sakes that we understand precisely which ones we use and when we use them.

In Section 1.3.3, we discussed unique existence. The next theorem is our first uniqueness proof. property A7 guarantees that every $a \in \mathbb{R}$ has *an* additive inverse $-a$ for which $a + -a = 0$. This is an axiom. But nothing about the axioms A1–A22 precludes (at first glance) that a real number might have more than one additive inverse. The next theorem shows that, in fact, $a \in \mathbb{R}$ has a unique additive inverse. Once we prove this, we will then be able to talk about *the* additive inverse of a. Remember how we show that something is unique. Suppose there are two things that both have the property in question, and show that they must, in actuality, be the same.

Theorem 2.3.3. *The additive inverse of a real number is unique.*

Proof: Pick $a \in \mathbb{R}$. Suppose $b, c \in \mathbb{R}$ are both additive inverses of a. Then $a + b = 0$ and $a + c = 0$. We must show that $b = c$. Now since $a + b = 0$ and $a + c = 0$, then by assumption A1, the transitive property of equality, $a + b = a + c$. But by Theorem 2.3.1, $b = c$. Thus the additive inverse of a is unique. ∎

Now that we know $-a$ is unique, we can make the following important observation. For every $a \in \mathbb{R}$, $-a \in \mathbb{R}$ is *the* number such that $a + (-a) = 0$. Not only can you read this equation as saying that $-a$ is the additive inverse of a, you can also read it as saying that a is the additive inverse of $-a$. In other words, this observation is a proof of the following theorem.

Theorem 2.3.4. *For every* $a \in \mathbb{R}$, $-(-a) = a$.

The behavior of 0 yields the next theorem.

Theorem 2.3.5. $-0 = 0$.

Proof: Since -0 is the additive inverse of 0, by definition it satisfies $0 + (-0) = 0$. But for any $a \in \mathbb{R}, 0 + a = a$. Thus $0 + (-0) = -0$. Thus we have $0 + (-0) = 0$ and $0 + (-0) = -0$. By the transitive property of equality, $-0 = 0$. ∎

Here's another algebraic property of real numbers with a hint for its proof (Exercise 1). The first part of Theorem 2.3.6 is a statement about $-(ab)$, the additive inverse of ab. To show that $(-a)b = -(ab)$, what you want to do is show that the expression $(-a)b$ exhibits the behavior that the additive inverse of ab ought to exhibit; then Theorem 2.3.3 guarantees that $(-a)b$ is that unique inverse. That is, $-(ab) = (-a)b$. The paragraph before Theorem 2.3.4 might give you insight into this proof. Also, recall assumption A18.

Theorem 2.3.6. *If $a, b \in \mathbb{R}$, then*

1. $(-a)b = -(ab)$,

2. $(-a)(-b) = ab$.

Sometimes the proof of a theorem can yield a special result that is important in its own right, or at least might be an observation that needs to be made. A simple theorem that follows immediately from another more complex theorem is called a *corollary*.

Corollary 2.3.7. *If $b \in \mathbb{R}$, then $(-1)b = -b$.*

Proof: Let $a = 1$ in part 1 of Theorem 2.3.6. ∎

Corollary 2.3.8. *If $a, b \in \mathbb{R}$, then $-(a + b) = (-a) + (-b)$.*

Proof: By Corollary 2.3.7 and the distributive property,

$$-(a + b) = (-1)(a + b) = (-1)a + (-1)b = (-a) + (-b). \quad (2.45)$$

∎

Corollary 2.3.8 gives us the right to make a statement such as the following: "The additive inverse of a sum is the sum of the additive inverses." It is quite common in mathematics to investigate the truth or falsity of a statement of the form "The X of the Y is equal to the Y of the X." Sometimes it is blatantly false. For example, let $X =$ "square root" and let $Y =$ "sum" and you have the statement $\sqrt{a + b} = \sqrt{a} + \sqrt{b}$. Sometimes, however, such a statement is true, and you're very glad to know that it is.

Here are results similar to some of those in the preceding, but for multiplication. You'll prove these in Exercises 2–5.

Theorem 2.3.9. *If $ac = bc$ and $c \neq 0$, then $a = b$.*

Theorem 2.3.10. *The multiplicative inverse of $a \neq 0$ is unique.*

Theorem 2.3.11. *For all $a \neq 0$, $(a^{-1})^{-1} = a$.*

Theorem 2.3.12. *For all nonzero $a, b \in \mathbb{R}$, $(ab)^{-1} = a^{-1}b^{-1}$.*

Suppose we've chosen some nonzero $a \in \mathbb{R}$. First, let's take its additive inverse $-a$, then let's take the multiplicative inverse of that to have $(-a)^{-1}$. Now start with a again, and do the same two processes in reverse order. First, take the multiplicative inverse of a to have a^{-1}, then take the additive inverse of that to have $-(a^{-1})$. You probably expect that these two processes done in reverse order produce the same result, but if so, it must be demonstrated. Look closely at the proof of Theorem 2.3.13, for this kind of stunt can come in handy sometimes.

Theorem 2.3.13. *For all nonzero $a \in \mathbb{R}$, $(-a)^{-1} = -(a^{-1})$.*

Proof: Suppose $a \neq 0$. Then there exists $a^{-1} \in \mathbb{R}$, and by part 2 of Theorem 2.3.6,

$$1 = a \cdot a^{-1} = (-a)[-(a^{-1})]. \tag{2.46}$$

Since $a \neq 0$, then neither is $-a$, else $a = -(-a) = 0$ by Theorem 2.3.5. Thus there exists $(-a)^{-1} \in \mathbb{R}$, and we may multiply both sides of Eq. (2.46) by $(-a)^{-1}$ to have $(-a)^{-1} = -(a^{-1})$. ∎

2.3.2 Ordering of the real numbers

In Chapter 0, we discussed the trichotomy law (A16). The way we address the *sign* of real numbers is by assuming that every nonzero real number can, in a sense, be compared to zero in one way or another by the symbol $>$. If $a > 0$, we call a *positive*, and if $0 > a$ (or $a < 0$), we call a *negative*. Notationally, we write the positive and negative real numbers as $\mathbb{R}^+$ and $\mathbb{R}^-$, respectively. Assumptions A17 and A18 describe some assumed behaviors of addition and multiplication in $\mathbb{R}^+$, namely, that it is closed under addition and multiplication. By splitting $\mathbb{R}$ into these three pieces, that is, $\mathbb{R}^+$, $\{0\}$, and $\mathbb{R}^-$, we can then assign meaning to the statement $a > b$ by declaring $a > b \Leftrightarrow a - b > 0$. (See Definition 0.2.1.) Thus, anytime we see a statement $a > b$, it means precisely that $a - b > 0$. Similarly, if someone asks you to demonstrate that $a > b$, we can do it by showing $a - b > 0$. This definition, with Theorem 2.3.4, allows us to prove the following.

Theorem 2.3.14. *If $a > b$, then $-a < -b$.*

In the proof, we'll suppose $a > b$, to have $a - b > 0$. To arrive at $-a < -b$, which is the same as $-b > -a$, we must arrive by way of $-b - (-a) > 0$, for this is what the statement $-b > -a$ means. Thus, the heart of the proof will be to transform $a - b > 0$ into $-b - (-a) > 0$. Easy enough.

Proof: Suppose $a > b$. Then $a - b > 0$. But $a = -(-a)$, so that $-(-a) - b > 0$, which we may write as $-(-a) + (-b) > 0$. By commutativity, $(-b) + [-(-a)] > 0$, or $-b - (-a) > 0$, so that $-b > -a$. Hence, $-a < -b$. ∎

Corollary 2.3.15. *Suppose $c \in \mathbb{R}$. Then $c > 0$ if and only if $-c < 0$.*

Proof: If $c > 0$, then let $a = c$ and $b = 0$ in Theorem 2.3.14 to have $-c < -0 = 0$. Conversely, if $-c < 0$, then let $b = -c$ and $a = 0$ in Theorem 2.3.14 to have $-0 < -(-c)$, or $c > 0$. ∎

Assumption A15 states that $1 \neq 0$, which might seem silly, but actually is an essential assumption. You'll show in Exercise 8 that the assumption $1 = 0$ causes the entire set of real numbers to collapse to the single element set $\{0\}$. Thus, since $1 \neq 0$, then by the trichotomy law either $1 > 0$ or $1 < 0$. You'll show in Exercise 9 that $1 > 0$ is the only remaining possibility, for $1 < 0$ produces a contradiction.

2.3.3 Absolute value

Undoubtedly you are familiar with the *absolute value* of a real number x.

Definition 2.3.16. For $x \in \mathbb{R}$, we define $|x|$, the *absolute value* of x by

$$|x| = \begin{cases} x, & \text{if } x \geq 0; \\ -x, & \text{if } x < 0. \end{cases} \tag{2.47}$$

Absolute value is a very important measure of the *size* of a real number. We can make two observations right off the bat:

(N1) For every $x \in \mathbb{R}$, $|x| \geq 0$.

(N2) $|x| = 0$ if and only if $x = 0$.

Property N1 follows directly from Corollary 2.3.15. For if we pick $x \in \mathbb{R}$, then either $x \geq 0$ or $x < 0$. If $x \geq 0$, then $|x| = x \geq 0$. However, if $x < 0$, then $|x| = -x > 0$. Property N2 follows directly from Corollary 2.3.15 and the trichotomy law. The reason we point these out is that properties N1–N2 are two of the three defining properties of a *norm*, a very important term in analysis. Given a set (perhaps of numbers, functions, sets), a norm $|\cdot|$ is a measure of the size of its elements. There is a third property of a norm that $|x|$ has, and we'll see it in Theorem 2.3.22.

First, let's explore some of the simplest and most familiar behaviors of $|x|$. We'll prove a few, either wholly or in part, and leave some to you in the exercises. Absolute value proofs often involve multiple cases because $|x|$ is defined piecewise. The first one is really easy, so it's all yours (Exercise 13).

Theorem 2.3.17. *For all $x \in \mathbb{R}$, $|-x| = |x|$.*

Theorem 2.3.18. *Suppose $a \geq 0$. Then $|x| = a$ if and only if $x = \pm a$.*

Proof: Suppose $a \geq 0$.

($\Rightarrow$) Suppose $|x| = a$. If $x \geq 0$, then $x = |x| = a$. If $x < 0$, then $x = -|x| = -a$. In either case, $x = \pm a$.

($\Leftarrow$) Suppose $x = \pm a$. For the case $x = a$, we have that $x = a \geq 0$, so that $|x| = x = a$. For the case $x = -a$, we have that $x \leq 0$. If $x = 0$, then $a = 0$ also, and $|x| = 0 = a$. But if $x < 0$, then $|x| = -x = a$. In all these cases, $|x| = a$. ∎

We'll prove the $\Rightarrow$ direction of Theorem 2.3.19 here, and leave the $\Leftarrow$ direction to you in Exercise 16. It might seem weird that we're going to suppose $a \geq 0$ only to state in the proof that $a = 0$ is impossible. The reason is that we want consistency between the hypothesis conditions of Theorems 2.3.18–2.3.21.

Theorem 2.3.19. *Suppose $a \geq 0$. Then $|x| < a$ if and only if $-a < x < a$.*

Proof: Let $a \geq 0$.

($\Rightarrow$) Suppose $|x| < a$. Since $|x| \geq 0$, it must be that $0 \leq |x| < a$, so $a = 0$ is impossible. If $x \geq 0$, then

$$-a < 0 \leq |x| = x < a. \tag{2.48}$$

On the other hand, if $x < 0$, we may write $-|x| > -a$ to have

$$-a < -|x| = -(-x) = x < 0 < a. \tag{2.49}$$

In either case, we have $-a < x < a$. ∎

Let's apply Exercise 3i from Section 1.2 by defining the following statements:

$$p : |x| = a \quad q : x = \pm a \quad r : |x| < a \quad s : -a < x < a. \tag{2.50}$$

Since $(p \leftrightarrow q) \wedge (r \leftrightarrow s)$ is stronger than $(p \vee r) \leftrightarrow (q \vee s)$, we can combine Theorems 2.3.18 and 2.3.19 to have the following.

Corollary 2.3.20. *Suppose $a \geq 0$. Then $|x| \leq a$ if and only if $-a \leq x \leq a$.*

Corollary 2.3.20 should make your proof of the next theorem quick (Exercise 17).

Theorem 2.3.21. *Suppose $a \geq 0$. Then $|x| > a$ if and only if either $x > a$ or $x < -a$.*

Now that we have some practice at absolute value proofs, let's look at the third property of a norm, and show that $|x|$ has this important property.

Theorem 2.3.22 (N3: Triangle Inequality). *For all $x, y \in \mathbb{R}$,*

$$|x + y| \leq |x| + |y|. \tag{2.51}$$

The proof of Theorem 2.3.22 is left to you in Exercise 18. What makes it more complicated than the previous absolute value theorems is the presence of x, y, and $x + y$. The ways $|x|$, $|y|$, and $|x + y|$ are calculated depend on the signs of x, y, and $x + y$, respectively. Thankfully, some of the cases can be consolidated "without loss of generality." There will be a hint if you need it.

Another triangle-type inequality will prove to be important in Part II of this text. It can be proved in two lines from Theorem 2.3.22, if you can see how to apply it creatively (Exercise 19).

Theorem 2.3.23. *For all $x, y \in \mathbb{R}$, $|x - y| \geq |x| - |y|$.*

There is one more triangle-type inequality that we'll need in Section 4.6. The proof requires some sneaky application of several of the results we've shown so far. You'll tackle it in Exercise 20, with hints if you need them.

Theorem 2.3.24. *For all $x, y \in \mathbb{R}$, $||x| - |y|| \leq |x - y|$.*

EXERCISES

1. Prove Theorem 2.3.6: If $a, b \in \mathbb{R}$, then:
   (a) $(-a)b = -(ab)$.
   (b) $(-a)(-b) = ab$.[6]
2. Prove Theorem 2.3.9: If $ac = bc$ and $c \neq 0$, then $a = b$.
3. Prove Theorem 2.3.10: The multiplicative inverse of $a \neq 0$ is unique.
4. Using reasoning similar to the argument for Theorem 2.3.4, prove Theorem 2.3.11: For all $a \neq 0$, $(a^{-1})^{-1} = a$.
5. Prove Theorem 2.3.12: For all nonzero $a, b \in \mathbb{R}$, $(ab)^{-1} = a^{-1}b^{-1}$.
6. Prove the *principle of zero products:* If $ab = 0$, then either $a = 0$ or $b = 0$.[7]
7. Prove $(a + b)(c + d) = ac + ad + bc + bd$ for all $a, b, c, d \in \mathbb{R}$.
8. Suppose we replace assumption A15 with the assumption that $1 = 0$. Show that, with this assumption, there are no nonzero real numbers.
9. Prove the following.
   (a) If $a < b$, then $a + c < b + c$.
   (b) If $a < b$ and $b < c$, then $a < c$.
   (c) If $a > b$ and $b > c$, then $a > c$.
   (d) If $a < 0$ and $b < 0$, then $a + b < 0$.
   (e) If $a > 0$ and $b < 0$, then $ab < 0$.
   (f) If $a < 0$ and $b < 0$, then $ab > 0$.
   (g) If $a < b$ and $c > 0$, then $ac < bc$.
   (h) If $a < b$ and $c < 0$, then $ac > bc$.
   (i) If $0 < a < b$, then $a^2 < b^2$.
   (j) If $a < b < 0$, then $a^2 > b^2$.
   (k) $1 > 0$.
   (l) For $a \in \mathbb{R}$, write $a^2 = a \cdot a$. Show that for every $a \in \mathbb{R}$, $a^2 \geq 0$.
   (m) Explain why the equation $x^2 = -1$ has no solution $x \in \mathbb{R}$.

---

[6] Part a and Theorem 2.3.4 should make this one quick.
[7] See Exercise 3g from Section 1.2.

10. Prove the following:
    (a) 0^{-1} does not exist in $\mathbb{R}$.
    (b) If $a > 0$, then $a^{-1} > 0$.
    (c) If $a < 0$, then $a^{-1} < 0$.
    (d) $c > 1$ if and only if $0 < c^{-1} < 1$.[8]
    (e) If $a > 0$ and $c > 1$, then $a/c < a$.
    (f) If $a, b \in \mathbb{Z}$ and $ab = 1$, then $a = b = \pm 1$.[9]

11. If $a, b \in \mathbb{R} \setminus \{0\}$ and $a < b$, does it follow that $1/a < 1/b$? Use results from Exercises 9 and 10 to state and prove the relationship between $1/a$ and $1/b$ depending on the signs of a and b.

12. Prove that if $a < b$ are real numbers, then $a < a + b/2 < b$. (How do you know that $2 > 0$? What exactly is 2 anyway?)

13. Prove Theorem 2.3.17: For all $x \in \mathbb{R}$, $|-x| = |x|$.

14. Prove that for all $x \in \mathbb{R}$, $-|x| \leq x \leq |x|$.

15. Suppose $x, y \in \mathbb{R}$. Prove the following.
    (a) $|xy| = |x| |y|$.
    (b) If $x \neq 0$, then $|x^{-1}| = |x|^{-1}$.[10]
    (c) If $y \neq 0$, then $|x/y| = |x|/|y|$.

16. Prove the $\Leftarrow$ direction of Theorem 2.3.19: Suppose $a \geq 0$. Then $|x| < a$ if $-a < x < a$.

17. Prove Theorem 2.3.21: Suppose $a \geq 0$. Then $|x| > a$ if and only if either $x > a$ or $x < -a$.[11]

18. Prove Theorem 2.3.22: If $x, y \in \mathbb{R}$, then $|x + y| \leq |x| + |y|$.[12]

19. Prove Theorem 2.3.23: For all $x, y \in \mathbb{R}$, then $|x - y| \geq |x| - |y|$.[13, 14]

20. Prove Theorem 2.3.24: For all $x, y \in \mathbb{R}$, $||x| - |y|| \leq |x - y|$.[15]

21. There are two theorems and one exercise in this section besides Corollary 2.3.8 that can be worded as "The X of the Y is equal to the Y of the X." Find them, and state them in this form.

---

[8] To show $0 < c^{-1} < 1$, you must show the two inequalities $0 < c^{-1}$ and $c^{-1} < 1$ separately.
[9] There are five cases: $a = -1, a = 0, a = 1, a > 1$, and $a < -1$. Three will produce a contradiction.
[10] Use a technique like the proof of Theorem 2.3.13.
[11] See Exercise 3j from Section 1.2.
[12] The case $x \geq 0$ and $y < 0$ is, without any loss of generality, the same as $x < 0$ and $y \geq 0$. However, the case $x \geq 0$ and $y < 0$ generates two subcases, depending on the sign of $x + y$.
[13] One of the mathematician's most useful tricks is knowing when and how to add zero. Start with Theorem 2.3.22, and reassign the roles of x and y. Don't look at the next hint unless you have to.
[14] Start with $|x|$ by itself and add zero inside the absolute value.
[15] Apply Corollary 2.3.20 by writing the expression in Exercise 19 in two ways, once switching the roles of x and y.

2.4 The Principle of Mathematical Induction

Take another look at property A19, the well-ordering principle on page 8. In the world of mathematics, the well-ordering principle (WOP) is often taken as an axiom. In this section, we derive a theorem based on the WOP called the *principle of mathematical induction* (PMI). In high school, you might have done what is called *proofs by induction*, where you built an argument that was analogous to knocking down an infinite row of dominoes. First, you showed figuratively that you can knock the first domino down. Then you showed that if the nth domino falls, then so does the $(n+1)$st. This very, very important proof technique is useful when the theorem you're trying to prove has a form like one of these:

$$1 + 2 + 3 + \cdots + n = \sum_{k=1}^{n} k = \frac{n(n+1)}{2} \tag{2.52}$$

or perhaps

$$A \cup (B_1 \cap B_2 \cap \cdots \cap B_n) = (A \cup B_1) \cap (A \cup B_2) \cap \cdots \cap (A \cup B_n) \tag{2.53}$$

where the theorem makes a statement about a finite but unspecified n number of things, and you want to prove that the claim is true for any $n \in \mathbb{N}$.

You might have found Eq. (2.52) handy if you had been in grammar school with Carl Friedrich Gauss in the 1780s. Gauss, a very precocious child, showed amazing mathematical ability at a very early age. A somewhat embellished story goes that when Gauss was eight years old, it was raining one day during recess, and Internet access was down. His teacher needed to keep the children in the class busy for a while, so he told them to add up the first 100 natural numbers without their calculators. Gauss figured out how to get the result quickly in the following way. By writing the sum twice, once in reverse order, he added vertically, term by term:

$$\begin{array}{c}
1 + 2 + 3 + 4 + \cdots + 99 + 100 \\
100 + 99 + 98 + 97 + \cdots + 2 + 1 \\
\hline
\underbrace{101 + 101 + 101 + 101 + \cdots + 101 + 101}_{100 \text{ terms}}
\end{array}$$

Gauss observed that 100×101 is twice the desired result, so he quickly reported the result of 5050. If you perform a similar trick replacing 100 with an arbitrary n, you get Eq. (2.52).

Although Gauss' technique might seem sufficient as a proof, there is something a little disconcerting about making a claim that involves a "dot dot dot" in it. The PMI is a theorem derived from the WOP that eliminates this untidiness.

So what is the PMI? And what does the WOP have to do with it? Let's conduct a thought experiment to set it up, first in its standard form. Then we'll look at some variations.

2.4.1 The standard PMI

Suppose we consider a set S, which is assumed to be a subset of the natural numbers $\mathbb{N}$, and suppose S is known to have the following properties:

(I1) $1 \in S$

(I2) If $n \geq 1$ and $n \in S$, then $n + 1 \in S$.

The question we ask is "What, precisely, is S?" Now I1 says $1 \in S$, but then I2 applies to guarantee that, since $1 \in S$, then $1 + 1 = 2 \in S$. But then I2 applies again to guarantee that $2 + 1 = 3 \in S$, and so forth. Now $S \subseteq \mathbb{N}$ is assumed, and it appears that every natural number is also in S, so that $\mathbb{N} \subseteq S$. Thus it appears that $S = \mathbb{N}$. Fair enough. But this argument has its own "dot dot dot" and is a little flimsy.

Another way to look at this same argument is still a little open-ended, but comes closer to the actual proof as we'll present it. If it were true that $S \neq \mathbb{N}$, then since $S \subseteq \mathbb{N}$, it must be that $\mathbb{N} \not\subseteq S$. Thus there exists $k \in \mathbb{N}$ such that $k \notin S$. By I1, $k \neq 1$, so that $k - 1 \in \mathbb{N}$. By the contrapositive of I2, $k - 1 \notin S$ either. Therefore, $k - 1 \neq 1$, so that $k - 2 \in \mathbb{N}$, and since $k - 1 \notin S$, I2 implies $k - 2 \notin S$. This is where the "dot dot dot" comes in, and we have that $k, k - 1, k - 2, \ldots, \notin S$. But this runs head-on into the fact that $1 \in S$.

If we look to the WOP, then we can clean up the untidiness in these arguments and find in that assumption enough strength to provide a rigorous proof that any set with properties I1–I2 must, in fact, be $\mathbb{N}$. This is the principle of mathematical induction.

Theorem 2.4.1 (PMI). *Suppose $S \subseteq \mathbb{N}$ has the properties that*

(I1) $1 \in S$

(I2) *If $n \geq 1$ and $n \in S$, then $n + 1 \in S$.*

Then $S = \mathbb{N}$.

The proof of the PMI is a great example of proof by contradiction. If you'd like to try to prove it yourself, then take a glance at the hints[16,17,18,19,20] if you need to. Here is the proof.

Proof: Suppose $S \subseteq \mathbb{N}$ satisfies properties I1–I2. Suppose also that $S \neq \mathbb{N}$. Then either $S \not\subseteq \mathbb{N}$ or $\mathbb{N} \not\subseteq S$. But $S \subseteq \mathbb{N}$ is assumed, so $\mathbb{N} \not\subseteq S$. Thus there exists $n \in \mathbb{N}$ such that $n \notin S$. If we define $T = \mathbb{N} \setminus S = \mathbb{N} \cap S'$, then $T \subseteq \mathbb{N}$ and $n \in T$. Thus T is a nonempty subset of the natural numbers, which, by the WOP, contains a smallest element a. Now it is impossible that $a = 1$ because $1 \in S$. Thus $a > 1$, so that $a - 1 \in \mathbb{N}$. Furthermore, since a is the smallest element of T, it must be that $a - 1 \notin T$. Thus $a - 1 \in S$. But by I2, since $a - 1 \in S$, it follows that $a = (a - 1) + 1 \in S$. But if $a \in S$, then $a \notin T$. This is a contradiction. Thus there is no smallest element of T, which means that $T = \emptyset$. Therefore, $\mathbb{N} \subseteq S$, from which $S = \mathbb{N}$. ∎

[16] The theorem says $[(S \subseteq \mathbb{N}) \wedge I1 \wedge I2] \to (S = \mathbb{N})$. Suppose this is false.
[17] State $\mathbb{N} \not\subseteq S$ in $\exists$ form.
[18] Let $T = \mathbb{N} \setminus S$. What does the WOP say about T?
[19] If T has a smallest element a, what can you say about $a - 1$?
[20] But if $a - 1 \in S$, then what is true of a?

What does Theorem 2.4.1 have to do with proving the kinds of formulas in Eqs. (2.52) and (2.53)? Think of Eqs. (2.52) and (2.53) as statements about a natural number n. They say, in effect, that a certain formula or statement is true for all $n \in \mathbb{N}$. To be as general as possible, address such a statement by $P(n)$. Let's define S to be the set of all natural numbers n for which the statement $P(n)$ is true. The trick is to show that S has properties I1–I2. Then the PMI will allow us to conclude that $S = \mathbb{N}$, which is the same as saying $P(n)$ is true for all $n \in \mathbb{N}$. First, we show that $1 \in S$ by showing $P(1)$ is true. Then we show that $n \in S \Rightarrow n+1 \in S$ by supposing $P(n)$ is true, and using this to show that $P(n+1)$ is true. The assumption that $n \in S$ is called the *inductive assumption*, and the part of the proof where we show $n \in S \Rightarrow n+1 \in S$ is called the *inductive step*. Having shown that S has properties I1–I2, we can then conclude $S = \mathbb{N}$; that is, the $P(n)$ holds true for all $n \in \mathbb{N}$. Here's a sample theorem.

Theorem 2.4.2 (Sample). *For all $n \in \mathbb{N}$, $P(n)$.*

Proof: We use induction on $n \geq 1$.

(I1) $P(1)$.

(I2) Suppose $n \geq 1$ and $P(n)$. Then .... Thus, $P(n+1)$.

Therefore, by induction, $P(n)$ is true for all $n \in \mathbb{N}$. ∎

There will be some point in step I2 where you use the inductive assumption $P(n)$ to get you over the hump of showing $P(n+1)$. Study the two proofs by induction that follow, and notice where the inductive assumption is used.

Theorem 2.4.3. *For all $n \in \mathbb{N}$,*

$$\sum_{k=1}^{n} k = \frac{n(n+1)}{2}. \quad (2.54)$$

Proof: We use induction on $n \geq 1$.

(I1) For the case $n = 1$, we have $\sum_{k=1}^{1} k = 1$ and $1(1+1)/2 = 1$, so that Eq. (2.54) holds true for $n = 1$.

(I2) Suppose $n \geq 1$ and $\sum_{k=1}^{n} k = n(n+1)/2$. Then

$$\sum_{k=1}^{n+1} k = \sum_{k=1}^{n} k + (n+1) = \frac{n(n+1)}{2} + (n+1)$$
$$= \frac{n^2 + n + 2n + 2}{2} = \frac{n^2 + 3n + 2}{2} = \frac{(n+1)(n+2)}{2}.$$

Thus Eq. (2.54) holds for $n+1$.

By the PMI, it follows that Eq. (2.54) holds for all $n \in \mathbb{N}$. ∎

Before we prove Eq. (2.53), we need to make an observation first. In Exercise 6 of Section 2.1, you showed that

$$A \cup (B \cap C) = (A \cup B) \cap (A \cup C). \tag{2.55}$$

This is the special case of Eq. (2.53) where $n = 2$. We need this fact in proving the inductive step in the following.

Theorem 2.4.4. *Suppose A, B_1, B_2, ..., B_n are sets. Then*

$$A \cup (B_1 \cap B_2 \cap \cdots \cap B_n) = (A \cup B_1) \cap (A \cup B_2) \cap \cdots \cap (A \cup B_n). \tag{2.56}$$

Proof: To clean up the notation, we write Eq. (2.56) as

$$A \cup \left[\cap_{k=1}^{n} B_k \right] = \cap_{k=1}^{n} (A \cup B_k). \tag{2.57}$$

(I1) If $n = 1$, then $\cap_{k=1}^{1} B_k = B_1$, so that both sides of Eq. (2.57) are simply $A \cup B_1$.

(I2) Suppose $n \geq 1$ and Eq. (2.57) holds for n. Then

$$A \cup \left[\cap_{k=1}^{n+1} B_k \right] = A \cup \left[\left(\cap_{k=1}^{n} B_k \right) \cap B_{n+1} \right]$$
$$\stackrel{\text{by (2.55)}}{=} \left[A \cup \left(\cap_{k=1}^{n} B_k \right) \right] \cap (A \cup B_{n+1})$$
$$= \left[\cap_{k=1}^{n} (A \cup B_k) \right] \cap (A \cup B_{n+1}) = \cap_{k=1}^{n+1} (A \cup B_k).$$

Thus, Eq. (2.57) holds for $n + 1$.

By the PMI, Eq. (2.57) holds for all $n \in \mathbb{N}$. ∎

2.4.2 Variation of the PMI

There is a natural and useful variation of the standard PMI that results from making one slight change. In defining S on page 62, Section 2.4.1, there was nothing magical about making 1 the first element of S. We could have supposed that j is any *whole number* and that $S \subseteq \mathbb{W}$ has the following properties.

(J1) $j \in S$

(J2) If $n \geq j$ and $n \in S$, then $n + 1 \in S$.

By mimicking almost word for word the proof of Theorem 2.4.1, we would have a similar PMI rooted, if you will, at $j \in \mathbb{W}$.

Theorem 2.4.5 (PMI, Version 2). *Suppose $S \subseteq \mathbb{W}$ has properties J1–J2. Then $S = \{j, j + 1, j + 2, \ldots\} = \{n \in \mathbb{W} : n \geq j\}$.*

The reason we might want Theorem 2.4.5 is that a theorem involving $n \in \mathbb{W}$ might not be true for all n, but only *eventually* true, that is, true for $n \geq j$ for some $j \in \mathbb{W}$. For

example, the formula
$$n^2 < 2^n \qquad (2.58)$$
is true for $n \geq 5$. You will prove this in Exercise 12.

Let's present an example of this variation of proof by induction just to make the general technique of induction more familiar. For $x \in \mathbb{R}\setminus\{0\}$ and $n \in \mathbb{W}$, make the following definition:
$$\begin{aligned} x^0 &= 1, \\ x^{n+1} &= x^n \cdot x \quad \text{for } n \geq 0. \end{aligned} \qquad (2.59)$$

This is the standard way of defining exponentiation. Instead of saying something like $x^9 = x \cdots x$ nine times, we say $x^9 = x^8 \cdot x$. This might sound silly, but it's an example of what is called a *recursive definition*, where the initial case is defined concretely, and then later cases are defined in terms of earlier ones. This definition of exponentiation has some familiar behaviors. If $a, b \in \mathbb{R}\setminus\{0\}$ and $m, n \in \mathbb{W}$, then
$$a^m \cdot a^n = a^{m+n} \qquad (2.60)$$
$$(a^m)^n = a^{mn} \qquad (2.61)$$
$$(ab)^n = a^n b^n. \qquad (2.62)$$

Notice we're not allowing the exponents to be negative at this point. You'll address that in Exercise 7. Let's prove Eq. (2.60) here, and leave the others to you in Exercise 6.

Theorem 2.4.6. *Suppose $a \in \mathbb{R}\setminus\{0\}$ and $m, n \in \mathbb{W}$. Then $a^m \cdot a^n = a^{m+n}$.*

Proof: We prove by induction on $n \geq 0$ (thinking of m as fixed).

(J1) For the case $n = 0$, we have $a^m \cdot a^0 = a^m \cdot 1 = a^m = a^{m+0}$. Thus the result is true for $n = 0$.

(J2) Suppose $n \geq 0$ and $a^m \cdot a^n = a^{m+n}$. We show that $a^m \cdot a^{n+1} = a^{m+(n+1)}$.
$$\begin{aligned} a^m \cdot a^{n+1} &= a^m \cdot (a^n \cdot a) = (a^m \cdot a^n) \cdot a \\ &= a^{m+n} \cdot a = a^{(m+n)+1} = a^{m+(n+1)}. \end{aligned}$$

Thus by induction, $a^m \cdot a^n = a^{m+n}$ for all $a \in \mathbb{R}\setminus\{0\}$ and $m, n \in \mathbb{W}$. ∎

Another example of a recursive definition is the *factorial*.
$$\begin{aligned} 0! &= 1, \\ (n+1)! &= (n+1) \cdot n! \quad \text{for } n \geq 0. \end{aligned} \qquad (2.63)$$

In Exercise 13, you'll prove some formulas involving sums of factorials.

2.4.3 Strong induction

There is another way to build the set S from page 62, Section 2.4.1. Consider the following. Suppose $S \subseteq \mathbb{N}$ has the following properties.

(K1) $1 \in S$

(K2) If $n \geq 2$ and $1, 2, \ldots, n - 1 \in S$, then $n \in S$.

Then by yet another mimicking of the proof of Theorem 2.4.1, we could show that $S = \mathbb{N}$. This result is called the *strong principle of mathematical induction*, or SPMI.

Theorem 2.4.7 (SPMI). *Suppose $S \subseteq \mathbb{N}$ has the properties K1–K2 above. Then $S = \mathbb{N}$.*

Before we consider an example of the usefulness of the SPMI, let's look at how it appears to be different from the standard PMI. If we're required to prove a theorem involving $n \in \mathbb{N}$ and induction seems to be the way to go, we might find that regular induction does not provide us with a strong enough assumption to make the inductive leap. With either form of induction, we would still need to show that $1 \in S$. But to make the inductive step, regular induction would only allow us to assume $n \in S$ and require us to conclude $n + 1 \in S$ solely from this. On the other hand, strong induction allows us to assume that all of $1, 2, \ldots, n - 1 \in S$, and requires us to conclude $n \in S$ from this more extensive set of assumptions.

A natural number $p \geq 2$ is said to be *prime* if it has exactly two factors in $\mathbb{N}$. Thus its only natural number factors are 1 and p. A number that is not prime is called *composite*, and such a number n can be written in the form $n = ab$, where $a, b \in \mathbb{N}$ and $1 < a, b < n$. The following theorem addresses the factorization of natural numbers into a product of primes. We use a slight variation of the SPMI by rooting it at $n = 2$.

Theorem 2.4.8 (Fundamental Theorem of Arithmetic). *Every natural number $n \geq 2$ can be written as the product of primes, and this factorization is unique, except perhaps for the order in which the factors are written.*

Proof: We'll prove only the existence part now, and save uniqueness for Section 2.7. Since 2 is prime, the result is true for $n = 2$. Thus let $n \geq 3$ be given, and suppose all of $2, 3, \ldots, n - 1$ can be written as the product of primes. Now if n is itself prime, then its prime factorization is trivial, and the result is true. On the other hand, if n is composite, then there exist $a, b \in \mathbb{N}$ such that $1 < a, b < n$ and $n = ab$. By the inductive assumption, since $2 \leq a, b \leq n - 1$, both a and b can be written as the product of primes. That is,

$$a = p_1 p_2 \cdots p_s$$
$$b = q_1 q_2 \cdots q_t,$$

where all p_k and q_k are prime. Thus $n = p_1 \cdots p_s q_1 \cdots q_t$, and we have found a prime factorization for n. ∎

EXERCISES

1. Show that the smallest element of a nonempty subset of $\mathbb{W}$ is unique.

2. Prove the following sum formulas:

(a) $\sum_{k=1}^{n} k^2 = n(n+1)(2n+1)/6$
(b) $\sum_{k=1}^{n} (-1)^k k^2 = (-1)^n n(n+1)/2$
(c) $\sum_{k=1}^{n} k^3 = \left(\sum_{k=1}^{n} k\right)^2$
(d) $\sum_{k=1}^{n} (2k-1) = n^2$
(e) $\sum_{k=0}^{n} 2^k = 2^{n+1} - 1$
(f) $\sum_{k=1}^{n} 1/k(k+1) = n/n+1$

3. Suppose $n \in \mathbb{N}$ and $a, b_1, b_2, \ldots, b_n \in \mathbb{R}$. Show that

$$a \sum_{k=1}^{n} b_k = \sum_{k=1}^{n} (ab_k). \qquad (2.64)$$

4. Suppose $m, n \in \mathbb{N}$ and $a_1, a_2, \ldots, a_m, b_1, b_2, \ldots, b_n \in \mathbb{R}$. Show that

$$\left(\sum_{j=1}^{m} a_j\right)\left(\sum_{k=1}^{n} b_k\right) = \sum_{j=1}^{m} \left(\sum_{k=1}^{n} a_j b_k\right). \qquad (2.65)$$

If $m = n = 2$, this is the FOIL technique of multiplying two binomials.[21]

5. Suppose $a_1, a_2, \ldots, a_n \in \mathbb{R}$. Show that $\left|\sum_{k=1}^{n} a_k\right| \leq \sum_{k=1}^{n} |a_k|$.

6. Prove Eqs. (2.61) and (2.62) for $a, b \in \mathbb{R}\setminus\{0\}$ and $m, n \in \mathbb{W}$.

7. For $x \in \mathbb{R}\setminus\{0\}$ and $n \in \mathbb{N}$, define

$$x^{-n} = (x^{-1})^n. \qquad (2.66)$$

This exercise will show that the rules for exponents in Eqs. (2.60)–(2.62) hold for all $a, b \in \mathbb{R}\setminus\{0\}$ and $m, n \in \mathbb{Z}$.

(a) Show that $a^{-n} = (a^n)^{-1}$ for all $n \in \mathbb{W}$. That is, the expression a^{-n} behaves like the inverse of a^n, allowing us to write $a^{-n} \cdot a^n = 1$.[22]
(b) Show that $a^{-m} a^{-n} = a^{-m-n}$ for all $m, n \in \mathbb{W}$.
(c) Show that $a^m a^{-n} = a^{m-n}$ for all $m, n \in \mathbb{W}$.[23]
(d) Show that $(a^{-m})^{-n} = a^{mn}$ for all $m, n \in \mathbb{W}$.
(e) Show that $(a^m)^{-n} = a^{-mn}$ for all $m, n \in \mathbb{W}$.
(f) Show that $(a^{-m})^n = a^{-mn}$ for all $m, n \in \mathbb{W}$.
(g) Show that $(ab)^{-n} = a^{-n} b^{-n}$ for all $m, n \in \mathbb{W}$.

8. Suppose $a \in \mathbb{R}\setminus\{0\}$ and $n \in \mathbb{N}$. Prove the following:[24]

(a) $(-a)^{2n} = a^{2n}$
(b) $(-a)^{2n+1} = -a^{2n+1}$

9. Prove the following for $n \in \mathbb{N}$:

(a) If $0 \leq a < b$, then $a^n < b^n$.[25]

---
[21] This should be a one-liner with the help of Exercise 3.
[22] Only this part will require induction.
[23] Take a hint from $a^8 \cdot a^{-6} = a^2 \cdot a^6 \cdot a^{-6} = a^2 = a^{8-6}$ and $a^2 \cdot a^{-8} = a^2 \cdot a^{-2} \cdot a^{-6} = a^{-6} = a^{2-8}$.
[24] You can prove these without induction if you call on previous Exercises appropriately.
[25] In the inductive step, Exercise 9g from Section 2.3 takes care of the case $0 < a < b$. Consider the case $a = 0$ separately.

(b) If $x > 1$, then $x^n > 1$.
(c) If $a < b \leq 0$, then
  i. $a^{2n} > b^{2n}$.[26]
  ii. $a^{2n+1} < b^{2n+1}$.
(d) If $a < 0 < b$, then $a^{2n+1} < b^{2n+1}$.[27]

10. By Exercise 4, polynomial multiplication will reveal that
$$(1-x)(1+x+x^2+x^3+x^4) = 1 - x^5.$$
If $x \neq 1$, we may divide both sides through by $1-x$ to have
$$1+x+x^2+x^3+x^4 = \frac{1-x^5}{1-x}.$$
This algebraic observation suggests a general formula for the sum of powers of a real number $x \neq 1$. If $n \geq 0$, then
$$1+x+x^2+x^3+\cdots+x^n = \sum_{k=0}^{n} x^k = \frac{1-x^{n+1}}{1-x}. \tag{2.67}$$
Prove Eq. (2.67) with an induction argument rooted at $n = 0$.

11. Use Exercise 10 to prove the following factorization formula:[28]
$$a^{n+1} - b^{n+1} = (a-b)(a^n + a^{n-1}b + a^{n-2}b^2 + \cdots + a^2 b^{n-2} + ab^{n-1} + b^n). \tag{2.68}$$

12. If $n \geq 3$, then $n^2 = n \cdot n \geq 3n = 2n + n > 2n + 1$. Use this and induction to prove $2^n > n^2$ for all $n \geq 5$.

13. Prove the following for $n \geq 0$.
  (a) $\sum_{k=0}^{n} k/(k+1)! = 1 - 1/(n+1)!$
  (b) $\sum_{k=0}^{n} k \cdot k! = (n+1)! - 1$.

2.5 Equivalence Relations: The Idea of Equality

2.5.1 Analyzing equality

What does it mean to say that two things are *equal*? Consider the following:
$$8 = 8 \tag{2.69}$$
$$\frac{3}{8} = \frac{9}{24} \tag{2.70}$$
$$2.7999999\overline{9} = 2.8. \tag{2.71}$$

---

[26] Use the fact that $0 \leq -b < -a$ and Exercise 8.
[27] Use Exercise 9l from Section 2.3.
[28] Don't make another induction argument. Letting $x = a/b$ in Eq. (2.67) provides a good start.

2.5 Equivalence Relations: The Idea of Equality

Perhaps it seems silly even to ask what Eq. (2.69) means. After all, doesn't everyone know that something is always equal to itself? But what about Eq. (2.70)? You might be thinking that these two fractions are equal because cross multiplying yields $72 = 72$. Thus Eq. (2.70) is true because we can trade it for another equation comparable to Eq. (2.69). Then what about Eq. (2.71)? Do you remember the trick from junior high where you convert a repeating decimal to a ratio of integers? If you let $x = 2.799999\overline{9}$, then you can say

$$100x = 279.99999\overline{9}$$
$$10x = 27.99999\overline{9},$$

which by subtraction yields

$$90x = 252.000000\overline{0}$$
$$x = \frac{252}{90} = \frac{28}{10} = 2.8.$$

You claim that Eq. (2.71) is true because you can reshape one side into the other by the rules of algebra.

In general, we ask the following question. Given any set S, and two elements $x, y \in S$, what does it mean to say $x = y$? What are the fundamental properties of this thing we call *equality* that should apply in all contexts, regardless of whether the elements of S are numbers, functions, sets, and so forth? This section is designed to address this question. Once we understand these fundamental properties, we might even be able to use some imagination and create some entirely new types of equality.

Before we begin this process, we need to be aware that the symbol for equality can vary greatly from situation to situation. Depending on what kind of things x and y are, there are different symbols commonly used to denote that x and y are equal. For example,

$$x = y$$
$$x \equiv y$$
$$x \equiv_n y$$
$$x \sim y$$
$$x \cong y$$
$$x \leftrightarrow y$$
$$x \mathrel{R} y$$

are but a few. The statement $x \mathrel{R} y$ is read "x is related to y," and derives from a way of addressing equality in terms of a *relation*, which we'll investigate in Section 2.9. Since our goal is to address the features of equality that transcend context, let's choose one of these common symbols, say, $\equiv$, and stick with it.

Here are three properties of equality that we would probably expect to be true in any context. For a set S, we would expect that

(E1) $x \equiv x$ for all $x \in S$ (Reflexive property);

(E2) If $x \equiv y$, then $y \equiv x$ (Symmetric property);

(E3) If $x \equiv y$ and $y \equiv z$, then $x \equiv z$ (Transitive property).

Notice that these are the same properties of equality that we assumed for $\mathbb{R}$ in assumption A1 from Chapter 0. These three properties are what equality is all about in any context. They are so important that we make the following definition:

Definition 2.5.1. Let S be a nonempty set, and let "$x \equiv y$" be a statement for all $x, y \in S$. That is, for every $x, y \in S$, either $x \equiv y$ or $x \not\equiv y$ is true. Then $\equiv$ is said to be an *equivalence relation* on S if properties E1–E3 are true for all elements of S.

Example 2.5.2. Let C be the set of all cities on earth. Suppose for the sake of argument that there are no one-way roads. Define two cities $x, y \in C$ to be equivalent, $x \equiv y$, if it is possible to drive on roads *from* city x *to* city y. Is this definition of $\equiv$ an equivalence relation on C?

Solution: We must verify that properties E1–E3 hold.

(E1) Let $x \in C$. Since it is possible to drive from city x to city x on roads, then $x \equiv x$.

(E2) Suppose $x \equiv y$. Then it is possible to drive from x to y on roads. Since we've assumed no roads are one-way, then it is also possible to drive from y to x, since the same roads going from x to y can be traveled in the opposite direction. Thus, $y \equiv x$.

(E3) Suppose $x \equiv y$ and $y \equiv z$. Then it's possible to drive from x to y on roads, and it's possible to drive from y to z on roads. By beginning at x, driving to y, then to z, it's possible to drive from x to z on roads. Thus, $x \equiv z$.

Since $\equiv$ satisfies properties E1–E3, we have that $\equiv$ is an equivalence relation. ∎

Although Example 2.5.2 seems on the surface not to be particularly mathematical, it illustrates the general form for proving that some definition of $\equiv$ is an equivalence relation. It illustrates the process of defining a term and applying that definition in a new context, a process that should be getting clearer to you. Here is a sample definition and theorem to drive the point home. We use the expression $P(x, y)$ to represent some statement involving x and y.

Definition 2.5.3 (Sample). For set Y and for $x, y \in Y$, define $x \equiv y$ if $P(x, y)$.

Theorem 2.5.4 (Sample). *The equivalence $\equiv$ in Definition 2.5.3 is an equivalence relation on Y.*

Proof: We show that properties E1–E3 hold for $\equiv$ on Y.

(E1) Pick $x \in Y$. Then ... so that $P(x, x)$ is true. Thus $x \equiv x$.

(E2) Suppose $x \equiv y$. Then $P(x, y)$. Thus ..., so that $P(y, x)$. Therefore, $y \equiv x$.

(E3) Suppose $x \equiv y$ and $y \equiv z$. Then $P(x, y)$ and $P(y, z)$. Then ..., so that $P(x, z)$. Thus $x \equiv z$.

Since $\equiv$ satisfies properties E1–E3, we have that $\equiv$ defines an equivalence relation on Y. ∎

Example 2.5.5. (A very important example). For $x, y \in \mathbb{Z}$, define $x \equiv_6 y$ if there exists $k \in \mathbb{Z}$ such that $x - y = 6k$. This is to be read, "x is congruent to y, modulo 6 (or mod 6)." This definition says that $x \equiv_6 y$ if $x - y$ is divisible by 6. Show that $\equiv_6$ defines an equivalence relation on $\mathbb{Z}$.

Solution: Details are left to you in Exercise 1. Here's the skeleton.

(E1) Pick $x \in \mathbb{Z}$, and let $k = ?$. Then $x - x = 0 = 6k$.

(E2) Suppose $x \equiv_6 y$. Then there exists $k_1 \in \mathbb{Z}$ such that $x - y = 6k_1$. Let $k_2 = \ldots$.

(E3) Suppose $x \equiv_6 y$ and $y \equiv_6 z$. Then there exist $k_1, k_2 \in \mathbb{Z}$ such that .... ∎

Example 2.5.5 is particularly important in algebra and pervades Part III of this text. The number 6 was chosen arbitrarily, of course. We can discuss $\equiv_n$ for any $n \in \mathbb{N}$, defining $x \equiv_n y$ if and only if there exists $k \in \mathbb{Z}$ such that $x - y = nk$. Notationally, there are two other common ways of writing this form of equivalence among integers. They are

$$x \equiv y \pmod{n} \tag{2.72}$$

$$x \equiv y \quad (n). \tag{2.73}$$

Example 2.5.6. Suppose we have an equilateral triangle, and we write the numbers $\{1, 2, 3\}$, one on each corner. Call one such writing of these numbers an *assignment*. How many assignments are there? For two assignments x and y, define $x \equiv y$ if it is possible to convert x into y by rotating and/or flipping the triangle over. Does $\equiv$ define an equivalence relation on the set of all assignments?

Solution: First, since there are three corners and three numbers to assign, we can count six assignments. (Any one of three numbers can be written on the top corner, then any one of two remaining numbers can be written on the lower left corner, and the only remaining number is written on the lower right corner. This yields $3 \times 2 \times 1 = 6$ assignments.) We verify that $\equiv$ satisfies E1–E3.

(E1) By rotating the triangle 360° and flipping it six times, we see that $x \equiv x$.

(E2) Suppose $x \equiv y$. Then it is possible to convert x into y by some combination of rotation and/or flip. By exactly reversing the process of converting x to y, we can convert y to x. Thus $y \equiv x$.

(E3) Suppose $x \equiv y$ and $y \equiv z$. Then x can be converted into y, and y can be converted into z by rotations and/or flips. By converting x into y, then y into z in succession, it is possible to convert x into z. Thus $x \equiv z$.

Since $\equiv$ satisfies E1–E3, $\equiv$ defines an equivalence relation on the set of all assignments. ∎

Example 2.5.7. Let S be the set of all school buildings in the United States. Say that two schools x and y are equivalent, that is, $x \equiv y$, if the height of the flagpole in front of school x is within one foot of the height of the flagpole in front of school y. Does this definition of equivalence constitute an equivalence relation?

Solution: Although E1 and E2 are satisfied, E3 is not. We can show that the statement "$x \equiv y$ and $y \equiv z$ implies $x \equiv z$" is a false statement. What is the logic behind this demonstration?[29] We demonstrate the existence of a counterexample to transitivity where $x \equiv y$ and $y \equiv z$, but $x \not\equiv z$. Let the heights of the flagpoles in front of schools x, y, and z be 29, 29.6, and 30.3 ft., respectively. Then, clearly, $x \equiv y$ and $y \equiv z$, but $x \not\equiv z$. Since transitivity fails, $\equiv$ is not an equivalence relation. ∎

2.5.2 Equivalence classes

In Example 2.5.2, it might have occurred to you that defining two cities to be equivalent as we did takes all cities on earth and lumps them together into groups that are mutually accessible from each other. For example, all cities on the north island of New Zealand are equivalent to each other because there is (presumably) a network of roads connecting them all. Furthermore, no city outside the north island of New Zealand is equivalent to any city on the island. If it were, say, for example, by some newly constructed bridge between Auckland, NZ, and Santiago, Chile, then Santiago would be equivalent to all the cities on the north island of New Zealand. Furthermore, all cities accessible from Santiago would also become equivalent to all cities on the north island of New Zealand. It's sort of like the molecules in two drops of water. Either the two drops do not touch at all, or, if they do, they instantly merge into one drop.

This illustrates that there is a lot of strength in the properties E1–E3. One very important feat that an equivalence relation defined on S performs is that it completely splits S up into very nice, nonempty, nonoverlapping subsets. In general, the splitting of a set into nonoverlapping subsets is called *partitioning*. When an equivalence relation is defined on a set S, it naturally partitions S into subsets where elements of the same subset are all equivalent to each other, and elements of different subsets are not equivalent. Each subset of this sort is called an *equivalence class*. First, we'll define a partition and look at an example. Then we'll see how an equivalence relation gives rise to a partition.

Definition 2.5.8. Suppose S is a set, and $\mathcal{F} = \{A\}$ is a family of subsets of S. Then $\mathcal{F}$ is said to be a *partition* of S if

(P1) $A \neq \emptyset$ for all $A \in \mathcal{F}$;

(P2) If $A, B \in \mathcal{F}$, and $A \cap B \neq \emptyset$, then $A = B$; and

(P3) $\bigcup_{A \in \mathcal{F}} A = S$.

---

[29] What is the negation of $(p \wedge q) \to r$?

Let's illustrate with a specific example before we discuss general properties of partitions. Just remember, a partition of a given set S is nothing but a family of subsets of S with some special properties.

Example 2.5.9. For $\mathbb{N}_{10}$, consider the following four families of subsets:

$$\begin{aligned}
\mathcal{F}_1 &= \{\{1\}, \{2, 3, 5, 7\}, \{4, 6, 8, 10\}, \{9\}\} \\
\mathcal{F}_2 &= \{\emptyset, \{1, 2, 3, 4\}, \{5, 6, 7, 8\}, \{9, 10\}\} \\
\mathcal{F}_3 &= \{\{1\}, \{2, 4, 6, 8, 10\}, \{3, 6, 9\}, \{5, 10\}, \{7\}\} \\
\mathcal{F}_4 &= \{\{\text{Primes in } \mathbb{N}_{10}\}, \{\text{Composites in } \mathbb{N}_{10}\}\}.
\end{aligned} \quad (2.74)$$

Of these four families, only $\mathcal{F}_1$ is a partition of $\mathbb{N}_{10}$. Notice how $\mathcal{F}_1$ satisfies properties P1–P3. First, no set in $\mathcal{F}_1$ is empty. Second, they are all disjoint. If we choose any two of them and they have a nonempty intersection, then they are the same set. Third, forming the union across $\mathcal{F}_1$ produces $\mathbb{N}_{10}$.

Notice how the other families fail to be a partition of $\mathbb{N}_{10}$. Since a partition of a set must contain only nonempty subsets, $\mathcal{F}_2$ fails to be a partition of $\mathbb{N}_{10}$. In $\mathcal{F}_3$, there exist two distinct sets with nonempty intersection. Finally, the union across $\mathcal{F}_4$ is not $\mathbb{N}_{10}$ for 1 is neither prime nor composite.

In the work we'll do in what follows, there will be some program by which we construct subsets of a given S in order to create a partition. To verify that this program does indeed partition S, we'll need to show that each of the properties P1–P3 is satisfied on $\mathcal{F}$.

(P1) The program for forming the subsets of S must always generate nonempty subsets. Thus if we choose an arbitrary set in the family, we must be able to find some element of S in the chosen set.

(P2) This property says that if A and B overlap at all ($A \cap B \neq \emptyset$), then they really are the same set, so that no $x \in S$ can be in more than one distinct set $A \in \mathcal{F}$. Thus if we suppose there exists some $z \in A \cap B$, we must be able to show $A = B$.

(P3) This property says that the generated sets in the family completely exhaust all elements of S. Naturally, the program for forming the subsets of S should be defined so that it creates only subsets of S. Then Exercise 5 from Section 2.2 guarantees that the $\subseteq$ part of P3 is satisfied. What remains to be shown is $\supseteq$, which amounts to showing that for every $x \in S$ there exists some A in the family $\mathcal{F}$ such that $x \in A$.

Example 2.5.10. For C in Example 2.5.2, choose a city $x \in C$ and define A_x to be the set of all cities accessible from city x. Show that $\mathcal{F} = \{A_x : x \in C\}$ is a partition of C.

Solution: We verify that properties P1–P3 are satisfied on $\mathcal{F}$.

(P1) Pick any $A \in \mathcal{F}$. Then there exists $x \in C$ for which $A = A_x$. Since city x is accessible from itself, we have that $x \in A$, so that $A \neq \emptyset$.

(P2) Pick $A_x, A_y \in \mathcal{F}$, and suppose $A_x \cap A_y \neq \emptyset$. Then there exists $z \in A_x \cap A_y$. We show that $A_x = A_y$. First, pick $a \in A_x$. Then city a is accessible from city

x, and since $z \in A_x$, z is accessible from x. (Draw a picture!) Furthermore, since $z \in A_y$, it follows that z is accessible from y. Thus, a is accessible from y by traveling $y \to z \to x \to a$, so that $a \in A_y$ and $A_x \subseteq A_y$. By an identical argument in the reverse direction $A_y \subseteq A_x$. Thus, $A_x = A_y$.

(P3) Since $A_x \subseteq C$ for all $x \in C$, Exercise 5 from Section 2.2 implies that $\cup_{x \in C} A_x \subseteq S$. To show $\supseteq$, pick any $y \in S$. We must find some set in $\mathcal{F}$ that contains y. But $y \in A_y$, so this is clearly true. Thus $y \in \cup_{x \in C} A_x$, and $S \subseteq \cup_{x \in C} A_x$. ∎

In showing P3 in Example 2.5.10, notice there is some redundancy in forming the union $\cup_{x \in C} A_x$. For example, Wellington, another city on the north island of New Zealand, is in $A_{\text{Wellington}}$, but it's also in A_{Auckland}. Thus dumping all cities in each A_x together to form $\cup_{x \in C} A_x$ means Wellington is tossed in more than once. But that's all right. All we care about is that every city is tossed in at least once.

Now let's see how an equivalence relation on a nonempty set S is tied to a partition of S. We'll work our way into a theorem, tying them together by way of some examples. First, a definition.

Definition 2.5.11. Suppose $\equiv$ defines an equivalence relation on a nonempty set S. For an element $x \in S$, define

$$[x] = \{y \in S : y \equiv x\}.$$

That is, $[x]$ is the set of all elements of S that are equivalent to x. This subset of S is called the *equivalence class* of x.

Example 2.5.12. From Example 2.5.2, the equivalence class of Auckland is the set of all cities accessible from Auckland, that is, all cities on the north island of New Zealand. It is the same as the equivalence class of Wellington.

Example 2.5.13. For equivalence mod 6 from Example 2.5.5, given any $x \in \mathbb{Z}$,

$$[x] = \{y \in \mathbb{Z} : y - x = 6k, \text{ for some } k \in \mathbb{Z}\}.$$

Thus, $[x]$ is the set of all integers y of the form $y = 6k + x$, where k can be any integer. (See Exercise 7.)

Example 2.5.14. From Example 2.5.6, there are six assignments, and a given assignment can be altered into any of the remaining five by rotations and/or flips. Thus there is only one equivalence class in the set of assignments.

In all the preceding examples of equivalence relations, the equivalence classes seem to be a basis of a partition of the set. This is no coincidence, of course. Theorem 2.5.15 states that properties E1–E3 can be used to demonstrate that the family of equivalence classes satisfies P1–P3. Most details are left to you in Exercise 8.

Theorem 2.5.15. *Suppose $\equiv$ defines an equivalence relation on a set S. Define $\mathcal{F} = \{[x] : x \in S\}$. That is, $\mathcal{F}$ is the family of all equivalence classes in S. Then $\mathcal{F}$ is a partition of S.*

Proof: We verify that properties P1–P3 from Definition 2.5.8 are satisfied.

(P1) No element of $\mathcal{F}$ is empty. For if we pick some $[x] \in \mathcal{F}$, then since $x \equiv x$, it follows that $x \in [x]$. Thus $[x] \neq \emptyset$.

(P2) Pick $[x], [y] \in \mathcal{F}$ and suppose $[x] \cap [y] \neq \emptyset$....

(P3) Since every set in $\mathcal{F}$ is defined to be a subset of S, we have $\cup_{x \in S}[x] \subseteq S$ by Exercise 5 from Section 2.2. Thus we must show $\supseteq$. Pick $y \in S$.... ∎

Theorem 2.5.15 is very important for the following reason. If some form of equivalence $\equiv$ is defined on a set S, and it can be verified that this definition is an equivalence relation, then Theorem 2.5.15 guarantees that S is automatically partitioned into equivalence classes by this definition of $\equiv$. We can then think of any element of a particular equivalence class as being representative of all elements in its class. Furthermore, we can think of $\equiv$ in the same way we think of equality, where two equivalent elements are, in some sense, interchangeable. The partitioning of S into equivalence classes lumps the elements of S into categories where one element can be replaced with any other element in its category, within limits, of course. In Section 2.6 we'll illustrate this phenomenon on the rational numbers, where we'll investigate the familiar definition of equality of two fractions. Then we'll show how the binary operations of addition and multiplication on $\mathbb{Q}$ are defined so that different representative elements of equivalence classes are indeed interchangeable.

EXERCISES

1. Prove that $\equiv_6$ in Example 2.5.5 is an equivalence relation.

2. Show that Definition 2.1.2 defines an equivalence relation on the set of all sets.

3. Let $\mathcal{F}$ be the family of all nonempty sets, and define $A \equiv B$ if $A \cap B \neq \emptyset$. Is $\equiv$ an equivalence relation on $\mathcal{F}$?

4. Let S be the set of four-letter words in the *Random House Unabridged Dictionary*. For two words $x, y \in S$ define $x \equiv y$ if it's possible to construct a sequence of words in S, beginning with x and changing one letter at a time to create another word in S at each step, ending up at y. (A sequence of one word is considered valid.) For example BABY $\equiv$ POOP because of the sequence BABY, BABE, BARE, BARN, BURN, BURP, BUMP, PUMP, POMP, POOP. Show $\equiv$ is an equivalence relation on S.

5. Is $\leq$ an equivalence relation on $\mathbb{R}$? Prove or disprove.

6. Is $\neq$ an equivalence relation on $\mathbb{R}$? Prove or disprove.

7. How many equivalence classes are there in Example 2.5.13? Describe them by listing some of their elements.

8. Complete the proof of Theorem 2.5.15 by showing that the set of equivalences of an equivalence relation $\equiv$ satisfies properties P2 and P3.[30]

[30]Take some hints from Example 2.5.10.

9. Consider a square in the xy plane, turned diagonally so that its corners are on the axes. Define an assignment of $\{1, 2, 3, 4\}$ as in Example 2.5.6. How many assignments are there? For two assignments x and y, define $x \equiv y$ if x can be converted into y by some combination of rotation about its center point and/or flipping the square over from top to bottom. How many equivalence classes are there? List their elements.

2.6 Equality, Addition, and Multiplication in $\mathbb{Q}$

As an illustration of the concept of equivalence relation, we want to take a look at equality, addition, and multiplication in the integers and see how they give rise to equality, addition, and multiplication in the rationals. This section is effectively a dissection of some of the real number properties assumed in Chapter 0, so a few prefacing words are in order about what we are trying to accomplish here.

First, let's consider the whole numbers. Beginning with the set $\{0, 1\}$, which by assumption A15 is indeed a two-element set, we calculate all possible sums of its elements. We know from the behavior of 0 that $0 + 0 = 0$ and $1 + 0 = 0 + 1 = 1$. What is not immediately clear is $1 + 1$. Is it possible that $1 + 1 = 0$ or $1 + 1 = 1$? In Exercise 1, you'll show that certain axioms and properties of $\mathbb{R}$ prevent either of these from being possible. Since $1 + 1 \notin \{0, 1\}$, it might be helpful and simpler to devise a new symbol, such as maybe 2, to denote $1 + 1$. This is just the beginning of a process of building the whole numbers as a set of *successors* of previously defined whole numbers. The assumptions used to build them are called *Peano's postulates*. Because they are formulated without reference to the set of real numbers as a context, Peano's postulates and their implications generate the set of whole numbers in a somewhat more complicated way than we describe here. One of the assumptions is that 0 is not a successor to any whole number, so that the sequence of successors, which we call $1, 2, 3, \ldots$, never circles back around to 0.

Once the set of whole numbers has been constructed, we extend it to the integers by tossing in the additive inverses of all the whole numbers. These additive inverses are in fact new elements not already in $\mathbb{W}$, except for $-0 = 0$. For if any whole number $a \geq 1$ had $b \in \mathbb{W}$ as its additive inverse, then $a + b = 0$ would imply that 0 is the successor to $a + b - 1$.

In this section we want to give ourselves the standard assumptions concerning equality and well-definedness of addition and multiplication on the integers, just as we did for $\mathbb{R}$ in properties A1, A2, and A8 from Chapter 0. For now, however, we assume these properties *only* for the integers. Specifically, using the symbol $=_\mathbb{Z}$ to represent equality on the set of integers, let's assume the following:

(Z1) For every $n \in \mathbb{Z}$, $n =_\mathbb{Z} n$.

(Z2) If $m, n \in \mathbb{Z}$ and $m =_\mathbb{Z} n$, then $n =_\mathbb{Z} m$.

(Z3) If $m, n, p \in \mathbb{Z}$, $m =_\mathbb{Z} n$ and $n =_\mathbb{Z} p$, then $m =_\mathbb{Z} p$.

Let's also restate assumptions A2 and A8, that addition and multiplication are well defined on $\mathbb{R}$, in terms that apply only to the integers. To be clear that addition and multiplication are being performed only on integers, let's use $+_\mathbb{Z}$ and $\times_\mathbb{Z}$ to represent these operations.

(Z4) Addition and multiplication are well defined on the integers. That is, if $m, n, p, q \in \mathbb{Z}$, $m =_\mathbb{Z} n$, and $p =_\mathbb{Z} q$, then $m +_\mathbb{Z} p =_\mathbb{Z} n +_\mathbb{Z} q$ and $m \times_\mathbb{Z} p =_\mathbb{Z} n \times_\mathbb{Z} q$.

In what follows, we want to use assumptions Z1–Z4 on $\mathbb{Z}$ to derive similar results on $\mathbb{Q}$. We'll also use the other assumptions from Chapter 0 concerning the behavior of addition and multiplication, but only for the integers. To be clear, we'll use the new notation from assumptions Z1–Z4 to show that the properties are being applied only to the integers. With these things in mind, we're ready to dissect equality, addition, and multiplication in $\mathbb{Q}$ and see how they derive from the same concepts in $\mathbb{Z}$.

2.6.1 Equality in $\mathbb{Q}$

In Section 0.2 we defined the set of rational numbers as

$$\mathbb{Q} = \{p/q : p, q \in \mathbb{Z}, q \neq 0\}.$$

There is nothing inherent in the definition of $\mathbb{Q}$ that suggests how you would address equality of two elements p/q and r/s. In fact, there is more than one useful way to define equality on $\mathbb{Q}$, but we will look only at the familiar one.

What is the familiar meaning of the statement $p/q = r/s$?[31] It certainly does not mean that $p = r$ and $q = s$. We'll think of the equality of two rational numbers as being defined in terms of equality of the integers ps and qr. That is, if we use the symbol $=_\mathbb{Q}$ to mean equality of two rational numbers, then the standard *definition* of rational equality is

$$\frac{p}{q} =_\mathbb{Q} \frac{r}{s} \quad \Leftrightarrow \quad p \times_\mathbb{Z} s =_\mathbb{Z} q \times_\mathbb{Z} r. \tag{2.75}$$

Notice what we've done here. We've defined $=_\mathbb{Q}$ in terms that use only $=_\mathbb{Z}$ and $\times_\mathbb{Z}$. Notice also the positions of p, q, r, s in Eqs. (2.75). These positions must be retained when we translate between $p/q =_\mathbb{Q} r/s$ and $p \times_\mathbb{Z} s =_\mathbb{Z} q \times_\mathbb{Z} r$.

Now we need to verify that $=_\mathbb{Q}$ has the properties we would want any definition of equality to have (E1–E3) by assuming Z1–Z4.

Theorem 2.6.1. *The relation $=_\mathbb{Q}$ is an equivalence relation on $\mathbb{Q}$.*

Proof: Most of the details are left to you in Exercise 2.

(E1) $=_\mathbb{Q}$ is reflexive. Pick $p/q \in \mathbb{Q}$. Then since multiplication of integers is commutative, we have that $p \times_\mathbb{Z} q =_\mathbb{Z} q \times_\mathbb{Z} p$. Thus by the definition of $=_\mathbb{Q}$ in Eq. (2.75), $p/q =_\mathbb{Q} p/q$.

(E2) $=_\mathbb{Q}$ is symmetric. Suppose $p/q =_\mathbb{Q} r/s$. Then .... Thus $r/s =_\mathbb{Q} p/q$.

(E3) $=_\mathbb{Q}$ is transitive. Suppose $p/q =_\mathbb{Q} r/s$ and $r/s =_\mathbb{Q} t/u$. Then .... Thus $p/q =_\mathbb{Q} t/u$. ∎

Thanks to Theorem 2.6.1, it follows from Theorem 2.5.15 that $\mathbb{Q}$ is partitioned into equivalence classes. Presumably, you have an idea of what these are. First, all expressions

---

[31] Think in terms of cross multiplication.

of the form $0/q$ where $q \neq 0$ are equivalent to each other. Conversely, if $p_1/q_1 =_\mathbb{Q} 0/q_2$, then $p_1 = 0$. For the case of nonzero numerators, if $p/q = r/s$, then we may find a common denominator qs, and write $p/q = ps/qs$ and $r/s = qr/qs$. By the way we defined equality of fractions, the numerators ps and qr are equal. Therefore, if p/q is a fraction in fully reduced form, the equivalence class of p/q can be thought of as all fractions that can be reduced by standard cancellation to p/q. We're then always free to address rational numbers in a form where the fraction has been fully reduced.

2.6.2 Well-defined $+$ and $\times$ on $\mathbb{Q}$

Now that we've shown standard equality of rational numbers is an equivalence relation, we can define a form of addition of rational numbers, which we'll denote $+_\mathbb{Q}$, and show that it is well defined. What definition of $+_\mathbb{Q}$ should we use?[32] We'll use the definition,

$$\frac{p}{q} +_\mathbb{Q} \frac{r}{s} = \frac{ps +_\mathbb{Z} qr}{qs}. \tag{2.76}$$

Observe that Eq. (2.76) will always produce a "legitimate" result: that is, the fact that $p/q, r/s \in \mathbb{Q}$ ensures that $q, s \neq 0$. Consequently, $qs \neq 0$ by the principle of zero products. Thus Eq. (2.76) fits the required form for elements of $\mathbb{Q}$ because it has an integer numerator and nonzero integer denominator, and we can say that $+_\mathbb{Q}$ is closed on $\mathbb{Q}$. Here's the theorem you'll prove in Exercise 3.

Theorem 2.6.2. *Addition on $\mathbb{Q}$ as defined in Eq. (2.76) is well defined. That is, if $p/q, r/s, t/u, v/w \in \mathbb{Q}$, and if $p/q =_\mathbb{Q} r/s$ and $t/u =_\mathbb{Q} v/w$, then $p/q +_\mathbb{Q} t/u =_\mathbb{Q} r/s +_\mathbb{Q} v/w$.*

In the same way you showed that $=_\mathbb{Q}$ is an equivalence relation by assuming that $=_\mathbb{Z}$ is an equivalence relation, you'll show $+_\mathbb{Q}$ is well defined by assuming $+_\mathbb{Z}$ and $\times_\mathbb{Z}$ are well defined. Notice from Eq. (2.76) that anything of the form $0/s$ functions as the additive identity in $\mathbb{Q}$, for

$$\frac{p}{q} + \frac{0}{s} = \frac{ps + q \cdot 0}{qs} = \frac{ps}{qs} = \frac{p}{q}. \tag{2.77}$$

Also, notice $(-p)/q$ and $p/(-q)$, which are equivalent, function as $-(p/q)$.

Multiplication of two rational numbers is defined in the following way:

$$\frac{p}{q} \times_\mathbb{Q} \frac{t}{u} = \frac{pt}{qu}. \tag{2.78}$$

Proving multiplication is well defined is easier than for addition (Exercise 4).

Theorem 2.6.3. *Multiplication on $\mathbb{Q}$ as defined in Eq. (2.78) is well defined. That is, if $p/q, r/s, t/u, v/w \in \mathbb{Q}$, and if $p/q =_\mathbb{Q} r/s$ and $t/u =_\mathbb{Q} v/w$, then $p/q \times_\mathbb{Q} t/u =_\mathbb{Q} r/s \times_\mathbb{Q} v/w$.*

Notice anything of the form p/p functions as the multiplicative identity, and $(p/q)^{-1} = q/p$, as long as $p \neq 0$.

---

[32]What can serve as a common denominator? What will the numerator therefore be?

EXERCISES

1. Show that $1 + 1 = 0$ and $1 + 1 = 1$ are impossible, given the assumptions and properties of $\mathbb{R}$ we have already addressed.[33]

2. Complete the proof of Theorem 2.6.1, that $=_\mathbb{Q}$ from page 77 is an equivalence relation, by showing it has properties E2 and E3.

3. Prove Theorem 2.6.2: If $p/q, r/s, t/u, v/w \in \mathbb{Q}$, and if $p/q =_\mathbb{Q} r/s$ and $t/u =_\mathbb{Q} v/w$, then $p/q +_\mathbb{Q} t/u =_\mathbb{Q} r/s +_\mathbb{Q} v/w$.

4. Prove Theorem 2.6.3: If $p/q, r/s, t/u, v/w \in \mathbb{Q}$, and if $p/q =_\mathbb{Q} r/s$ and $t/u =_\mathbb{Q} v/w$, then $p/q \times_\mathbb{Q} t/u =_\mathbb{Q} r/s \times_\mathbb{Q} v/w$.

2.7 The Division Algorithm and Divisibility

A theory of real numbers is a fascinating progression of ideas beginning with the set $\{0, 1\}$ and developing through the famous subsets of $\mathbb{R}$: $\mathbb{W} \subset \mathbb{Z} \subset \mathbb{Q} \subset \mathbb{R}$. In Section 2.3 we investigated properties of $\mathbb{R}$ that follow from assumptions A1–A22, but we did not look into any unique properties of certain subsets of $\mathbb{R}$. This text is hardly designed to be a thorough treatment of the real numbers, for the real numbers are an intricate set with properties characteristic of some of the most complex mathematical structures. However, some characteristics of the integers, rationals, and irrationals are basic and representative of the more abstract mathematical structures that you will become very familiar with in time. In this section, we investigate a few of the basic properties of the integers.

2.7.1 Even and odd integers; the division algorithm

One way to assign meaning to the words *even integer* and *odd integer* is the following:

Definition 2.7.1. If $n \in \mathbb{Z}$, we say that n is *even* if there exists $k \in \mathbb{Z}$ such that $n = 2k$. We say that n is *odd* if there exists $k \in \mathbb{Z}$ such that $n = 2k + 1$.

Theorem 2.7.2. *Suppose $m, n \in \mathbb{Z}$ are both even. Then mn is even.*

Proof: Suppose $m, n \in \mathbb{Z}$ are both even. Then there exist $k_1, k_2 \in \mathbb{Z}$ such that $m = 2k_1$ and $n = 2k_2$. Let $l = 2k_1k_2$, which is an integer. Then

$$mn = (2k_1)(2k_2) = 4k_1k_2 = 2(2k_1k_2) = 2l. \tag{2.79}$$

Thus mn is even. ∎

---

[33] Consider the trichotomy law and cancellation of addition.

Corollary 2.7.3. *If $n \in \mathbb{Z}$ is even, then n^2 is even.*

Proof: Let $m = n$ in Theorem 2.7.2. ∎

The proof of the following similar theorem is left to you in Exercise 1. Corollary 2.7.5 will prove useful in Section 2.8.

Theorem 2.7.4. *Suppose $m, n \in \mathbb{Z}$ are both odd. Then mn is odd.*

Corollary 2.7.5. *If $n \in \mathbb{Z}$ is odd, then n^2 is odd.*

At this point, you may be thinking that Definition 2.7.1 addresses the only two possible situations that can happen in the integers. After all, every integer is either even or odd, isn't it? Well, maybe, but how do you know that? How do you know that every integer can be written either in the form $2k$ or $2k + 1$, but not both? If you've sensed the need to justify the fact that every integer is either even or odd, but not both, then you're catching onto the game of mathematics.

Ever since elementary school, you've been familiar with the idea of dividing an integer b by another integer $a > 0$ to produce a *quotient q* and *remainder r*. One way you could express the results of your division calculation in an equation is to write

$$b = aq + r, \tag{2.80}$$

where, you hope, $0 \leq r < a$. For example, if $a = 12, b = 88$, we can write $88 = 12 \times 7 + 4$, and if $a = 6, b = -13$, we have $-13 = 6 \times (-3) + 5$. One nice thing about the form of Eq. (2.80) is that every number involved is an integer. What resembles the division of integer by integer is written without having to resort to the rational numbers. The following theorem says that the existence of a quotient q and a remainder $0 \leq r < a$ are guaranteed. Even more, they are unique. The theorem is a surprisingly useful one.

Theorem 2.7.6 (Division Algorithm). *Let $a, b \in \mathbb{Z}$, where $a > 0$. Then there exist unique $q, r \in \mathbb{Z}$ such that $b = aq + r$ and $0 \leq r < a$.*

We're going to provide part of the proof here, the existence of q and r for the case $b \geq 0$. In Exercise 5, you'll show existence for $b < 0$ by exploiting the existence of q and r for $-b > 0$ and then you'll show uniqueness. The technique we'll use in showing the existence for $b \geq 0$ will call on the WOP of $\mathbb{W}$, and is similar to a technique you'll use later when you prove Theorem 2.7.13. For Theorem 2.7.6, we use the set

$$S = \{b - aq : q \in \mathbb{Z}, \ b - aq \geq 0\}, \tag{2.81}$$

which is merely the set of all whole numbers you can generate by subtracting integer multiples of a from b. For example, if $a = 12$ and $b = 104$, then

$$S = \{\ldots, 128, 116, 104, 92, 80, 68, 56, 44, 32, 20, 8\}. \tag{2.82}$$

The last element of S is the result of letting $q = 8$. Notice that it is the smallest element of S, and is strictly less than 12. By constructing the corresponding S according to Eq. (2.81) for an arbitrary $a > 0$ and $b \geq 0$, we can apply the WOP to S, then show that this smallest element is the value of r we want. Here's the proof of Theorem 2.7.6, and in as much detail as we promised.

Proof: Let $a, b \in \mathbb{Z}$, where $a > 0$. First, we consider the case $b \geq 0$. Define the set S as in Eq. (2.81). By definition, $S \subseteq \mathbb{W}$, and since $b \geq 0$, we may let $q = 0$ to see that $b \in S$, so that $S \neq \emptyset$. By the WOP, S has a smallest element, which we may denote r, and we also note that r is of the form $b - aq$ for some $q \in \mathbb{Z}$. Thus, we have that $b = aq + r$, where $r \geq 0$, and we must show that $r < a$. Suppose $r \geq a$. Then

$$r = b - aq > b - aq - a = b - a(q+1) = b - aq - a = r - a \geq 0, \quad (2.83)$$

and therefore $b - a(q+1)$ is an element of S that is strictly smaller than r. This contradicts our choice of r and so it must be that $r < a$.

According to Exercise 5, the requisite values of q and r exist for $b < 0$, and q and r are unique. ∎

From Section 2.5 we know that $81 \equiv_7 25$ because $81 - 25 = 56 = 8 \times 7$. Applying the division algorithm to $a = 7$ and $b_1 = 81$, then to $a = 7$ and $b_2 = 25$, we have $81 = 7 \times 11 + 4$ and $25 = 7 \times 3 + 4$. Therefore, 81 and 25 have the same remainder when divided by 7 according to the division algorithm. This illustrates the following theorem, which says merely that x and y differ by a multiple of n if and only if they have the same remainder when divided by n (Exercise 6).

Theorem 2.7.7. *Suppose $x, y \in \mathbb{Z}$ and $n \in \mathbb{N}$. Then $x \equiv_n y$ if and only if x and y have the same remainder when divided by n according to the division algorithm.*

2.7.2 Divisibility in $\mathbb{Z}$

Motivated by Theorem 2.7.6, if $a, b \in \mathbb{Z} \setminus \{0\}$, and there exists $k \in \mathbb{Z}$ such that $b = ak$, we say that *a divides b* or that *a is a divisor of b*, and we write $a \mid b$. If $a \mid b$ where $a \notin \{\pm 1, \pm b\}$, we call a a *proper* divisor of b. First, here are some really easy theorems about divisibility that you'll prove in Exercises 8–11.

Theorem 2.7.8. *If $a \in \mathbb{Z} \setminus \{0\}$, then $a \mid a$.*

Theorem 2.7.9. *If $a, b, c \in \mathbb{Z} \setminus \{0\}$ such that $a \mid b$ and $b \mid c$, then $a \mid c$.*

Theorem 2.7.10. *If $a \mid b$ and $a \mid c$, then for all $m, n \in \mathbb{Z}$, $a \mid (mb + nc)$.*

An expression of the form $mb + nc$ is called a *linear combination* of b and c. Theorem 2.7.10 says that if a divides both b and c, then it divides any linear combination of them.

Theorem 2.7.11. *If $a \mid b$ and $b \mid a$, then $a = \pm b$.*

Some important and useful results in algebra stem from what we call the *greatest common divisor* (gcd) of two nonzero integers a and b. One practical way you might try to find the gcd of two integers is to break them down into their prime factorizations (Theorem 2.4.8 even though we didn't prove uniqueness there), then see how many 2's, 3's, 5's, etc. you can extract from both in order to construct the gcd. This might work well practically, but: 1) its logical basis in the work we've done so far is uncertain at best; and 2) it is not particularly useful as leverage in our later theorems. Instead, we define gcd in terms of two criteria, with one motivated by the word *common* and the other motivated by the word *greatest*. Then we show that such a thing exists uniquely and can be written in a somewhat surprising way. So that we will have uniqueness of $\gcd(a, b)$, we're going to insist that any integer we might be inclined to call a gcd must be positive.

Definition 2.7.12. Suppose $a, b \in \mathbb{Z}\setminus\{0\}$, and suppose $g \in \mathbb{Z}^+$ has the following properties:

(D1) $g \mid a$ and $g \mid b$.

(D2) If h is any positive integer with the properties that $h \mid a$ and $h \mid b$, then it must be that $h \mid g$ also.

Then g is called a *greatest common divisor* of a and b, and is denoted $\gcd(a, b)$.

Some remarks about Definition 2.7.12 are in order. First, nothing about the definition of $\gcd(a, b)$ (or of any term for that matter) guarantees that any such thing exists. It merely lays out criteria by which we declare some positive integer to be $\gcd(a, b)$. Second, property D1 clearly states that anything you want to call $\gcd(a, b)$ will in fact be a common divisor of a and b. However, we need to explain and perhaps justify our choice of property D2 as the other criterion, for it might seem a bit unnatural. If property D2 is supposed to be some way of describing what it means for g to be greatest of all the common divisors of a and b, you might think a more natural way to say it would be:

(D2') If h is any positive integer with the properties that $h \mid a$ and $h \mid b$, then it must be that $h \leq g$.

Well, we could do it that way, but we don't and this is why: It all centers around how you decide to measure greatness. In $\mathbb{Z}^+$, we can measure relative greatness by $\leq$. But as you'll see in your later work in algebra, there are other more abstract settings where we discuss divisibility, then ask about a gcd, but we do not necessarily have a notion of $\leq$ to use as our criterion for greatness. For example, if you want to talk about one polynomial dividing another polynomial and then ask about their gcd, it would not seem to mean much to say that some common divisor of theirs is greatest of all, in the sense of $\leq$. After all, what could $f \leq g$ mean for two functions? Not that we cannot *give* meaning to $\leq$ in that context, but the point is that it's better to develop a criterion for greatness of a divisor that stays within the context of divisibility rather than relying on some external measure like $\leq$, which might or might not already exist. All this is to say that property D2 is our measure of greatness among all common divisors of a and b. If g is what we're going to call $\gcd(a, b)$, it has the property that any positive integer h that comes down the pike that also has property D1 cannot, in some sense, be greater than g. Our way of saying h is no greater than g is that $h \mid g$.

Therefore, given $a, b \in \mathbb{Z} \setminus \{0\}$, how do we even know that there exists some $g \in \mathbb{Z}^+$ having properties D1–D2? And if such a $g \in \mathbb{Z}^+$ does exist, how many can there be? The following theorem claims that gcd(a, b) exists uniquely. Furthermore, hidden inside its proof is some additional information about gcd(a, b) that will come in handy later. Finally, here it is. Only a few of the details are required of you in Exercise 15.

Theorem 2.7.13. *Suppose $a, b \in \mathbb{Z}\setminus\{0\}$. Then there exists a unique $g \in \mathbb{Z}^+$ having properties D1–D2.*

Proof: Pick $a, b \in \mathbb{Z}\setminus\{0\}$ and define

$$S = \{ma + nb : m, n \in \mathbb{Z}, ma + nb > 0\}. \tag{2.84}$$

That is, S is the set of all *positive* linear combinations of a and b. First, S is not empty, for depending on the signs of a and b, we may let $m, n = \pm 1$ to produce some $ma + nb > 0$. By the WOP, S contains a smallest element g, which may be written in the form $g = m_0 a + n_0 b$ for some $m_0, n_0 \in \mathbb{Z}$. With Exercise 15, g has properties D1–D2, and if $g_1, g_2 \in \mathbb{Z}$ both have properties D1–D2, then $g_1 = g_2$. ∎

Look at the serendipity in the proof of Theorem 2.7.13. The gcd of a and b can be written in the form $ma + nb$ for some $m, n \in \mathbb{Z}$, and the smallest such positive expression is in fact the gcd. This can be particularly helpful if gcd$(a, b) = 1$. If gcd$(a, b) = 1$, then a and b are said to be *relatively prime*. Immediately we can see that if a and b are relatively prime, then there exist $m, n \in \mathbb{Z}$ such that $ma + nb = 1$. Furthermore, if it is possible to find a linear combination of a and b that equals 1, then this linear combination must be the smallest element of S in Eq. (2.84), hence gcd$(a, b) = 1$.

In Section 2.4, we defined $p \in \mathbb{N}$ to be prime if it has exactly two divisors in $\mathbb{N}$, namely 1 and p. Thus, if $a \in \mathbb{Z}\setminus\{0\}$ and $p \in \mathbb{N}$ is prime, then gcd(a, p) is either 1 or p, because these are the only divisors of p. If gcd$(a, p) = p$, then p satisfies D1, so that $p \mid a$. With this, we have proved the following:

Theorem 2.7.14. *If $a \in \mathbb{Z}$ and $p \in \mathbb{N}$ is prime, then either a and p are relatively prime or $p \mid a$.*

Theorem 2.7.14 will help you prove the following in Exercise 17:

Theorem 2.7.15. *If $a, b \in \mathbb{Z}$, $p \in \mathbb{N}$ is prime, and $p \mid ab$, then either $p \mid a$ or $p \mid b$.*

As an immediate corollary, by letting $a = b$ in Theorem 2.7.15, we have the following:

Corollary 2.7.16. *If $p \mid a^2$, then $p \mid a$.*

Then with an induction argument in Exercise 18, you can prove the following:

Theorem 2.7.17. *Suppose $a_1, a_2, \ldots, a_n \in \mathbb{Z}$ and $p \in \mathbb{N}$ is prime. Suppose also that $p \mid a_1 a_2 \cdots a_n$. Then there exists some k $(1 \leq k \leq n)$ such that $p \mid a_k$.*

The last thing we will do in this section is return to the prime factorization of $n \in \mathbb{N}$ whose existence we proved in Theorem 2.4.8 to extend it to include something like uniqueness.

Theorem 2.7.18. *The prime factorization of $n \geq 2$ is unique, except perhaps for the order in which the factors are written.*

Proof: Choose $n \in \mathbb{N}$, and suppose $n = p_1 p_2 \cdots p_k$ and $n = q_1 q_2 \cdots q_l$ are two ways of writing n as a product of primes. We show by induction on k that $l = k$ and, with some possible reordering of the q_i, that $p_i = q_i$ for all $1 \leq i \leq k$.

If $k = 1$, then $n = p_1$, so that n is prime. Thus $l = 1$ and $p_1 = q_1$. Therefore suppose $k \geq 2$, and suppose that any factorization of a natural number into $k - 1$ primes is unique up to order of the factors. Since

$$p_1 p_2 \cdots p_k = q_1 q_2 \cdots q_l, \qquad (2.85)$$

then $p_k \mid q_1 q_2 \cdots q_l$. By Theorem 2.7.17, there exists some j, where $1 \leq j \leq l$, such that $p_k \mid q_j$. Since q_j is prime, $p_k = q_j$. Reordering the right-hand side of Eq. (2.85) by switching q_l and q_j, then canceling p_k and q_l from Eq. (2.85), we have factorizations of the natural number n/p_k into $k - 1$ and $l - 1$ primes. Since $p_k \geq 2$, we have $n/p_k < n$. Applying the inductive assumption to n/p_k, we conclude that $k - 1 = l - 1$, and with some possible reordering of $q_1, \ldots, q_{l-1}$, we have $p_i = q_i$ for $1 \leq i \leq l - 1$. Thus $l = k$, and the factorization of n into k primes is unique up to the order of the factors. ∎

EXERCISES

1. Prove Theorem 2.7.4: Suppose $m, n \in \mathbb{Z}$ are both odd. Then mn is odd.

2. Prove that the sum of two even integers is even.

3. Prove that the sum of two odd integers is even.

4. Prove that the sum of an even and an odd integer is odd.

5. This exercise finishes the proof of the division algorithm, Theorem 2.7.6. Suppose $a, b \in \mathbb{Z}$, and $a > 0$.

    (a) Suppose $b < 0$. Use the result already demonstrated on $-b > 0$ to prove the existence of $q, r \in \mathbb{Z}$ such that $b = aq + r$ and $0 \leq r < a$.[34]

    (b) Now that you know $b = aq + r$, where $0 \leq r < a$ is possible for all $a, b \in \mathbb{Z}$ with $a > 0$, show the uniqueness of q and r.[35]

6. Prove Theorem 2.7.7: Suppose $x, y \in \mathbb{Z}$ and $n \in \mathbb{N}$. Then $x \equiv_n y$ if and only if x and y have the same remainder when divided by n according to the division algorithm.

7. Suppose a and b are integers such that $a \equiv_3 b \not\equiv_3 0$. Show that $ab \equiv_3 1$.[36]

---

[34] Write $-b = aq_1 + r_1$ and use two cases, $r_1 = 0$ and $0 < r_1 < a$, to find the desired q and $0 \leq r < a$ for b.

[35] Suppose $q_1, q_2, r_1, r_2 \in \mathbb{Z}$ are such that $b = aq_1 + r_1$ ($0 \leq r_1 < a$) and $b = aq_2 + r_2$ ($0 \leq r_2 < a$). If you suppose $q_1 \neq q_2$, then you may assume without any loss of generality that $q_2 > q_1$. How does this produce a contradiction?

[36] From Theorem 2.7.7, a and b will have the same nonzero remainder upon division by 3. Consider both cases.

8. Prove Theorem 2.7.8: If $a \in \mathbb{Z}\setminus\{0\}$, then $a \mid a$.

9. Prove Theorem 2.7.9: If $a, b, c \in \mathbb{Z}$ such that $a \mid b$ and $b \mid c$, then $a \mid c$.

10. Prove Theorem 2.7.10: If $a \mid b$ and $a \mid c$, then for all $m, n \in \mathbb{Z}$, $a \mid (mb + nc)$.

11. Prove Theorem 2.7.11: If $a \mid b$ and $b \mid a$, then $a = \pm b$.[37]

12. Use induction to show that $3 \mid (4^n - 1)$ for all $n \in \mathbb{N}$.

13. Show that if $n \in \mathbb{Z}$ is odd, then $8 \mid (n^2 - 1)$.[38]

14. Show that if $a, b, c \in \mathbb{Z}$ are all odd, then there are no rational solutions x to the equation $ax^2 + bx + c = 0$.[39]

15. Prove the following parts of Theorem 2.7.13.

    (a) If $g = m_0 a + n_0 b$ is the smallest element of S as defined in Eq. (2.84), then $g \mid a$. (The proof that $g \mid b$ is identical.)[40]

    (b) If h is any positive integer with the properties that $h \mid a$ and $h \mid b$, then it must be that $h \mid g$.

    (c) If g_1 and g_2 both have properties D1–D2, then $g_1 = g_2$.[41]

16. Construct a parallel to Definition 2.7.12 of the term $\gcd(a_1, a_2, \ldots, a_n)$. State and prove a parallel to Theorem 2.7.13.[42]

17. Prove Theorem 2.7.15: If $a, b \in \mathbb{Z}$ and $p \mid ab$, then either $p \mid a$ or $p \mid b$.[43]

18. Prove Theorem 2.7.17: Suppose $a_1, a_2, \ldots, a_n \in \mathbb{Z}$ and $p \in \mathbb{N}$ is prime. Suppose also that $p \mid a_1 a_2 \cdots a_n$. Then there exists some k ($1 \le k \le n$) such that $p \mid a_k$.

19. Suppose $p \in \mathbb{N}$ is prime and $a \in \mathbb{Z}$. Show that $a^2 \equiv_p 1$ if and only if $a \equiv_p \pm 1$.

2.8 Roots and irrational numbers

In this section, we return to the real numbers to investigate a few more of its properties. First, we investigate $\sqrt[n]{x}$ for $n \in \mathbb{N}$ and $x \in \mathbb{R}$. This provides a perfect opportunity for a presentation of one of the most famous theorems of all time: the fact that $\sqrt{2}$ is not rational. Alas, you will provide the proof.

---

[37] See Exercise 10f from Section 2.3.
[38] If $k \in \mathbb{Z}$, then either k or $k + 1$ is even.
[39] If $x = p/q$ is a rational solution, then you may assume p and q are not both even. Multiply the equation through by q and apply earlier results of this section to show that $ap2 + bpq + cq^2$ is always odd.
[40] If $g \nmid a$, then the division algorithm produces a contradiction by generating an element of S that is smaller than g.
[41] Theorem 2.7.11 should come in handy.
[42] Relax. Induction won't be necessary. The definition and steps of the proof can be done using universal and existential quantifiers.
[43] See Exercise 3g from Section 1.2 for the umpteenth time.

2.8.1 Roots of real numbers

In Section 0.2, assumption A22 states that for every real number $x > 0$ and $n \in \mathbb{N}$, there is a real number y solution to the equation $y^n = x$. That is, nth roots of positive real numbers exist in $\mathbb{R}$. In this section, we address the existence and possible uniqueness of solutions to $y^n = x$ for all $x \in \mathbb{R}$. We'll not make any claims that you are not already familiar with, but we do need to prove them. Most of the proof is left to you in Exercise 1, with a thorough outline provided.

You might be wondering why assumption A22 uses the language of equation solving ($y^n = x$) to address roots of real numbers and only makes slight reference to the expression $\sqrt[n]{x}$. To do otherwise is to put the cart before the horse. We begin by letting the expression $\sqrt[n]{x}$ mean *any* solution y to the equation $y^n = x$. To say $y = \sqrt[n]{x}$ is to say $y^n = x$. However, there could be a problem because such a solution $y \in \mathbb{R}$ might not exist if $x < 0$. Furthermore, possible confusion could result because there might be more than one such y. In this case, we need to decide which solution of $y^n = x$ we will declare to be the unambiguous value of $\sqrt[n]{x}$. By the end of this section, then, the expression $\sqrt[n]{x}$ will be clearly and uniquely defined for appropriate $x \in \mathbb{R}$ and $n \in \mathbb{N}$.

Another point to make here is that assumption A22 is not a standard axiomatic assumption in a rigorous study of the real numbers. It actually follows from the Least Upper Bound axiom. In this text, we accept it without question. Here is the main theorem concerning roots of real numbers.

Theorem 2.8.1. *Let $n \in \mathbb{N}$. Then the following hold:*

(X1) *The equation $y^n = 0$ has the unique solution $y = 0$.*

(X2) *Concerning even roots:*

    (a) *If $x > 0$, the equation $y^{2n} = x$ has precisely two distinct solutions in $\mathbb{R}$, and these are additive inverses of each other.*

    (b) *If $x < 0$, the equation $y^{2n} = x$ has no solution in $\mathbb{R}$.*

(X3) *Concerning odd roots, the equation $y^{2n+1} = x$ has a unique solution for every $x \in \mathbb{R}$.*

Part X3 of Theorem 2.8.1 overlooks the case $n = 1$, which is acceptable because it's trivial. Property X1 should be clear, for certainly $0^n = 0$, and if $y^n = 0$, then $y = 0$ from the principle of zero products. Thus there is no other value of $\sqrt[n]{0}$. See Exercise 1 for the completion of claims X2–X3.

With Theorem 2.8.1, we can now introduce the notation $\sqrt[n]{x}$ and define it unambiguously.

Definition 2.8.2. If $n \in \mathbb{N}$, then $\sqrt[2n+1]{x}$ is defined to be the unique solution y of the equation $y^{2n+1} = x$. If $x \geq 0$, the expression $\sqrt[2n]{x}$ is defined to be the unique, nonnegative solution y of the equation $y^{2n} = x$.

Theorem 2.8.1 and Definition 2.8.2 lead us to the following principles of algebraic manipulation: For $x \geq 0$, $y^{2n} = x$ if and only if $y = \pm\sqrt[2n]{x}$. Similarly, for all $x \in \mathbb{R}$, $y^{2n+1} = x$ if and only if $y = \sqrt[2n+1]{x}$.

2.8.2 Existence of irrational numbers

Up to this point, we have said virtually nothing about *irrational numbers*, that is, real numbers that are not rational. You've worked with many numbers that are irrational, including π, e, and most numbers of the form $\sqrt[n]{x}$. You've also been taught that irrational numbers have a decimal representation that does not terminate and does not fall into a pattern of repetition. However, you might never have seen a definitive argument that any particular number is irrational. Here we address this. But first, a little history.

Some of the intuitive properties of $\mathbb{R}$ that we likely take for granted were not a part of the thinking of the ancient Greeks, most notably the Pythagoreans, that very secret order of thinkers who produced some amazingly sophisticated mathematics. One such intuitive notion we probably have is that the set of real numbers is a sort of continuum, a set that can be visualized in terms of the ordered points on a straight line, where rationals and irrationals are spread up and down like strewn grains of salt and pepper, leaving no holes in the continuum. The Greeks had no concept of irrational numbers at first, and now is a good time to touch on some of their views. Here is a very brief description of what they meant.

To the Greeks, numbers were best visualized in terms of lengths of segments, areas of rectangles, and volumes of rectangular solids, all of which were *constructible* according to certain rules. They had very sophisticated techniques for constructing them using the only two geometric shapes they considered perfect, straight line segments (crudely constructible with a straight-edge) and circles (crudely constructible with a compass). In Euclid's all-important compilation of the best of Greek mathematics (a five-volume set called *Elements*), he demonstrated amazingly sophisticated mathematical results, such as the theorem attributed to Pythagoras, using only some assumptions about circles and line segments.

With these techniques of construction, it is possible to imagine in the following way the drawing, if you will, of a line segment whose length is any positive rational number. Beginning with a line segment whose length is arbitrarily declared to be one unit, it is possible to construct segments representing 2, 3, 4, etc. by extending the given segment with a straight-edge, then twirling a compass around to tack the measured unit length onto itself end to end. It is also fairly easy to take a segment of length n and use similar triangles to construct a line segment of length $1/n$. (See Fig. 2.3.) With another technique by which segments of length a and b can be used to construct a segment of length ab, all the positive rational lengths are constructible. (See Exercise 2.) Easy enough, right?

The fact that segments of arbitrarily long length, say, $n \in \mathbb{N}$, could be constructed meant that segments of arbitrarily short length $1/n$ could also be constructed by the reciprocation technique. Such a short line segment could then, in theory, be added to itself as many times as needed to produce a segment of any arbitrary length m/n. The Greeks erroneously assumed the converse—that a segment of *any* constructible length could be constructed by imagining it to be a finite sum of very short segments of the form $1/n$. Slap any segment down onto an imaginary piece of paper using techniques of construction. How long is it? It is some whole multiple of a (possibly very short) segment that can be constructed by reciprocating a segment of length $n \in \mathbb{N}$. In modern language, what number is represented by the length of a segment? Always a rational one.

Now suppose we are given two segments of arbitrary lengths. If both of them can be visualized in this way, what sort of relationship must exist between these two segments? In

1. Construct segment *AB* of length *n*.
2. Construct segment *AC* of length 1.
3. Construct segment *BC*.
4. Construct segment *FE* parallel to *BC*.
5. Segment *AE* has length 1/*n*.

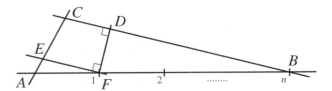

Figure 2.3 Constructing length $1/n$ using similar triangles.

the same way that we would find a common denominator of two fractions m/n and p/q, our ancient mathematicians would say that there is a single segment, perhaps very short, that can be attached to itself a finite number of times to produce each of the two given segments. In our language, $1/nq$ can be added to itself mq times to produce m/n and np times to produce p/q. The term that is used to describe this assumed relationship between two segments is that they are *commensurable*. The Greeks believed that all constructible segments are commensurable, which is equivalent to our believing that all real numbers are rational.

There was a problem with this and the Greeks eventually figured it out. Take a segment of length one, then construct another segment of length one at one endpoint and at a right angle to this first segment. Then sketch the hypotenuse. This hypotenuse line segment is obviously constructible. We just described how to do it easily. Furthermore, by the Pythagorean theorem, its length l satisfies $l^2 = 2$. The problem is that it is not commensurable with some other constructible segments. That is, l is not rational. This is the amazing result that the Pythagoreans discovered, and it created no small crisis. Given that the theorem traditionally named after Pythagoras was already known, it boiled down to this fact: Since "the sum of the [areas of the] squares on the legs of a right triangle is equal to the [area of the] square on the hypotenuse" (see Fig. 2.4), then, as we would write it, $\sqrt{2}$ is a constructible length. Therefore, $\sqrt{2}$ is commensurable with 1, or, in our language, $\sqrt{2}$ must be writable in the form p/q, where $p, q \in \mathbb{Z}$. The traditional proof that this is impossible is attributed to Aristotle. And, by the way, this is the one you'll discover in Exercise 3, perhaps with a few hints.

Theorem 2.8.3. $\sqrt{2}$ *is irrational.*

A few words are in order about irrational numbers. Let's work backwards from Theorem 2.8.3. Theorem 2.8.3 says $\sqrt{2}$ is not rational, while Theorem 2.8.1 says $\sqrt{2}$ is real. Thus, irrational numbers do exist in $\mathbb{R}$. However, Theorem 2.8.1 is based on assumption A22, which we have said very little about. As we said, assumption A22 is not an axiom of the real numbers. It can be proved from the Least Upper Bound axiom,

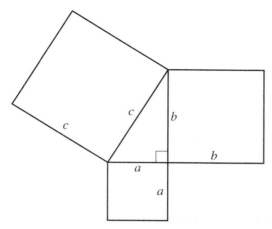

Figure 2.4 A sketch of the Pythagorean theorem.

which is a standard axiom of $\mathbb{R}$. What we discover is that the irrational numbers owe their existence to the Least Upper Bound axiom.

EXERCISES

1. This exercise leads you through the proofs of claims X2–X3 from Theorem 2.8.1. We begin with assumption A22, assuming that for every $x > 0$ and $n \in \mathbb{N}$, the equation $y^n = x$ has *some* solution $y \in \mathbb{R}$.

    (a) Existence in X2(a): Let $n \in \mathbb{N}$ and $x > 0$. Suppose $y_0 \in \mathbb{R}$ is a solution to $y^{2n} = x$. (Notice $y_0 \neq 0$.) Show that $-y_0$ is also a solution to $y^{2n} = x$.[44]

    (b) Nonexistence in X2(b): Let $n \in \mathbb{N}$ and $x < 0$. Explain why there is no $y \in \mathbb{R}$ such that $y^{2n} = x$.

    (c) Existence in X3 for $x < 0$: Let $n \in \mathbb{N}$ and $x < 0$. Prove that there exists a solution $y \in \mathbb{R}$ to $y^{2n+1} = x$.[45]

    (d) Nonexistence in X3 of solutions of opposite sign: Let $n \in \mathbb{N}$, and let y_0 be a solution to $y^{2n+1} = x$. Prove that $x > 0$ if and only if $y_0 > 0$.

    (e) Uniqueness in X2(a) and X3 (positive case): If $y_1 > 0$ and $y_2 > 0$ satisfy $y_1^n = x$ and $y_2^n = x$, then $y_1 = y_2$.[46]

    (f) Uniqueness in X3 (negative case): If $y_1 < 0$ and $y_2 < 0$ satisfy $y_1^{2n+1} = x$ and $y_2^{2n+1} = x$, then $y_1 = y_2$.[47]

2. Devise a technique similar to that described in Fig. 2.3 by which a segment of length ab can be constructed from segments of length a and b.

---

[44] Exercise 8 from Section 2.4 should come in handy.
[45] Use the fact that $-x > 0$.
[46] Use the factorization formula in Exercise 11 from Section 2.4.
[47] Apply part (e) to $-y_1$ and $-y_2$.

3. Prove Theorem 2.8.3: $\sqrt{2}$ is irrational. (Naturally, you'll want to assume it is rational and arrive at a contradiction. Here are some hints if you need them.[48,49,50,51])

4. The proof of Theorem 2.8.3 can be easily generalized to demonstrate that $\sqrt{p}$ is irrational for any prime $p \in \mathbb{N}$. Explain how your proof of Theorem 2.8.3 can be adapted into a proof of this more general claim.[52]

5. Suppose $a, b, c, d \in \mathbb{Q}$ and $a + b\sqrt{2} = c + d\sqrt{2}$. Does it follow that $a = c$ and $b = d$?[53]

6. Prove that the sum of a rational and an irrational must be irrational.[54]

7. Prove that the product of a nonzero rational and an irrational must be irrational.

8. Suppose $x, y \geq 0$. Show that if $x < y$, then $\sqrt{x} < \sqrt{y}$.[55]

2.9 Relations in General

There is another way to construct an equivalence relation that starts at a place different from that in Definition 2.5.1. A *relation* is a set construction that puts all kinds of element comparisons such as equality, less than ($<$), divisibility, and subset inclusion into one mathematical idea. It's also a way of linking elements of two different sets together, and is a context in which functions can be defined. In this section, we'll define a relation and look at examples and special kinds of relations. First, however, we define the *Cartesian product* of two sets A and B by

$$A \times B = \{(a, b) : a \in A, b \in B\}, \tag{2.86}$$

the set of ordered pairs where the first term in the pair is an element of A and the second is an element of B. You might have seen the Cartesian plane written as $\mathbb{R} \times \mathbb{R}$, or more succinctly as $\mathbb{R}^2$. A *relation from A to B* is defined to be any subset of $A \times B$.

Example 2.9.1. For $A = \{1, 2, 3\}$ and $B = \{11, 12, 13, 21, 22, 23, 31, 32, 33\}$, the set

$$R = \{(1, 11), (2, 21), (2, 22), (3, 31), (3, 32), (3, 33)\} \tag{2.87}$$

is a relation from A to B.

If a relation involves subsets of $\mathbb{R}$, we can represent it graphically as a set of points in the xy-plane, just as in high school algebra.

---

[48] Write $\sqrt{2} = m/n$, where you can assume no common factors between m and n.
[49] Square both sides and look at the contrapositive of Corollary 2.7.5.
[50] If m^2 is even then, ... What does this mean? Cancel out a 2.
[51] So n^2 is even. What contradiction does this cause?
[52] Corollary 2.7.16 might help.
[53] If $b \neq d$, then what must be true of $\sqrt{2}$?
[54] See Exercise 3h from Section 1.2.
[55] Exercise 9i from Section 2.3 should make this quick.

Example 2.9.2. $\{(x, y) : y > |x| + 1\}$ is a relation from $\mathbb{R}$ to $\mathbb{R}^+$ (Fig. 2.5).

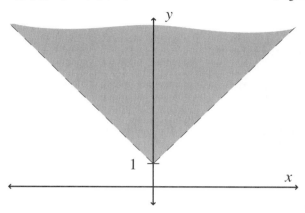

Figure 2.5 The relation $y > |x| + 1$.

In this section, we're going to delve only into subsets of $A \times A$, which we call a *relation on A* instead of a relation from A to A.

Example 2.9.3. Let $A = \{1, 2, 3, 4, 5, 6\}$. Then
$$R = \{(1, 3), (1, 5), (2, 4), (2, 6), (3, 5), (4, 6)\} \tag{2.88}$$
is a relation on A.

You might think there's nothing particularly exciting about a relation, for any collection of ordered element pairs qualifies as one. What makes the idea take shape and become mathematically important is that we can lay out certain criteria by which pairs are included in the relation, and these different criteria might have particular properties that make interesting statements about A.

Example 2.9.4. Define a relation $R \subseteq \mathbb{Q} \times \mathbb{Q}$ by $R = \{(p/q, r/s) : ps = qr\}$. This relation consists of all pairs of rational numbers that are equal, as we defined equality in Eq. (2.75). Thus, $(3/8, -30/-80)$, $(0/2, 0/12) \in R$, while $(2/5, 5/3) \notin R$.

Example 2.9.5. $R = \{(x, y) : x \leq y\}$ defines a relation on $\mathbb{R}$ (Exercise 1).

Example 2.9.6. $R_1 = \{(a, b) : a < b\}$ and $R_2 = \{(a, b) : a \mid b\}$ are relations on $\mathbb{Z}$.

Example 2.9.7. For a set A, $R = \emptyset$ and $R = A \times A$ are relations on A.

The *power set* of a given set A is defined to be the family of all subsets of A. For example, if $A = \{1, 2, 3\}$, the power set of A is
$$\{\emptyset, \{1\}, \{2\}, \{3\}, \{1, 2\}, \{1, 3\}, \{2, 3\}, \{1, 2, 3\}\}. \tag{2.89}$$

We're saving a discussion of the number of elements in a set until Chapter 3, but notice that A has three elements and the power set of A has eight elements. This motivates the notation 2^A to mean the power set of A.

Example 2.9.8. For $A = \{1, 2, 3\}$, define a relation on 2^A by $R = \{(A_1, A_2) : A_1 \subseteq A_2\}$. Thus $(\{1\}, \{1\}), (\emptyset, A) \in R$, but $(A, \{2\}), (\{1, 2\}, \emptyset) \notin R$.

In Examples 2.9.4–2.9.8, a particular pair (x, y) from the set is included in the relation if some stated property $P(x, y)$ is true. That is, $(x, y) \in R$ if x and y are *related* according to some criterion. We give names to some relations when the criterion for inclusion of ordered pairs in R has certain properties. Here's an example.

Definition 2.9.9. Suppose $R \subseteq A \times A$ has the following properties:

(E1) $(x, x) \in R$ for all $x \in A$ (Reflexive property);

(E2) If $(x, y) \in R$, then $(y, x) \in R$ (Symmetric property);

(E3) If $(x, y), (y, z) \in R$, then $(x, z) \in R$ (Transitive property).

Then R is called an *equivalence relation* on A.

Notice how E1–E3 say the same things here as E1–E3 from Definition 2.5.1, but in different language. Instead of saying that $x \equiv x$ for all $x \in A$ is a true statement, we say that all ordered pairs of the form (x, x) are in the relation. Instead of saying that the truth of $x \equiv y$ implies the truth of $y \equiv x$, now we say that the inclusion of the ordered pair (x, y) in the relation is always accompanied by the inclusion of (y, x). And instead of saying $x \equiv y$ and $y \equiv z$ implies $x \equiv z$, we say that the presence of (x, y) and (y, z) in the relation is always accompanied by the presence of (x, z). Thus an equivalence relation on A is a subset of $A \times A$ with some special properties that motivate a partition of A.

Any time we define a term that describes how elements of a set may or may not compare to each other, we can use that basis of comparison as a criterion for inclusion in a relation. For example, the symbol $\leq$ denotes a way that two real numbers relate or compare to each other. In Example 2.9.5, we used the statement $x \leq y$ as a criterion for inclusion in a relation. The relation defined by $\leq$ has some special properties that motivate a definition of another type of relation. Since it's common to write xRy to mean $(x, y) \in R$, and say that x is *related to* y instead of saying (x, y) is an element of the relation, we'll use this somewhat more efficient notation in the next definition and example.

Definition 2.9.10. Suppose R is a relation on a set A with the following properties:

(O1) xRx for all $x \in A$ (Reflexive property);

(O2) If xRy and yRx, then $x = y$ (Antisymmetric property);

(O3) If xRy and yRz, then xRz (Transitive property).

Then R is called an *order relation* on A, or a *partial ordering* of A.

Example 2.9.11. Show that the relation in Example 2.9.5 is an order relation on $\mathbb{R}$.

Solution: We show that R has properties O1–O3.

(O1) Since $x = x$ for all $x \in \mathbb{R}$, we have that $x \leq x$, so that $x R x$ for all $x \in \mathbb{R}$.

(O2) Suppose xRy and yRx. Then $x \leq y$ and $y \leq x$, so that $x = y$.

(O3) Suppose xRy and yRz. Then $x \leq y$ and $y \leq z$. Now if either $x = y$ or $y = z$, then $x \leq z$ by substitution. If, on the other hand, $x < y$ and $y < z$, then $x < z$ by Exercise 9 from Section 2.3. In either case, xRz.

Since R has properties O1–O3, it defines an order relation on $\mathbb{R}$. ∎

We can be even more efficient with our language and notation than we were in Example 2.9.11 by dispensing with the statement xRy and saying simply that $\leq$ defines an order relation on $\mathbb{R}$. That would make the preceding verification look like this:

Solution: We show that $\leq$ has properties O1–O3.

(O1) Clearly $x \leq x$ for all $x \in \mathbb{R}$.

(O2) If $x \leq y$ and $y \leq x$, then $x = y$.

(O3) Suppose $x \leq y$ and $y \leq z$. If either $x = y$ or $y = z$, then clearly $x \leq z$. If, on the other hand, $x < y$ and $y < z$, then $x < z$ from Exercise 9 in Section 2.3. In either case, $x \leq z$.

Since $\leq$ has properties O1–O3, it defines an order relation on $\mathbb{R}$. ∎

Verifying the claim in the next example will be quick if you reference the applicable results from Section 2.1 (Exercise 3).

Example 2.9.12. If A is a set, then $\subseteq$ defines an order relation on 2^A.

For the set $\{1, 2, 3\}$, the order relation defined by $\subseteq$ can be illustrated with a *directed graph* as in Fig. 2.6. The arrow from $\{1\}$ to $\{1, 2\}$ indicates that the former is related to

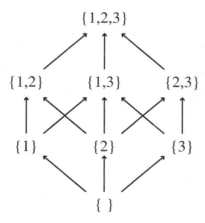

Figure 2.6 Digraph of the partial order relation $\subseteq$ on $2^{\{1,2,3\}}$.

the latter by $\subseteq$. Notice that $\{1, 2, 3\}$ is reachable from $\{1\}$ by a *directed path* by way of either $\{1, 2\}$ or $\{1, 3\}$, and either of these directed paths indicates that $\{1\} \subseteq \{1, 2, 3\}$.

You'll verify the claim of the next example in Exercise 4, then in Exercise 5 you'll sketch a directed graph that illustrates divisibility on a given subset of $\mathbb{Z}$.

Example 2.9.13. Divisibility defines an order relation on $\mathbb{Z}$.

Another type of relation is reminiscent of $<$.

Definition 2.9.14. Suppose R is a relation on a set A with the following properties:

(S1) $x \not R x$ for all $x \in A$ (Irreflexive property);

(S2) If $x R y$, then $y \not R x$ (Asymmetric property);

(S3) If $x R y$ and $y R z$, then $x R z$ (Transitive property).

Then R is called a *strict order relation* on A, or a *strict ordering* of A.

Example 2.9.15. Show that $<$ defines a strict order relation on $\mathbb{Z}$.

Solution: We show that $<$ satisfies properties S1–S3.

(S1) Since $x - x = 0$, we have that $x - x \not> 0$, so $x \not< x$ for all $x \in \mathbb{Z}$.

(S2) Suppose $x < y$. Then $y - x > 0$, so that $y - x \not< 0$. Thus $y \not< x$.

(S3) From Exercise 9 from Section 2.3, if $x < y$ and $y < z$, then $x < z$. ■

To this point we made no more than a passing reference in Chapter 0 to strict subset inclusion $\subset$. Saying that $A_1 \subset A_2$ creates a compound statement. First, if $x \in A_1$, then $x \in A_2$. As well, there exists $x \in A_2$ such that $x \notin A_1$. In Exercise 6, you'll show that $\subset$ defines a strict order relation on 2^A.

One notable difference between $\leq$ and divisibility on $\mathbb{Z}$ is that any pair of integers we choose is comparable in either one way or the other according to $\leq$. That is, for all $a, b \in \mathbb{Z}$, either $a \leq b$ or $b \leq a$. For divisibility, however, there are many pairs of integers that are not comparable at all. For example, neither $6 \mid 10$ nor $10 \mid 6$ is true. This distinction motivates yet another type of relation.

Definition 2.9.16. Suppose R is a relation on a set A with the following properties:

(T1) $x R x$ for all $x \in A$ (Reflexive property);

(T2) If $x R y$ and $y R x$, then $x = y$ (Antisymmetric property);

(T3) If $x R y$ and $y R z$, then $x R z$ (Transitive property);

(T4) For all $x, y \in A$, either $x R y$ or $y R x$.

Then R is called a *total order relation* on A, or a *total ordering* of A.

Notice that properties T1–T3 are the same as O1–O3, so a total order relation is a special kind of order relation where every pair of elements is comparable in one way or another.

Example 2.9.17. On $\mathbb{R}$, the relation $\leq$ is a total ordering.

Example 2.9.18. For $A = \{1, 2, 3\}$, $\subseteq$ does not define a total order of 2^A, for $\{1, 2\} \not\subseteq \{2, 3\}$ and $\{2, 3\} \not\subseteq \{1, 2\}$.

One of our assumptions from Chapter 0 is the WOP of $\mathbb{W}$. To say that a is the smallest element of a nonempty $S \subseteq \mathbb{W}$ is to say that $a \leq x$ for all $x \in S$. The last type of relation we define is a generalization of the WOP of $\mathbb{W}$.

Definition 2.9.19. Suppose R is a relation on a set A with the following properties:

(W1) $x R x$ for all $x \in A$ (Reflexive property);

(W2) If $x R y$ and $y R x$, then $x = y$ (Antisymmetric property);

(W3) If $x R y$ and $y R z$, then $x R z$ (Transitive property);

(W4) If S is any nonempty subset of A, then there exists $a \in S$ such that $a R x$ for all $x \in S$.

Then R is called a *well order* on A, or a *well ordering* of A.

Properties W1–W3 are the same as O1–O3, so a well ordering of A is an order relation. Property W4 adds the feature that every nonempty subset of A contains what we might call a least element. The antisymmetric property implies that such a least element is unique (Exercise 8). You'll show in Exercise 9 that W4 implies T4, so that a well ordering is a total ordering. However, a total ordering is not necessarily a well ordering. For example, $\leq$ is a total ordering of $\mathbb{Z}$ that is not a well ordering, as is illustrated by $\mathbb{Z}$ itself. If $a \in \mathbb{Z}$ were a least element of $\mathbb{Z}$, then $a - 1$ is an integer for which $a \leq a - 1$ is false. Since W4 implies T4, but not vice versa, property W4 is stronger than T4.

Just because a given set with an order relation fails to be a well ordering, it does not mean that the set cannot be well ordered by some other relation. In fact, one of the most notable results in modern set theory is that *any* set can be well ordered. Ernst Zermelo demonstrated this in 1904, using an axiom of set theory that we have said nothing about so far in this text. The *axiom of choice* is a somewhat mysterious axiom of set theory that is simple to state, but not often addressed at this level of the mathematical game, at least not without some caveats and near apologies. It says "Given any family $\mathcal{F}$ of mutually disjoint nonempty sets, there is a set S that contains a single element from each set in $\mathcal{F}$." Thus S can be thought of as the result of having chosen a representative element from each set in $\mathcal{F}$. In the axiom of choice there is enough strength to demonstrate that for any set, there exists a relation A that is a well ordering. One way to see how $\mathbb{Z}$ can be well ordered is to list its elements as $\langle 0, -1, 1, -2, 2, -3, 3, \ldots \rangle$ and define $x R y$ if x does not come after y in this listing. Ordering $\mathbb{Z}$ in this way, every nonempty subset has a least element.

EXERCISES

1. For each of the relations on $\mathbb{R}$, sketch the set of included points in the xy plane:
   (a) $\{(x, y) : x \leq y\}$
   (b) $\{(x, y) : x^2 + y \leq 4\}$
   (c) $\{(x, y) : |x| < 1\}$
   (d) $\{(x, y) : x^2 + y^2 \geq 1\}$

2. List all elements of the power set of the following sets:
   (a) $\emptyset$
   (b) $\{1\}$
   (c) $\{1, 2\}$
   (d) $\{1, 2, 3, 4\}$

3. Show that for any set A, $\subseteq$ defines an order relation on 2^A.

4. Show that divisibility defines an order relation on $\mathbb{Z}$.

5. Sketch a directed graph to illustrate the order relation of divisibility on $\{1, 2, 3, 4, 6, 8, 9, 12, 18, 24, 27, 36, 54\}$.

6. Show that for any set A, $\subset$ defines a strict order relation on 2^A.

7. For $a, b \in \mathbb{Z}$, write $a <_{\mathbb{Z}} b$ to mean $a < b$ as in Example 2.9.15. For $p/q, r/s \in \mathbb{Q}$ where $0 <_{\mathbb{Z}} q$ and $0 <_{\mathbb{Z}} s$, define $p/q <_{\mathbb{Q}} r/s$ if $ps <_{\mathbb{Z}} qr$. Use the fact that $<_{\mathbb{Z}}$ is a strict order relation on $\mathbb{Z}$ to show that $<_{\mathbb{Q}}$ is a strict order relation on $\mathbb{Q}$.

8. Show that the least element of a well-ordered set is unique.

9. Show that a well ordering is a total ordering by showing that property W4 implies T4.

CHAPTER 3

Functions

3.1 Definitions and Terminology

Second only to sets, *functions* are likely the most important mathematical concept. In reality, functions can be defined solely in terms of sets, and some authors take that approach. So it's arguable that sets are the heart of all mathematics. In this section, we define the term *function* and study some examples, and then we expand on some important terminology.

3.1.1 Definition and examples

Definition 3.1.1. Given two nonempty sets A and B, a *function* f is a rule, or set of instructions, by which each element of A is paired with exactly one element of B. A is called the *domain*, and is denoted Dom f. B is called the *codomain*. Notationally, if f is a function from A to B, we write $f : A \to B$. If $a \in A$ is paired by f with $b \in B$, we write $f(a) = b$ or $a \mapsto b$. We say that b is the image of a, or that a *maps* to b, and a is a *pre-image* of b. The subset of B consisting of the images of all elements of A is called the *range*, and is denoted Rng f. (See Fig. 3.1 for a basic sketch.)

Another way to define a function is in terms of a relation, that is, a subset of $A \times B$. Instead of imagining elements of A being associated with elements of B via the input-output imagery of Definition 3.1.1, it's possible to define a function as a set of pairs (a, b), where $a \in A$ and $b \in B$, and with some additional restrictions we'll mention in what follows. We also use the term *mapping* to refer to any pairing of the elements of two sets. Thus a function is also a mapping with some special restrictions.

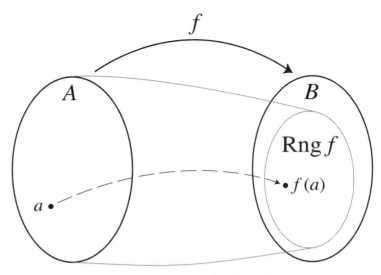

Figure 3.1 Schematic of a function: $f : A \to B$.

Let's present some clarifying thoughts and examples to give us a handle on Definition 3.1.1. First, for a mapping $f : A \to B$ to be a function, *every* $a \in A$ must have an image $b \in B$. That is:

(F1) For every $a \in A$, there exists $b \in B$ such that $f(a) = b$.

In addition to property F1, the rule defining f must produce a unique $f(a)$ for all $a \in A$. Mathematically, we say that f is *well defined* if $f(a)$ is unique for every $a \in A$. How can we say mathematically that the image of $a \in A$ must be unique? (See Fig. 3.2 for a sketch of what must not happen.) One way to say that f is well defined is to say

(F2) If $b_1, b_2 \in B$ are such that $f(a) = b_1$ and $f(a) = b_2$, then $b_1 = b_2$.

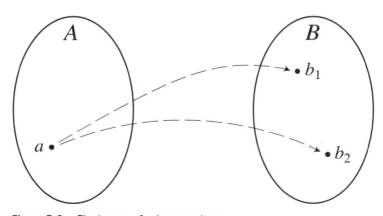

Figure 3.2 The image of a is not unique.

At first glance, F2 sounds like a statement that would be true for any mapping. After all, how could b_1 ever be different from b_2 when all you have to do is apply the transitive property of equality to the two equations $f(a) = b_1$ and $f(a) = b_2$? Unfortunately, this way of expressing the uniqueness of $f(a)$ doesn't reveal possible problems that can prevent f from being well defined. Here are two examples that illustrate how a mapping can fail to be well defined.

Example 3.1.2. It's possible that the rule defining $f : A \to B$ can be ambiguous. For example, let $f : \mathbb{R}^+ \cup \{0\} \to \mathbb{R}$ be defined in the following way. For $x \in \mathbb{R}^+ \cup \{0\}$, define $f(x)$ to be a solution y to the equation $y^2 = x$. You cannot doubt that this is a set of instructions for generating $f(x)$ from x, that the generated $f(x)$ is in $\mathbb{R}$, and that it works for every x in the domain. However, the rule is ambiguous for every $x \neq 0$. Letting $x = 25$, $y_1 = 5$, and $y_2 = -5$, we have demonstrated the existence of $y_1, y_2 \in \mathbb{R}$ where $f(x) = y_1$, $f(x) = y_2$, and $y_1 \neq y_2$.

Example 3.1.2 illustrates a possible pitfall you might run into with the notation $f(x)$ if you haven't verified that f is well defined. Unless you know that f is well defined, you might find yourself using an expression like $f(25)$ as one name for more than one thing. Make sure that the set of instructions detailing how $f(x)$ is to be determined doesn't produce more than one value for any x in the domain.

Here's another example of how a mapping can fail to be well defined, this time because a single element in the domain might have more than one distinct name.

Example 3.1.3. Define $f : \mathbb{Q} \to \mathbb{Z}$ in the following way. For $p/q \in \mathbb{Q}$, define $f(p/q) = p$. That is, the image of a rational number written in a standard form of integer over integer is defined to be the numerator. Unlike Example 3.1.2, there is no ambiguity concerning what $f(p/q)$ is. The problem here is that the standard definition of equality in the rationals ($=_{\mathbb{Q}}$, as discussed in Section 2.6.1) lumps a lot of different expressions of the form p/q into the same equivalence class. As a specific example, although $2/5 = 6/15$, $f(2/5) \neq f(6/15)$. The fact that one domain element can be addressed by more than one name causes problems, because the image of $x \in \mathbb{Q}$ depends on the form it's written in. Thus f is not well defined because we have exhibited $a \in \mathbb{Q}$ and $b_1, b_2 \in \mathbb{Z}$ where $f(a) = b_1$, $f(a) = b_2$, but $b_1 \neq b_2$.

Example 3.1.3 illustrates a good way to restate property F2. If a_1 and a_2 are two names for the same thing, that is if $a_1 = a_2$, then the rule must produce the same functional value for a_1 and a_2. That is, $f(a_1) = f(a_2)$.

(F2) If $a_1, a_2 \in A$ and $a_1 = a_2$, then $f(a_1) = f(a_2)$.

In the language of relations, properties F1 and F2 translate to the following. If a relation $R \subseteq A \times B$ is a function, property F1 says that, for all $a \in A$, there must exist $b \in B$ such that $(a, b) \in R$. Property F2 says that if $(a, b_1), (a, b_2) \in R$, then $b_1 = b_2$.

Suppose someone gives us two sets A and B, and a rule f for pairing elements of A with elements of B. If we are asked to verify that f is a function, then we must verify that F1 and F2 hold.

Example 3.1.4. Show that $f : \mathbb{R}\setminus\{1\} \to \mathbb{R}$ defined by

$$f(x) = \frac{x^2 + 2}{x - 1}$$

is a function.

Solution:

(F1) Pick any $x \in \mathbb{R}\setminus\{1\}$. Then by properties A3 and A9 (closure in $\mathbb{R}$ of addition and multiplication, respectively), $x^2 + 2$ and $x - 1$ exist in $\mathbb{R}$. Furthermore, because $x \neq 1$, then $x - 1 \neq 0$, so that $(x - 1)^{-1} \in \mathbb{R}$ by property A13 (existence of multiplicative inverses). Finally, $(x^2 + 2) \cdot (x - 1)^{-1} \in \mathbb{R}$ by closure of multiplication. Thus $f(x)$ exists in $\mathbb{R}$.

(F2) Pick $x_1, x_2 \in \mathbb{R}\setminus\{1\}$ and suppose $x_1 = x_2$. Then by properties A2 and A8 (addition and multiplication in $\mathbb{R}$ are well defined), it follows that $x_1^2 + 2 = x_2^2 + 2$ and $x_1 - 1 = x_2 - 1$. By the uniqueness of multiplicative inverses, $(x_1-1)^{-1} = (x_2-1)^{-1}$. Finally, applying property A8 again, $(x_1^2+2)/(x_1-1) = (x_2^2 + 2)/(x_2 - 1)$, so that $f(x_1) = f(x_2)$. ∎

Example 3.1.4 illustrates a broad result that is fairly easy to see using only the example. If f is a rule that can be written in the form of polynomial over polynomial, that is, if

$$f(x) = \frac{a_n x^n + a_{n-1}x^{n-1} + \cdots + a_1 x + a_0}{b_m x^m + b_{m-1}x^{m-1} + \cdots + b_1 x + b_0}, \tag{3.1}$$

where $a_i, b_j \in \mathbb{R}$ ($1 \leq i \leq n, 1 \leq j \leq m$), then, assuming we avoid values of x that make the denominator zero, f is a (well-defined) function from an appropriate subset of $\mathbb{R}$ to $\mathbb{R}$. Here's another way to say it. Since $+$ and $\times$ are well defined on $\mathbb{R}$, and additive and multiplicative inverses are unique, then any mapping whose rule is built up from the operations of $+, -, \times,$ and $\div$, and whose domain avoids any occurrence of 0^{-1} will be well defined.

We can go even further. Thanks to Theorem 2.8.1 and Definition 2.8.2, we have ensured that $\sqrt[n]{x}$ is well defined for all $x \in \mathbb{R}$ if n is odd, and for all $x \geq 0$ if n is even. Therefore, we can say that any real-valued mapping whose rule is built up algebraically with $+, -, \times, \div, \square^n$, and $\sqrt[n]{\ }$, and whose domain is a subset of $\mathbb{R}$ that avoids 0^{-1} and even roots of negative numbers will be well defined. For example,

$$f(x) = 5 + x + \sqrt[3]{\frac{1 + \sqrt{x^3 - 1/(x-3)}}{x^5 - 1 + x^{-5}}} \tag{3.2}$$

is a (well-defined) function whose domain is some hideous subset of $\mathbb{R}$. Functions built up as is f in Eq. (3.2) are called *algebraic*.

There is a simple function that will come in handy in Section 3.3. In its simplicity it illustrates the sorts of things we have to show when verifying that a mapping is a function and when showing a function has certain properties.

Example 3.1.5. Let $n \in \mathbb{N}$ and $m \in \mathbb{Z}$, and write $\mathbb{N}_n = \{1, 2, \ldots, n\}$ and $\mathbb{N}_n^m = \{1 + m, 2 + m, \ldots, n + m\}$. Define a translation mapping $T : \mathbb{N}_n \to \mathbb{N}_n^m$ by $T(x) = x + m$. Show T is a function.

Solution: We show that T satisfies properties F1 and F2.

(F1) Pick $x \in \mathbb{N}_n$. Since $1 \leq x \leq n$, it follows that $1 + m \leq x + m \leq n + m$, or $1 + m \leq T(x) \leq n + m$. Thus $T(x) \in \mathbb{N}_n^m$.

(F2) Pick $x_1, x_2 \in \mathbb{N}_n$ and suppose $x_1 = x_2$. Then since addition in $\mathbb{R}$ is well defined, $T(x_1) = x_1 + m = x_2 + m = T(x_2)$. Thus T is well defined on $\mathbb{N}_n$. ∎

3.1.2 Other terminology and notation

There is a wealth of terminology surrounding functions, and we present some of it now. As you read the informal introductions to the terms, you should try to construct formal definitions yourself for practice. After you've tried to do it yourself, then read the definitions provided.

First we ask what meaning we would like to assign to the statement $f = g$. The standard definition is the following.

Definition 3.1.6. Two functions f and g are said to be equal if they have the same domain and codomain, and $f(a) = g(a)$ for all a in the domain.

It's a quick mental exercise to see that Definition 3.1.6 is an equivalence relation. For example, to show E1, we note that f has the same domain and codomain as itself, and $f(a) = f(a)$ for all a in the domain. Showing E2 and E3 is equally easy.

If $f : A \to B$ is a function, we would like to create some notation to denote the image, not of a single element, but of a subset of the domain $A_1 \subseteq A$. The notation we use is $f(A_1)$. Clearly, it is a subset of the range. How should we define $f(A_1)$ for some $A_1 \subseteq A$? To put it another way, under what conditions is $y \in f(A_1)$? Here's the definition:

Definition 3.1.7. Suppose $f : A \to B$ is a function and $A_1 \subseteq A$. Then the *image of A_1*, denoted $f(A_1)$, is defined by

$$f(A_1) = \{y \in B : (\exists x \in A_1)(y = f(x))\}. \tag{3.3}$$

Thus $y \in f(A_1)$ if and only if there exists $x \in A_1$ such that $y = f(x)$.

With the notation of Definition 3.1.7, we can then rigorously define the range of $f : A \to B$ by

$$\text{Rng } f = f(A) = \{y \in B : (\exists x \in A)(y = f(x))\}. \tag{3.4}$$

Property F1 says that a function f must map *every* element of A to some element of B. Property F2 says that the path from $a \in A$ to $f(a) \in B$ must not have a fork in the road, as was illustrated in Fig. 3.2. Two other terms relate to the possibility that a function $f : A \to B$ might have analogous features when viewed, as it were, in the reverse direction. Analogous to every $a \in A$ having an image, perhaps every $b \in B$ has

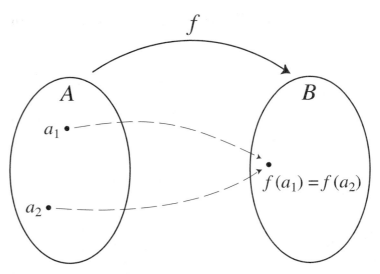

Figure 3.3 This function is not one-to-one.

a pre-image. Also, analogous to the uniqueness of the image of $a \in A$, perhaps every $b \in \text{Rng } f$ has a unique pre-image. A function with the former feature is called *onto*. A function with the latter feature is called *one-to-one*. In a one-to-one function, the forbidden merging of the path from different elements of A to the same image in B is illustrated in Fig. 3.3.

So how shall we define the terms *onto* and *one-to-one*? Think about it before you read the definitions that follow.

Definition 3.1.8. Suppose $f : A \to B$ is a function. Then f is said to be *onto* provided for every $b \in B$, there exists $a \in A$ such that $f(a) = b$. That is,

$$f \text{ is onto} \Leftrightarrow (\forall y \in B)(\exists x \in A)(y = f(x)). \tag{3.5}$$

As a notational shorthand, we write $f : A \xrightarrow{\text{onto}} B$. An onto function is also called a *surjection*.

Example 3.1.9. Show that $f : \mathbb{R} \to \mathbb{R}$ defined by $f(x) = x^3 - 1$ is a surjection.

Solution: Memories from pre-calculus resurface, and we recall that the graph of a polynomial function of odd degree extends all the way up and down the y-axis, so that everything in the codomain $\mathbb{R}$ has a pre-image. But how do we prove f is onto? We have to choose an arbitrary $y \in \mathbb{R}$ and work backwards to find some $x \in \mathbb{R}$ that maps to it. If $y \in \mathbb{R}$ is chosen arbitrarily, what does x need to be so that $y = f(x)$ will be true? Assuming you found it, here's a cleaned up proof.

Pick $y \in \mathbb{R}$. Let $x = \sqrt[3]{y+1}$. Then x is, in fact, a real number (Theorem 2.8.1), and

$$f(x) = f(\sqrt[3]{y+1}) = [\sqrt[3]{y+1}]^3 - 1 = y + 1 - 1 = y. \tag{3.6}$$

Thus f is onto. ∎

Example 3.1.10. Show that the function T from Example 3.1.5 is a surjection.

Solution: Pick $y \in \mathbb{N}_n^m$. Let $x = y - m$. Since $1 + m \leq y \leq n + m$, we have that $1 \leq x \leq n$, so that $x \in \mathbb{N}_n$. Also, $T(x) = x + m = y$. Thus T is onto. ∎

Before you read the definition of one-to-one, think about uniqueness of a pre-image and try to come up with your own definition.

Definition 3.1.11. Suppose $f : A \to B$ is a function. Then f is said to be one-to-one provided $[f(a_1) = f(a_2)] \to [a_1 = a_2]$. Or, if you prefer,

$$f \text{ is one-to-one} \Leftrightarrow (\forall a_1, a_2 \in A)([f(a_1) = f(a_2)] \to [a_1 = a_2]). \tag{3.7}$$

As a notational shorthand, we write $f : A \xrightarrow{1\text{-}1} B$. A one-to-one function is also called an *injection*.

Example 3.1.12. Show that the function T from Example 3.1.5 is an injection.

Solution: Pick $x_1, x_2 \in \mathbb{N}_n$ and suppose that $T(x_1) = T(x_2)$. Then $x_1 + m = x_2 + m$, so that $x_1 = x_2$ by cancellation. Thus T is one-to-one. ∎

If $f : A \to B$ is both one-to-one and onto, we may write $f : A \xrightarrow[\text{onto}]{1\text{-}1} B$. A one-to-one, onto function is also called a *bijection*. A bijection from A to B is said to put the sets A and B into a *one-to-one correspondence*.

Perhaps the simplest example of a one-to-one correspondence is the *identity function* $i : A \to A$ defined by $i(a) = a$ for all $a \in A$. The need for this ho-hum function arises more often than you might think, and the fact that it is a one-to-one, onto function from a set A to itself must be demonstrated. You'll do that in Exercise 5.

3.1.3 Three important theorems

The following theorems are not the prettiest ones you'll ever see in your mathematical life, but they're not the ugliest either. They're not complicated in principle, but proving them requires slavish attention to some rather minute details. If nothing else, they illustrate the inescapable fact that laying the groundwork for later, more elegant results sometimes requires you to muddle your way through preliminary theorems whose proofs are sticky and might involve multiple cases. You'll prove the first of these theorems in Exercise 10. It comes in handy in proving the other two.

Theorem 3.1.13. *Suppose* $f_1 : A_1 \xrightarrow[onto]{1\text{-}1} B_1$ *and* $f_2 : A_2 \xrightarrow[onto]{1\text{-}1} B_2$ *are functions, and that* $A_1 \cap A_2 = B_1 \cap B_2 = \emptyset$. *Define* $f : A_1 \cup A_2 \to B_1 \cup B_2$ *by*

$$f(x) = \begin{cases} f_1(x), & \text{if } x \in A_1; \\ f_2(x), & \text{if } x \in A_2. \end{cases}$$

Then f is a one-to-one, onto function from $A_1 \cup A_2$ to $B_1 \cup B_2$.

Theorem 3.1.13 says that if you're given two one-to-one, onto functions defined from disjoint domains to disjoint codomains, then it's possible to paste together the domains, paste together the codomains, and then define a single function between these new sets that is also one-to-one and onto.

We'll prove the next theorem here, for it's a little sticky and will prove to be a good warm-up for proving the third.

Theorem 3.1.14. *Let $n \in \mathbb{N}$, and let $S \subseteq \mathbb{N}_n$ be any nonempty set. Then there exists a natural number $m \leq n$ and some $f : S \to \mathbb{N}_m$ where f is a one-to-one, onto function.*

Proof: We prove by induction on $n \geq 1$.

(I1) If $n = 1$, then $S = \mathbb{N}_n = \{1\}$. Letting $m = 1$ we may define $f : S \to \mathbb{N}_m$ by $f(1) = 1$, which is clearly a one-to-one, onto function.

(I2) Suppose $n \geq 1$ and that the result is true for any nonempty subset of $\mathbb{N}_n$. Let S be any nonempty subset of $\mathbb{N}_{n+1}$, and consider the following three cases.

If $S = \mathbb{N}_{n+1}$, then we may let $m = n + 1$ and $f : S \to \mathbb{N}_m$ be the identity function.

If $S \neq \mathbb{N}_{n+1}$ and $n + 1 \notin S$, then $S \subseteq \mathbb{N}_n$. Thus, the inductive assumption applies, and there exists $m \leq n$ and a function $f : S \xrightarrow[onto]{1\text{-}1} \mathbb{N}_m$.

If $S \neq \mathbb{N}_{n+1}$ and $n + 1 \in S$, then there exists $k \in \mathbb{N}_n$ such that $k \notin S$. Let $T = S \cup \{k\} \setminus \{n + 1\}$. (Draw a picture!) Then $T \subseteq \mathbb{N}_n$, and the inductive assumption applies to yield $m \leq n$ and a function $f : T \xrightarrow[onto]{1\text{-}1} \mathbb{N}_m$. Define $g : S \to \mathbb{N}_m$ by

$$g(x) = \begin{cases} f(x), & \text{if } x \in T \setminus \{k\}; \\ f(k), & \text{if } x = n + 1. \end{cases} \tag{3.8}$$

Defining $A_1 = T \setminus \{k\}$, $B_1 = \{n + 1\}$, $A_2 = \mathbb{N}_m \setminus \{f(k)\}$, and $B_2 = \{f(k)\}$, we may apply Theorem 3.1.13 to conclude that $g : S \to \mathbb{N}_m$ is one-to-one and onto. ∎

The last theorem is a sort of nonexistence theorem whose importance will become clear in Section 3.3. You'll prove it in Exercise 11.

Theorem 3.1.15. *Suppose $m, n \in \mathbb{N}$ and $m < n$. Then there is no function $f : \mathbb{N}_n \xrightarrow[onto]{1\text{-}1} \mathbb{N}_m$.*

EXERCISES

1. What does it mean to say that a function is not onto?

2. What does it mean to say that a function is not one-to-one?

3. Here are several pairs of sets.

    (a) $\{1, 2, 3, 4\}$ and $\{a, b, c\}$
    (b) $\{1, 2, 3, 4\}$ and $\mathbb{N}$
    (c) $\mathbb{Z}$ and $\mathbb{Z}$
    (d) $\mathbb{Z}$ and $\mathbb{N}$
    (e) $\mathbb{R}$ and $\mathbb{R}$

    For each of the preceding pairs of sets, find four functions $\{f_1, f_2, f_3, f_4\}$ from the first set to the second set with the following properties, if such functions are possible.

    (i) f_1 is one-to-one but not onto.
    (ii) f_2 is onto but not one-to-one.
    (iii) f_3 is both one-to-one and onto.
    (iv) f_4 is neither one-to-one nor onto.

    Be prepared to provide as much explanation as necessary to support your claim.

4. Prove that the function $f : \mathbb{R} \to \mathbb{R}$ defined by $f(x) = x^3 - 1$ is one-to-one.

5. Show that the identity mapping $i : A \to A$ is a one-to-one function from $A \neq \emptyset$ onto itself.

6. Suppose $f : A \to B$ is a function and $A_1, A_2 \subseteq A$. Prove the following, or disprove it by providing a counterexample. If the claim is not true, is at least one direction of subset inclusion true?

    (a) $f(A_1 \cap A_2) = f(A_1) \cap f(A_2)$.[1]
    (b) $f(A_1 \cup A_2) = f(A_1) \cup f(A_2)$.

7. From the parts of Exercise 6 that are valid, state analogous theorems for a family of sets $\mathcal{F} = \{A\}$.

8. This exercise is a return to Exercise 6a, which we hope you have already solved. Suppose $f : A \xrightarrow{1\text{-}1} B$. Show that $f(A_1 \cap A_2) = f(A_1) \cap f(A_2)$.

9. In this exercise we want to consider functions $f : S \to \mathbb{R}$ where $S \subseteq \mathbb{R}$ and f is of the form

$$f(x) = \frac{ax + b}{cx + d}, \tag{3.9}$$

where c and d are not *both* zero. Clearly since f is algebraic, it is well defined on whatever subset of $\mathbb{R}$ prevents $cx + d = 0$.

---

[1] See Exercise 8 if you have to.

(a) Under what conditions (i.e., restrictions on a, b, c, d) will f be one-to-one?[2]

(b) Given that the restriction you discovered in part (a) holds, f is a one-to-one function. Consider the two cases $c = 0$ and $c \neq 0$.

  i. Show that if $c = 0$, then f is defined on all of $\mathbb{R}$ and is onto.

  ii. Show that if $c \neq 0$, then there exist $x_0, y_0 \in \mathbb{R}$ such that f is defined on $\mathbb{R} \setminus \{x_0\}$ and f is onto $\mathbb{R} \setminus \{y_0\}$.

10. Prove Theorem 3.1.13: Suppose $f_1 : A_1 \xrightarrow[\text{onto}]{1\text{-}1} B_1$ and $f_2 : A_2 \xrightarrow[\text{onto}]{1\text{-}1} B_2$ are functions, and that $A_1 \cap A_2 = B_1 \cap B_2 = \emptyset$. Define $f : A_1 \cup A_2 \to B_1 \cup B_2$ by

$$f(x) = \begin{cases} f_1(x), & \text{if } x \in A_1; \\ f_2(x), & \text{if } x \in A_2. \end{cases}$$

Then f is a one-to-one, onto function from $A_1 \cup A_2$ to $B_1 \cup B_2$.

11. Prove Theorem 3.1.15 by induction on $m \geq 1$ in the following way:

(a) Show that there is no such function for the case $m = 1$.

(b) Use the contrapositive to prove the inductive step: If there exists $f : \mathbb{N}_{n+1} \xrightarrow[\text{onto}]{1\text{-}1} \mathbb{N}_{m+1}$ for some $1 \leq m < n$, then there exists $g : \mathbb{N}_n \xrightarrow[\text{onto}]{1\text{-}1} \mathbb{N}_m$.[3]

3.2 Composition and Inverse Functions

One way to think of a function is as a linking of a set A to a set B with certain restrictions. If A is linked to B by a function f, and B is linked to C by a function g, then we might want to look into how to link A to C via B, using f and g in succession. *Composition* is the term we use when combining functions in this way. It might remind you a little of the transitive property, whereby some relationship between a and b, in conjunction with a relationship between b and c, yields the same relationship between a and c. Then, reminiscent of the symmetric property, whereby $a = b$ implies $b = a$, we study the possibility of using $f : A \to B$ to link B back to A by reversing the function f. Such a reverse linking is called the *inverse* of f.

3.2.1 Composition of functions

Suppose $f : A \to B$ and $g : B \to C$ are functions. We want to define a new mapping $h : A \to C$ that uses the rules of f and g together as in Fig. 3.4.

---

[2] If the statement $f(x_1) = f(x_2) \Rightarrow x_1 = x_2$ is going to be a true statement, what must be assumed about a, b, c, d to make it so? Play with the statement $f(x_1) = f(x_2)$.
[3] If $f(n+1) = m+1$, then defining g should be easy. Otherwise, $f(n+1) = k$ for some $1 \leq k \leq m$ and $f(l) = m+1$ for some $1 \leq l \leq n$. Define g to map l to k.

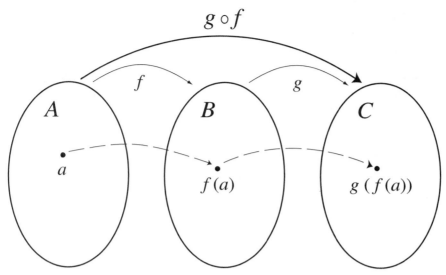

Figure 3.4 Composition: $(g \circ f) : A \to C$.

Definition 3.2.1. If $f : A \to B$ and $g : B \to C$ are functions, we define the *composition* $g \circ f : A \to C$ to be the mapping defined by $(g \circ f)(a) = g(f(a))$ for all $a \in A$.

Example 3.2.2. Consider the functions $f : \mathbb{R} \to \mathbb{R}^+ \cup \{0\}$ defined by $f(x) = x^2$ and $g : \mathbb{R}^+ \cup \{0\} \to \mathbb{R}^+ \cup \{0\}$ defined by $g(x) = \sqrt{x}$. Then $(g \circ f)(x) = \sqrt{x^2}$. This is another way to define $|x|$.

Notice we did not presume to say that $g \circ f$ is a function in Definition 3.2.1. We need to use the fact that f and g are themselves functions to show that properties F1 and F2 hold for $g \circ f$ (Exercise 1).

Composition should not be new to you. You used it extensively in calculus, and the chain rule is the technique you used to differentiate a function that was composed of other functions. The questions we want to address here involve relationships between f, g, and $g \circ f$ based on their individual characteristics. For example, if f and g are one-to-one, does it follow that $g \circ f$ is one-to-one? What about onto? What if you know something about f and $g \circ f$? Can you conclude anything about g?

Given f, g, and $g \circ f$, and given the characteristics of one-to-one and onto, let's put them together in all possible combinations and consider the following six possible theorems. Given $f : A \to B$ and $g : B \to C$:

(Q1) If f and g are onto, then $g \circ f$ is onto.

(Q2) If f and $g \circ f$ are onto, then g is onto.

(Q3) If g and $g \circ f$ are onto, then f is onto.

(Q4) If f and g are one-to-one, then $g \circ f$ is one-to-one.

(Q5) If f and $g \circ f$ are one-to-one, then g is one-to-one.

(Q6) If g and $g \circ f$ are one-to-one, then f is one-to-one.

Some of the questions Q1–Q6 are true, and some are false. In fact, some might be true even if one of the hypothesis conditions is omitted. To try to prove or disprove them, a picture might be helpful. Counterexamples can be constructed easily if you let simple pictures with simple sets inspire them. We'll prove that Q4 is true and then you'll attack the rest in Exercise 2.

Theorem 3.2.3. *If $f : A \to B$ and $g : B \to C$ are one-to-one functions, then $g \circ f : A \to C$ is one-to-one.*

Proof: Suppose f and g are one-to-one, and suppose $(g \circ f)(a_1) = (g \circ f)(a_2)$. Thus $g[f(a_1)] = g[f(a_2)]$. Since g is one-to-one, we have $f(a_1) = f(a_2)$. Similarly, since f is one-to-one, it follows that $a_1 = a_2$. Thus $g \circ f$ is one-to-one. ∎

3.2.2 Inverse functions

If $f : A \to B$ is a function, then the rule linking A to B is well defined on all of A. We now ask a question from the perspective of B. Given $f : A \to B$, can we find a function $g : B \to A$ whose rule, in effect, is the exact reversal of the rule for f? That is, if $f(8) = -2$, then we want $g(-2) = 8$. It might occur to you right away that f will need to be a certain kind of function in order for the g derived from it to be a function itself. For example, what feature will f need to have in order for g to be defined on all of B? The answer: f must be onto. What feature will f need in order for g to be well defined? The answer: f must be one-to-one. Such a function g, if one exists, is called an *inverse* of f. Here are a definition and a theorem addressing its unique existence.

Definition 3.2.4. Suppose $f : A \to B$ is a function. Then we say that a function $g : B \to A$ is an *inverse* of f if $(g \circ f)(a) = a$ for all $a \in A$ and $(f \circ g)(b) = b$ for all $b \in B$. Such a function g is generally denoted f^{-1}.

Theorem 3.2.5. *Given a function $f : A \xrightarrow[\text{onto}]{1\text{-}1} B$, there exists a unique inverse function $f^{-1} : B \xrightarrow[\text{onto}]{1\text{-}1} A$.*

The proof of this theorem has several parts to it. We'll get it started here, though you'll supply the details in Exercise 4. To define the rule for $g : B \to A$, we must pick $b \in B$ and explain how to find some $a \in A$ that we're going to call $g(b)$. So for $b \in B$, define $g(b)$ to be any solution a to the equation $f(a) = b$. Showing g has property F1 is to show that the equation $f(a) = b$ has *some* solution in A. Showing g has property F2 is to show that $f(a) = b$ has a unique solution. And you're on your way.

In Section 3.1, we exploited the notation of functions to discuss the image of a set $f(A_1)$, where $A_1 \subseteq A$. Similarly, we can slightly abuse the notation f^{-1} to talk about the *pre-image* of a set $B_1 \subseteq B$. Even though the function f^{-1} might not exist because f might fail to be one-to-one or onto, we can still talk about all elements of the domain that map to the elements of B_1.

Definition 3.2.6. Given a function $f : A \to B$, and $B_1 \subseteq B$, we define the set $f^{-1}(B_1)$, the *pre-image* of B_1 by

$$f^{-1}(B_1) = \{a \in A : f(a) \in B_1\}. \tag{3.10}$$

Thus $x \in f^{-1}(B_1)$ if and only if $f(x) \in B_1$. If $B_1 = \{b\}$ has only one element, we generally write $f^{-1}(b)$ instead of $f^{-1}(\{b\})$.

The notation $f^{-1}(b)$ instead of $f^{-1}(\{b\})$ for a pre-image set is technically inaccurate, but it's a common abuse of notation. If f is one-to-one, so that f^{-1} exists as a function, then $f^{-1}(b)$ is simply the *element* of A that maps to b. In this case, we don't generally think of $f^{-1}(b)$ as a pre-image subset of the domain that contains this one element. However, if nothing is known about the existence of f^{-1} as a function, then $f^{-1}(b)$ should be thought of as the set of all domain elements that map to b by f. Naturally this set could be empty, or contain any number of elements.

Example 3.2.7. Consider $f : \mathbb{R} \to \mathbb{R}$ defined by $f(x) = x^2$. Determine $f^{-1}(4)$, $f^{-1}([0, 9])$, and $f^{-1}([-2, -1])$.

Solution: The first one is easy, for $f^{-1}(4) = \{\pm 2\}$. Without going into any detail for the others, we know that $0 \leq f(x) \leq 9$ if and only if $x \in [-3, 3]$. Thus $f^{-1}([0, 9]) = [-3, 3]$. Also, since $f(x) \geq 0$ for all $x \in \mathbb{R}$, $f^{-1}([-2, -1]) = \emptyset$. ∎

Given $A_1 \subseteq A$ and $B_1 \subseteq B$, we might want to address the following statements:

$$f^{-1}(f(A_1)) \stackrel{?}{=} A_1, \tag{3.11}$$

$$f(f^{-1}(B_1)) \stackrel{?}{=} B_1. \tag{3.12}$$

It turns out that neither of these statements is true, but in each case, one direction of subset inclusion is true. Example 3.2.7 can suggest counterexamples to demonstrate how subset inclusion fails in the other direction. See Exercise 7.

EXERCISES

1. Given functions $f : A \to B$ and $g : B \to C$, show that $g \circ f : A \to C$ from Definition 3.2.1 is a function.

2. Prove or disprove the remaining statements from Q1–Q6 on pages 107–108. Of those that are true, which have hypothesis conditions that can be relaxed?

3. For the false statements in Q1–Q6, replace one hypothesis condition with another that will make the statement true. Prove.

4. Prove Theorem 3.2.5 in the following way: Let $f : A \xrightarrow[\text{onto}]{1\text{-}1} B$ be given. Define a mapping $g : B \to A$ by letting $g(b)$ be any solution a to the equation $f(a) = b$. Show g is a one-to-one, onto function from B to A with the following steps:

   (a) Show that g is defined on all of B (F1).
   (b) Show that g is well defined (F2).
   (c) Show that g is onto.
   (d) Show that g is one-to-one.
   (e) Show that $(g \circ f)(a) = a$ for all $a \in A$ and $(f \circ g)(b) = b$ for all $b \in B$.
   (f) Show that g is unique: Suppose g_1 and g_2 are two functions from B to A that are one-to-one, onto and satisfy $(g_1 \circ f)(a) = (g_2 \circ f)(a) = a$ for all $a \in A$ and $(f \circ g_1)(b) = (f \circ g_2)(b) = b$ for all $b \in B$. Show that $g_1 = g_2$ by showing $g_1(b) = g_2(b)$ for all $b \in B$.[4]

5. For non-empty sets A and B, define $A \equiv B$ if there exists a function $f : A \xrightarrow[\text{onto}]{1\text{-}1} B$. Show that $\equiv$ defines an equivalence relation on the family of all non-empty sets.

6. Suppose $f : A \to B$ is a function. Prove the following:

   (a) If $A_1 \subseteq A_2 \subseteq A$, then $f(A_1) \subseteq f(A_2)$.
   (b) If $B_1 \subseteq B_2 \subseteq B$, then $f^{-1}(B_1) \subseteq f^{-1}(B_2)$.

7. Suppose $f : A \to B$ is a function, $A_1 \subseteq A$ and $B_1 \subseteq B$.

   (a) Prove the following:
      i. $A_1 \subseteq f^{-1}(f(A_1))$.
      ii. $f(f^{-1}(B_1)) \subseteq B_1$.
   (b) Provide examples to show that the reverse subset inclusions from part (a) are false.
   (c) Determine condition(s) under which the reverse subset inclusions from part (a) would be true. Prove your claims.

8. Let $f : A \to B$ be a function and $B_1 \subseteq B$. Show $f^{-1}(B_1') = [f^{-1}(B_1)]'$.

3.3 Cardinality of Sets

What would you have in mind by saying that a set has n elements? Or given two sets, what would you mean by saying they are the same *size*, that is, they have the same *number of elements*? The term we use to denote the number of elements in a set is *cardinality*. The question of cardinality is a bit stickier than you might imagine at first. Consider the following sets:

$$A = \{-5, -2, 1, 4, 7, 10\} \quad \text{and} \quad B = \{b, d, h, p, f, l\}. \tag{3.13}$$

[4]All you need are $(f \circ g_1)(b) = (f \circ g_2)(b) = b$ and that f is one-to-one.

What would you say is the cardinality of A? Would you be inclined to say that A and B have the same cardinality? What about

$$C = \{\ldots, -4, -2, 0, 2, 4, \ldots\} \quad \text{and} \quad D = \{\ldots, -10, -5, 0, 5, 10, \ldots\}?$$

What would you say is the cardinality of C? Would you say C and D have the same cardinality? What about $\mathbb{Z}$ and $\mathbb{Q}$? $\mathbb{Q}$ and $\mathbb{R}$? If a set is finite in cardinality, whatever that might mean, then saying it has cardinality n ought to be a pretty straightforward term to define. Also, if two sets are finite, then saying they have the same cardinality ought to suggest a pretty natural relationship between them. However, when sets are not finite, things get a little more complicated, as we'll see.

3.3.1 Finite sets

When you looked at A and B in Eq. (3.13), you probably counted six elements of A, then used that as a basis for saying A has cardinality six. Doing the same for B, you then said A and B have the same cardinality. To count elements in this way is precisely the motivation behind the following definition.

Definition 3.3.1. Let A be a set, and suppose there exists $n \in \mathbb{N}$ such that A can be placed into one-to-one correspondence with $\mathbb{N}_n = \{1, 2, 3, \ldots, n\}$. That is, there exists $n \in \mathbb{N}$ and some $f : \mathbb{N}_n \xrightarrow[\text{onto}]{1\text{-}1} A$. Then we say that A is a *finite* set, and that it has *cardinality* n, which we write $|A| = n$. For the empty set, we make a special definition, writing $\mathbb{N}_0 = \emptyset$, and we define the *empty mapping* $f : \mathbb{N}_0 \to \emptyset$ in order to say that $\emptyset$ is finite and has cardinality zero.

Defining $f : \mathbb{N}_0 \to \emptyset$ as an empty mapping does not really jibe with Definition 3.1.1, for in our definition of function, we required that domains and codomains be nonempty. However, there's no reason we cannot extend the definition of function to include this case, as long as we make sure that properties F1 and F2 apply to it. And they do. In fact, the empty mapping meets all the requirements for being a one-to-one, onto function because all the requirements are statements that involve the universal quantifier. For example, the empty mapping is onto, for if it were not, then there would exist some $y \in \emptyset$ which has no pre-image. But no such y exists, so the empty mapping is onto. This illustrates a strange way that a statement involving the universal quantifier (or if-then) can be true. Statements of the form $(\forall x \in \emptyset)(P(x))$ are always true. For example, every human being on Mars has three legs. If a statement is written in if-then form, and if the hypothesis condition cannot ever be true, then regardless of the conclusion, the implication statement is true.

Example 3.3.2. Show that the set B in Eq. (3.13) has cardinality 6.

Solution: Define $f : \mathbb{N}_6 \to B$ in the following way: Let $f(1) = $ b; $f(2) = $ d; $f(3) = $ h; $f(4) = $ p; $f(5) = $ f; $f(6) = $ l. Since f is one-to-one and onto B, we've shown that $|B| = 6$. ∎

Example 3.3.3. By the identity mapping, $|\mathbb{N}_n| = n$.

Definition 3.3.1 introduces two terms concerning the size of some, but not all sets: *finiteness* and *cardinality*. Regardless of whether a set A is empty, to say that it is finite is to mean that it has some $n \in \mathbb{W}$ associated with it, called its cardinality, which derives from the existence of some function $f : \mathbb{N}_n \to A$. Conversely, to say that $|A| = n$, where $n \in \mathbb{W}$, is to say that A is finite. One other thing to notice is that $\emptyset$ is the only set with cardinality zero. For if $A \neq \emptyset$, then no $f : \mathbb{N}_0 \to A$ can be onto. If A is not finite, then we can still talk about its cardinality, but not in terms of functions $f : \mathbb{N}_n \xrightarrow[\text{onto}]{1\text{-}1} A$.

With Example 3.3.2, we have earned the right to make a statement such as "B has six elements," and validated our inclination to walk our fingers over the elements of a finite set and count them, so to speak. We must be careful, though. The existence of $f : \mathbb{N}_n \xrightarrow[\text{onto}]{1\text{-}1} A$ is a basis for declaring the cardinality of A to be n. But how do we know that there is not some other $m \neq n$ for which there exists $g : \mathbb{N}_m \xrightarrow[\text{onto}]{1\text{-}1} A$? If there were such an m, then cardinality would not be well defined, for $|A|$ could be two different numbers. With some of the theorems from Sections 3.1 and 3.2, your proof of the following theorem in Exercise 1 should be quick.

Theorem 3.3.4. *The cardinality of a finite set A is well defined. That is, if $|A| = m$ and $|A| = n$, then $m = n$.*

Exercise 5 from Section 3.2 allows us to say that the family of all sets is partitioned into equivalence classes based on the existence of one-to-one, onto functions between them. Now we can see what some of these equivalence classes are. Definition 3.3.1 effectively defines cardinality for some sets based on whether they are in the equivalence class of $\mathbb{N}_n$ for some $n \in \mathbb{W}$. Furthermore, by Theorem 3.3.4, that n is unique. So if $m \neq n$, then $\mathbb{N}_m$ and $\mathbb{N}_n$ are representative elements of different equivalence classes. With this, we have proved the following.

Theorem 3.3.5. *Suppose A is a finite set, and B is any set. Then A and B have the same cardinality if and only if there exists a bijection from A to B.*

The proof of Theorem 3.3.6 will require you to call on a lot of the results from Sections 3.1 and 3.2. Just remember that all results must be justified in terms of the existence of certain one-to-one, onto functions.

Theorem 3.3.6. *If $|A| = m$, $|B| = n$, and $A \cap B = \emptyset$, then $|A \cup B| = m + n$.*

Theorem 3.3.6 says that if there exist $f_1 : \mathbb{N}_m \xrightarrow[\text{onto}]{1\text{-}1} A$ and $f_2 : \mathbb{N}_n \xrightarrow[\text{onto}]{1\text{-}1} B$, then you can construct some $g : \mathbb{N}_{m+n} \xrightarrow[\text{onto}]{1\text{-}1} A \dot\cup B$. You'll do this in Exercise 2. Once you've defined g, showing it is a one-to-one, onto function from $\mathbb{N}_{m+n}$ to $A \dot\cup B$ should amount to little more than calling on previous exercises you've done. You'll prove the next two theorems in Exercises 3 and 5.

Theorem 3.3.7. *If $|B| = n$ and $A \subseteq B$, then A is finite and satisfies $|A| \leq n$.*

3.3 Cardinality of Sets

Corollary 3.3.8. *Suppose $|A| = n$ and C is any set. Then $|A \cap C| \leq |A|$.*

Proof: Since $A \cap C \subseteq A$, the result is immediate from Theorem 3.3.7. ∎

Theorem 3.3.9. *If $|A| = m$ and $|B| = n$, then $|A \cup B| \leq m + n$.*

3.3.2 Infinite sets

The following definition shouldn't be too surprising.

Definition 3.3.10. If a set is not finite, it is said to be *infinite*.

In terms of the definition of finite set, what is characteristic of an infinite set?[5] A set A is infinite if, for every $n \in \mathbb{W}$ and every $f : \mathbb{N}_n \to A$, either f is not one-to-one, or f is not onto. In reality, if there were some $f : \mathbb{N}_n \xrightarrow{\text{onto}} A$ that is not one-to-one, we could create a function $f_1 : \mathbb{N}_m \xrightarrow[\text{onto}]{1\text{-}1} A$ for some $m < n$ that culls out any repetition in f. Here's how. First create a subset $S \subseteq \mathbb{N}_n$ so that $f : S \to A$ is one-to-one in the following way. For each $a \in A$, take a single element of $f^{-1}(a)$, and let S consist of all these chosen pre-images.[6] Then apply Theorem 3.1.14 to conclude the existence of $g : S \xrightarrow[\text{onto}]{1\text{-}1} \mathbb{N}_m$ for some $m < n$. Then $f_1 = f \circ g^{-1}$ is a one-to-one, onto function from $\mathbb{N}_m$ to A. The upshot is that a set A is infinite if for every $n \in \mathbb{N}$ and $f : \mathbb{N}_n \to A$, f is not onto. Loosely speaking, there are not enough elements in $\mathbb{N}_n$ for any $n \in \mathbb{W}$ to tag all the elements of A. No matter how you might try to count them exhaustively using only the elements of $\mathbb{N}_n$ for some n, you'll never count them all.

Before we get into the interesting results of infinite sets, let's point out where it will lead. Strangely, just because two sets are both infinite, it does not follow that they have the same cardinality, in the sense that there is a one-to-one correspondence between them. Some infinite sets are actually bigger than others. This makes for some real surprises, and motivates us to discuss different *orders* of infinity. In fact, it's possible to generate an infinite sequence of infinite sets $A_1, A_2, \ldots$, where $|A_n|$ is a higher order of infinity than $|A_{n-1}|$. It's mind boggling. Not only is there more than one size of infinity, but there are infinitely many infinities. We'll look at only two. Here's our first infinity.

Definition 3.3.11. Suppose A is an infinite set. Then A is said to be *countably infinite* if there exists a function $f : \mathbb{N} \xrightarrow[\text{onto}]{1\text{-}1} A$, and we say that A has cardinality $\aleph_0$ (the Hebrew letter aleph). If A is finite or countably infinite, we say that A is *countable*. If A is not countable, we say that it is *uncountable*.

By calling on the identity mapping $i : \mathbb{N} \to \mathbb{N}$, we see that $\mathbb{N}$ is countably infinite, and the family of all countably infinite sets is precisely the equivalence class of $\mathbb{N}$ from the equivalence definition in Exercise 5 from Section 3.2. Let's look at some other countably infinite sets and some theorems about them.

---
[5] Negate Definition 3.3.1.
[6] See the discussion of the *axiom of choice* on page 95.

Theorem 3.3.12. $\mathbb{Z}$ *is countable.*

You'll prove Theorem 3.3.12 in Exercise 7. Your task is to construct a function $f : \mathbb{N} \xrightarrow[\text{onto}]{1\text{-}1} \mathbb{Z}$ by declaring $f(1)$, $f(2)$, $f(3)$, etc., Once you've developed such a function, see if you can come up with an explicit formula for $f(n)$. Specifically, you might find a way to define the one-to-one correspondence as

$$f(n) = \begin{cases} ? & \text{if } n \text{ is even,} \\ ? & \text{if } n \text{ is odd,} \end{cases}$$

and then apply Exercise 10 from Section 3.1.

If we expect to stumble eventually onto an uncountable set, the rationals might seem to be the first one we would find. After all, in Exercise 12 from Section 2.8, we showed that $a < (a+b)/2 < b$, so that between any two rational numbers there is another rational number. Thus the rational numbers are strewn densely up and down the real number line, as opposed to the integers, which have plenty of room in between each one. Well, $\mathbb{Q}$ is countable. Theorem 3.3.13 starts with only the positive rationals and says that there are no more positive rationals than there are natural numbers. Even though $\mathbb{N} \subset \mathbb{Q}^+$, and even though rational numbers are densely scattered up and down the real line, it is possible to list elements of $\mathbb{Q}^+$ sequentially as $f(1)$, $f(2)$, $f(3)$, etc. to create a function $f : \mathbb{N} \xrightarrow[\text{onto}]{1\text{-}1} \mathbb{Q}^+$. Here's how.

Theorem 3.3.13. $\mathbb{Q}^+$ *is countable.*

Proof: Consider Fig. 3.5, which lists all elements of $\mathbb{Q}^+$. From this figure, we may define f in the following way. Starting in the upper left-hand corner of the table, define $f(1) = 1/1$. We then move down successive diagonals from upper right to

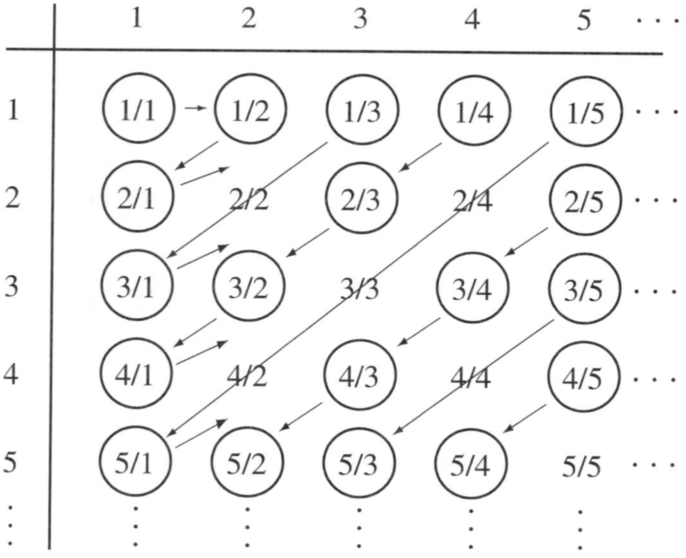

	1	2	3	4	5	...
1	1/1	1/2	1/3	1/4	1/5	...
2	2/1	2/2	2/3	2/4	2/5	...
3	3/1	3/2	3/3	3/4	3/5	...
4	4/1	4/2	4/3	4/4	4/5	...
5	5/1	5/2	5/3	5/4	5/5	...
⋮	⋮	⋮	⋮	⋮	⋮	

Figure 3.5 Ordering of $\mathbb{Q}^+$ to show countability.

lower left, defining $f(2) = 1/2$ and $f(3) = 2/1$ from the first diagonal. Moving down the next diagonal, we define $f(4) = 1/3$. But since $2/2 = 1/1$, we skip $2/2$ and define $f(5) = 3/1$. Continuing in this fashion,

$$f(6) = \frac{1}{4}, \quad f(7) = \frac{2}{3}, \quad f(8) = \frac{3}{2}, \quad f(9) = \frac{4}{1}, \quad f(10) = \frac{1}{5}, \quad f(11) = \frac{5}{1},$$

and so on, making sure f is one-to-one by skipping over rational numbers that are not in reduced form. This program guarantees that f is one-to-one and onto, so we have shown that $\mathbb{Q}^+$ is countable. ∎

Let's pause here for a moment and reflect on the proof of Theorem 3.3.13. We can think of finding a one-to-one correspondence $f : \mathbb{N} \to A$ as sequencing the elements of A as $\langle a_1, a_2, a_3, \ldots \rangle$ where every element of A is in the list exactly once. If such a listing of elements of A can be found, then you've shown A is countably infinite.

In tracing through the grid in the proof of Theorem 3.3.13, we ensured that $f : \mathbb{N} \to \mathbb{Q}^+$ is one-to-one by skipping over entries that were not in reduced form. If we had not done this skipping, the resulting f would still have been onto, but not one-to-one. Knowing a priori that $\mathbb{Q}^+$ is infinite, this failure of f to be one-to-one would not be particularly disconcerting. Loosely speaking, here's why. If there are enough natural numbers to tag all elements of $\mathbb{Q}$ with some repetition, then there ought to be enough to tag them without repetition. The next theorem gives us the freedom not to worry about this repetition, and will save us from some minor headaches later.

Theorem 3.3.14. *A nonempty set A is countable if and only if there exists a function $f : \mathbb{N} \xrightarrow{onto} A$.*

Proof:

($\Rightarrow$) Suppose A is nonempty and countable. Then by definition either A is finite or countably infinite. If A is countably infinite, then there exists $f : \mathbb{N} \xrightarrow[onto]{1\text{-}1} A$, and clearly there exists an onto function. So suppose A is finite. Then there exists $n \in \mathbb{N}$ and a function $f : \mathbb{N}_n \xrightarrow[onto]{1\text{-}1} A$. We may extend the domain of f to all of $\mathbb{N}$ and map every $k \geq n+1$ to any element of A we choose. Thus $f : \mathbb{N} \to A$ is onto.

($\Leftarrow$) Suppose there exists $f : \mathbb{N} \xrightarrow{onto} A$, and suppose A is not finite. (See Exercise 3g from Section 1.2.) The plan is to consider the sequence

$$f(1), f(2), f(3), f(4), \ldots \tag{3.14}$$

and select a *subsequence* that we can write as

$$g(1), g(2), g(3), g(4), \ldots,$$

where we effectively remove repetition from (3.14).

Let $g(1) = f(1)$. Let $g(2)$ be the first term in (3.14) different from $g(1)$. Let $g(3)$ be the first term of (3.14) different from $g(1)$ and $g(2)$, and so on. The fact that A is not finite means that a new, different term from (3.14) can always be found. We claim that the resulting $g : \mathbb{N} \to A$ is one-to-one and onto. Here's why.

Pick any $a \in A$. Because f is onto, there exists $n \in \mathbb{N}$ such that $f(n) = a$. Thus $S = \{n \in \mathbb{N} : f(n) = a\}$ is a nonempty subset of $\mathbb{N}$. By the WOP, S contains a smallest element k. According to the program by which g was constructed, $f(k)$ is chosen to be $g(j)$ for some $j \leq k$. Thus g is onto. Also, g is one-to-one because k is the unique smallest element of S and all other $f(n)$ for $n > k$ are skipped over in the defining of g. ∎

The convenience of Theorem 3.3.14 will become apparent in the following theorems. For example, given two countable sets, Theorem 3.3.15 allows you to show that the union is countable by constructing a mapping $f : \mathbb{N} \xrightarrow{\text{onto}} A \cup B$ without having to worry whether A and B are disjoint, or whether they are finite or countably infinite. The proofs of Theorems 3.3.15–3.3.17 are left as exercises. Since Theorem 3.3.14 does not apply to the empty set, it can only be invoked in proving Theorems 3.3.15–3.3.17 if we know that all sets involved are nonempty. But since the union across a family of sets is unchanged if all empty sets in the family are omitted, we're free to assume that all sets in the following theorems are nonempty.

Theorem 3.3.15. *If A and B are countable, then so is $A \cup B$.*

Theorem 3.3.16. *If $\{A_k\}_{k=1}^n$ is a finite family of countable sets, then $\cup_{k=1}^n A_k$ is countable.*

Theorem 3.3.17. *If $\{A_n\}_{n=1}^\infty$ is a countably infinite family of countable sets, then $\cup_{n=1}^\infty A_n$ is countable.*

By Theorem 3.3.16, it follows that $\mathbb{Q}$ is countable, for $\mathbb{Q} = \mathbb{Q}^+ \cup \mathbb{Q}^- \cup \{0\}$, and $\mathbb{Q}^-$ is countable in the same way that $\mathbb{Q}^+$ is.

If every $f : \mathbb{N} \to A$ fails to be onto, then A is uncountable. This is how we prove the following:

Theorem 3.3.18. *$[0, 1]$ is uncountable.*

Proof: Suppose $f : \mathbb{N} \to [0, 1]$ is *any* function whatsoever. We can show that f fails to be onto. From assumption A21, we may think of $[0, 1]$ as the set of all decimal representations of the form 0.XXXXX.... Now consider the following diagram:

$$f(1) = 0.a_1a_2a_3a_4\ldots$$
$$f(2) = 0.b_1b_2b_3b_4\ldots$$
$$f(3) = 0.c_1c_2c_3c_4\ldots$$
$$f(4) = 0.d_1d_2d_3d_4\ldots$$
$$f(5) = 0.e_1e_2e_3e_4\ldots$$
$$\vdots$$

We can show f is not onto by constructing a real number $x \in [0, 1]$ that is not in the list. Let $x = 0.x_1x_2x_3x_4\ldots$ where $x_1 \neq a_1$, $x_2 \neq b_2$, $x_3 \neq c_3$, etc. and no $x_k = 9$.

Then $x \in [0, 1]$, and x is different from every number in the range of f. Thus $[0, 1]$ is not countable. ∎

You'll prove Theorem 3.3.19 in Exercise 11.

Theorem 3.3.19. *If B is countable and $A \subseteq B$, then A is countable.*

Corollary 3.3.20. $\mathbb{R}$ *is uncountable.*

Proof: Since $[0, 1]$ is uncountable and $[0, 1] \subset \mathbb{R}$, Theorem 3.3.19 implies that $\mathbb{R}$ is uncountable. ∎

EXERCISES

1. Prove Theorem 3.3.4: If $|A| = m$ and $|A| = n$, then $m = n$.[7]
2. Prove Theorem 3.3.6: If $|A| = m$, $|B| = n$, and $A \cap B = \emptyset$, then $|A \cup B| = m + n$.[8]
3. Prove Theorem 3.3.7: If $|B| = n$ and $A \subseteq B$, then A is finite and satisfies $|A| \leq n$.[9]
4. Suppose $|A| = |B| = n$, and suppose $f : A \xrightarrow{1\text{-}1} B$ is any function. Show that f must be onto.[10]
5. Prove Theorem 3.3.9: If $|A| = m$ and $|B| = n$, then $|A \cup B| \leq m + n$.[11]
6. Prove that the union of a finite number of finite sets is finite.[12]
7. Prove Theorem 3.3.12: $\mathbb{Z}$ is countable.
8. Prove Theorem 3.3.15: If A and B are countable, then so is $A \cup B$.
9. Prove Theorem 3.3.16: If $\{A_k\}_{k=1}^n$ is a finite family of countable sets, then $\cup_{k=1}^n A_k$ is countable.
10. Prove Theorem 3.3.17: If $\{A_n\}_{n=1}^\infty$ is a countably infinite family of countable sets, then $\cup_{n=1}^\infty A_n$ is countable.[13]
11. Prove Theorem 3.3.19: If B is countable and $A \subseteq B$, then A is countable.[14]
12. Prove that the irrationals are uncountable.

---

[7] Take care of $\emptyset$ as a special case. Otherwise, if the f is false, it violates Theorem 3.1.15.
[8] Make use of Theorem 3.1.13 where the domain of h is $\mathbb{N}_{m+n}$ and the range is $A \cup B$. The function $T : \mathbb{N}_n \to \mathbb{N}_n^m$ from Example 3.1.5 should be helpful, too.
[9] If $f : \mathbb{N}_n \xrightarrow[\text{onto}]{1\text{-}1} B$, apply Theorem 3.1.14 to $f^{-1}(A)$. Construct $g : \mathbb{N}_m \xrightarrow{1\text{-}1} A$ by composition.
[10] What if f is not onto? By theorem 3.3.5, $|A| = |\text{Rng } f|$.
[11] See Exercise 7 from Section 2.1.
[12] Use induction and Theorem 3.3.9.
[13] Induction won't help here. Try a grid like that in the proof of Theorem 3.3.13.
[14] Use Theorem 3.3.14 to take care of $A \neq \emptyset$.

13. Let $a < b$ and $c < d$. Using interval notation $[a, b] = \{x : a \leq x \leq b\}$, show that $|[a, b]| = |[c, d]|$ by finding a function $f : [a, b] \xrightarrow[\text{onto}]{1\text{-}1} [c, d]$.

3.4 Counting Methods and the Binomial Theorem

How many ways are there to put together a meal from all the cafeteria offerings? To arrange your CD collection on your dresser? To name a baby boy from a list of family preferences? To choose a steering committee for your club? In this section, we're going to apply the cardinality results from Section 3.3 to answer these and other questions. We'll develop and apply results to finite, nonempty sets and functions that relate them, though some results will be more generally applicable. Because the main goal of this section is to gain exposure to some complex cardinality questions whose answers are really quite plausible and have very practical application, we'll sometimes talk through informal arguments instead of providing or requiring rigorous proof.

To simplify some of the language and notation, if A is a finite set and $f : \mathbb{N}_n \to A$ is a one-to-one, onto function, we'll address elements of A as $\{a_1, a_2, \ldots, a_n\}$, where $a_k = f(k)$ for $1 \leq k \leq n$. Since f is a function, every $k \in \mathbb{N}_n$ can be used to reference a unique $a_k \in A$. Since f is one-to-one, all the a_k are different, and since f is onto, every element of A can be addressed as a_k for some $k \in \mathbb{N}_n$.

In Section 2.9, we defined the *Cartesian product* of two sets A and B as

$$A \times B = \{(a, b) : a \in A, b \in B\}, \tag{3.15}$$

the set of ordered pairs whose first coordinate comes from A, and whose second coordinate comes from B. Assuming A and B each have a form of equality defined on them, which we temporarily write as $=_A$ and $=_B$, we can define equality in $A \times B$ by

$$(a_1, b_1) =_{A \times B} (a_2, b_2) \quad \text{if and only if} \quad a_1 =_A a_2 \text{ and } b_1 =_B b_2.$$

Assuming $=_A$ and $=_B$ are equivalence relations, we can show that $=_{A \times B}$ defines an equivalence relation. For property E1, if we pick $(a, b) \in A \times B$, then $a =_A a$ and $b =_B b$, so that $(a, b) =_{A \times B} (a, b)$. Showing properties E2 and E3 is equally easy (Exercise 1). From now on we won't bother with different notations for equality in the different sets, instead simply writing that $(a_1, b_1) = (a_2, b_2)$ if and only if $a_1 = a_2$ and $b_1 = b_2$.

3.4.1 The product rule

Our first theorem will serve you as the root of an induction argument when you prove Theorem 3.4.4.

Theorem 3.4.1. *Suppose $|A| = 1$ and $|B| = n \geq 1$. Then $|B| = |A \times B|$.*

We'll prove Theorem 3.4.1 by demonstrating a function $f : B \xrightarrow[\text{onto}]{1\text{-}1} A \times B$ to show that they're in the same cardinality class.

Proof: Write $A = \{a\}$ and $B = \{b_1, b_2, \ldots, b_n\}$. Define $f : B \to A \times B$ by $f(b_k) = (a, b_k)$. We show f is a one-to-one function from B onto $A \times B$. First, for any $b_k \in B$, $(a, b_k) \in A \times B$, so f has property F1. Also, if $b_j = b_k$, then $f(b_j) = (a, b_j) = (a, b_k) = f(b_k)$, so that f has property F2. If we suppose $f(b_j) = f(b_k)$, then $(a, b_j) = (a, b_k)$, which implies $b_j = b_k$, and f is one-to-one. Finally, if we pick any $(a, b_k) \in A \times B$, then $b_k \in B$ and $f(b_k) = (a, b_k)$, so that f is onto. Since we have found a one-to-one correspondence between B and $A \times B$, they are in the same cardinality class, and $|B| = |A \times B|$. ∎

Here's a theorem you'll prove in Exercise 2, and then an immediate corollary.

Theorem 3.4.2. *If A_1, A_2, and B are sets, then*

$$(A_1 \cup A_2) \times B = (A_1 \times B) \cup (A_2 \times B). \tag{3.16}$$

Furthermore, if $A_1 \cap A_2 = \emptyset$, then $(A_1 \times B) \cap (A_2 \times B) = \emptyset$.

Corollary 3.4.3. *If A_1, A_2, and B are finite sets and $A_1 \cap A_2 = \emptyset$, then $|(A_1 \cup A_2) \times B| = |A_1 \times B| + |A_2 \times B|$.*

Proof: By Theorems 3.3.6 and 3.4.2,

$$|(A_1 \dot\cup A_2) \times B| = |(A_1 \times B) \dot\cup (A_2 \times B)| = |A_1 \times B| + |A_2 \times B|. \tag{3.17}$$

∎

With Theorem 3.4.1 and Corollary 3.4.3, you have all you need to write an induction argument for the following (Exercise 3):

Theorem 3.4.4. *Suppose A and B are nonempty, finite sets. Then*

$$|A \times B| = |A| \times |B|. \tag{3.18}$$

We can form the Cartesian product of more than two sets, but not without giving ourselves a freedom that would make a rigorous set theorist a bit uncomfortable. We want $A \times B \times C$ to be the set of ordered triples (a, b, c) where $a \in A$, $b \in B$, and $c \in C$. However, to define $A \times B \times C$ in terms of the *binary* Cartesian product defined in Eq. (3.15), we need to associate either A and B, or B and C. This leaves us with $(A \times B) \times C$ or $A \times (B \times C)$, which, unfortunately, are different. As we defined the Cartesian product in Eq. (3.15),

$$(A \times B) \times C = \{((a, b), c) : a \in A, b \in B, c \in C\}, \tag{3.19}$$

but

$$A \times (B \times C) = \{(a, (b, c)) : a \in A, b \in B, c \in C\}. \tag{3.20}$$

Expressions of the form $((a, b), c)$ and $(a, (b, c))$ are not the same, though this doesn't create an insurmountable obstacle. Rather than trying to deal with the lack of associativity in the Cartesian product of sets, we'll make the following recursive definition, then give ourselves a freedom to ignore the associativity question.

Definition 3.4.5. Suppose $\{A_k\}$ is a family of sets indexed by $\mathbb{N}$. Then we define the Cartesian product recursively as

$$\prod_{k=1}^{1} A_k = A_1, \tag{3.21}$$

$$\prod_{k=1}^{n+1} A_k = \left(\prod_{k=1}^{n} A_k\right) \times A_{n+1} \quad \text{for } n \geq 1. \tag{3.22}$$

With Definition 3.4.5, elements of $\prod_{k=1}^{n} A_k$ for the first few values of n would look like this:

$$\begin{aligned}
\prod_{k=1}^{1} A_k &= \{a : a \in A_1\} \\
\prod_{k=1}^{2} A_k &= \{(a_1, a_2) : a_k \in A_k \text{ for } 1 \leq k \leq 2\} \\
\prod_{k=1}^{3} A_k &= \{((a_1, a_2), a_3) : a_k \in A_k \text{ for } 1 \leq k \leq 3\} \\
\prod_{k=1}^{4} A_k &= \{(((a_1, a_2), a_3), a_4) : a_k \in A_k \text{ for } 1 \leq k \leq 4\}.
\end{aligned} \tag{3.23}$$

Instead of writing elements of $\prod_{k=1}^{n} A_k$ this way, we'll give ourselves the freedom to address elements as $(a_1, a_2, \ldots, a_n)$, n-tuples where $a_k \in A_k$ for all $1 \leq k \leq n$. With this definition in hand, and another induction argument, you can prove the following (Exercise 4):

Theorem 3.4.6. Suppose $\{A_k\}_{k=1}^{n}$ is a finite family of finite sets. Then

$$\left|\prod_{k=1}^{n} A_k\right| = \prod_{k=1}^{n} |A_k|. \tag{3.24}$$

Here are some practical applications of Theorem 3.4.6.

Example 3.4.7. Your university cafeteria has the following menu for today's lunch:

Meats: Meatloaf, Chicken, Fishsticks
Starchy vegetables: Potatoes, Rice, Corn, Pasta
Green vegetables: Beans, Broccoli, Salad, Spinach
Breads: Rolls, Cornbread
Desserts: Chocolate cake, Pudding

If you choose one item from each category of the menu, how many different meals could you put together?

Solution: If we let A_1 be the set of meat offerings, A_2 the set of starchy vegetables, etc. then each potential complete meal is an element of $\prod_{k=1}^{5} A_k$. By Theorem 3.4.6, there are $3 \times 4 \times 4 \times 2 \times 3 = 288$ possible meals. ∎

In Example 3.4.8, counting the number of possible meals can be visualized in the following way. We have five empty slots to fill on our plate, where the first is to be filled with a choice of meat, the second with a choice of starchy vegetable, etc. Furthermore, the number of choices available to fill a particular slot is unaffected by the way any of the previous slots has been filled, or the way the remaining slots will be filled. Multiplying the number of ways to fill each slot illustrates the *product rule*. If an n-step process is such that the kth step can be done in a_k ways, and the number of ways each step can be done is unaffected by the choice made for any other step, then the total number of ways to perform all n steps is $\prod_{k=1}^{n} a_k$.

Example 3.4.8. Suppose you have five shirts, three pairs of pants, five pairs of socks, and two pairs of shoes, all of which match each other. How many different outfits can you put together, if an outfit consists of one shirt, one pair of pants, one pair of socks, and one pair of shoes?

Solution: If we let C_1, C_2, C_3, and C_4 be the shirts, pants, socks, and shoes, respectively, then $\prod_{k=1}^{4} C_k$ is the set of all distinct outfits. But $\left| \prod_{k=1}^{4} C_k \right| = 5 \times 3 \times 5 \times 2 = 150$, so there are 150 distinct outfits. ∎

Now suppose A and B are nonempty sets with $|A| = m$ and $|B| = n$. We can use Theorem 3.4.6 to calculate the number of functions from A to B in the following way. If we form the Cartesian product of m copies of B, then any ordered m-tuple from $\prod_{j=1}^{m} B$ can be thought of as a function $f : A \to B$, where the first coordinate of the m-tuple is $f(a_1)$, the second coordinate is $f(a_2)$, etc. Since the cardinality of $\prod_{k=1}^{m} B$ is n^m, we have the following:

Theorem 3.4.9. *If A and B are nonempty sets with $|A| = m$ and $|B| = n$, then there are n^m distinct functions from A to B.*

Theorem 3.4.9 motivates a general notation. Regardless of whether A or B is finite, we write B^A to mean the set of all functions from A to B. For example, $\mathbb{R}^{[0,1]}$ is the set of all functions from the interval $[0, 1]$ to $\mathbb{R}$. Here's an illustration of the practical use of Theorem 3.4.9.

Example 3.4.10. Suppose you toss a coin 10 times and observe the sequence of outcomes of heads (H) or tails (T). Each possible outcome of the ten tossings can be written as a 10-tuple of the form (H,H,H,T,H,H,T,T,H,T), each of which we can visualize as a function $f : \mathbb{N}_{10} \to \{H, T\}$. There are $2^{10} = 1{,}024$ such functions.

If a counting problem can be visualized as filling m labeled slots with any of n possible objects, where any one of the objects can be used with unlimited repetition, then Theorem 3.4.9 implies that there are n^m ways to do it.

Example 3.4.11. A password to your computer account must be precisely 7 alphanumeric characters, that is, a sequence of 7 characters taken from the 26 letters of the alphabet and the digits 0–9. Filling 7 ordered slots with any arrangement of these 36 characters reveals that there are $36^7 = 78{,}364{,}164{,}096$ passwords.

3.4.2 Permutations

Suppose we want to fill m slots with choices from n objects, but repetition of the objects is not allowed. Such an arrangement is called a *permutation of n objects taken m at a time*. We can think of such an assignment as a function from an m-element set to an n-element set, but the function must be one-to-one. In order for there to exist any such function, it must be that $m \leq n$. So assuming $|A| = m \leq n = |B|$, let's count the number of functions from A to B that are one-to-one. To see what the right answer ought to be, imagine making choices for $f(a_1)$, then $f(a_2)$, etc. For $f(a_1)$, there are n possible choices, then $n - 1$ choices for $f(a_2)$, $n - 2$ choices for $f(a_3)$, etc. If the product rule proves to be correct, there will be $n(n-1)(n-2)\cdots(n-m+1)$ functions. We can write this as

$$P(n, m) = n(n-1)(n-2) \cdots (n-m+1) = \frac{n!}{(n-m)!}. \tag{3.25}$$

Proving there are $P(n, m)$ one-to-one functions from an m-element set to an n-element set demonstrates that $P(n, m)$ is the number of permutations of n objects taken m at a time. A rigorous proof should be done by induction (Exercise 5).

Theorem 3.4.12. *Let A and B be nonempty sets with $|A| = m \leq n = |B|$. Then there are $P(n, m)$ distinct one-to-one functions from A to B.*

Example 3.4.13. The Fitzpatricks are expecting a baby boy any day now. Family members have strong opinions about what the child will be named. Suggested names are William, Warren, Benjamin, Fitzhugh, Chancellor, Millhouse, and Nebuchadnezzar. The parents want their son to have three given names, such as William Fitzhugh Millhouse Fitzpatrick. How many distinct names are there for the Fitzpatrick son?

Solution: We may let $A = \{1, 2, 3\}$ be the positions for the names to be chosen, and B the set of names. A complete name can be thought of as a function $f : A \xrightarrow{1\text{-}1} B$. There are $P(7, 3) = 7!/4! = 7 \cdot 6 \cdot 5 = 210$ such functions. ∎

If there are as many slots to fill as objects to place in them, then there are $P(n, n) = n!/(n-n)! = n!$ ways to do it. We call such an assignment a *permutation of n objects*.

Example 3.4.14. If you have a collection of eight CDs and want to arrange them in a row on top of your dresser, there are $P(8, 8) = 8! = 40{,}320$ ways to do it.

3.4.3 Combinations and partitions

Now let's suppose we're going to select m objects out of a supply of n, but instead of counting the number of ways that m of the n objects can be arranged, we want to count how many ways there are simply to choose m of the n objects, without considering the order of the objects as relevant. In other words, given a set of n elements, how many m-element subsets are there? When we create an m-element subset of an n-element set, we call it a *combination of n objects taken m at a time*.

One way to see how many m-element subsets exist from an n-element set is to finagle the answer from $P(n, m)$. Let S be the set of all the $P(n, m)$ arrangements of m elements of an n-element set, and partition S into subsets of arrangements that all contain precisely the same choices. For example, if $A = \{a, b, c, d\}$ and we are going to choose $m = 3$ of these four elements, then S contains $P(4, 3) = 4!/1! = 24$ permutations, and the family of subsets of S that we've described consists of:

$$S_1 = \{(a, b, c), (a, c, b), (b, a, c), (b, c, a), (c, a, b), (c, b, a)\},$$
$$S_2 = \{(a, b, d), (a, d, b), (b, a, d), (b, d, a), (d, a, b), (d, b, a)\},$$
$$S_3 = \{(a, c, d), (a, d, c), (c, a, d), (c, d, a), (d, a, c), (d, c, a)\},$$
$$S_4 = \{(b, c, d), (b, d, c), (c, b, d), (c, d, b), (d, b, c), (d, c, b)\}.$$
(3.26)

Thus, the $P(n, m)$ permutations of n objects taken m at a time can be partitioned into subsets, each of which consists of the $m!$ arrangements of an m-element subset of A. This means that there are $P(n, m)/m!$ m-element subsets of an n-element set. There are two common notations for the number of combinations of n objects taken m at a time:

$$C(n, m) = \binom{n}{m} = \frac{n!}{m!(n-m)!}. \tag{3.27}$$

Since $C(n, m)$ is the number of ways of choosing m out of n objects, it is also called "n choose m." Notice that $C(n, m) = C(n, n - m)$, so the number of ways to choose m out of n objects to serve some purpose is the same as the number of ways to choose $n - m$ of the objects not to serve. Most of the time we'll use $C(n, m)$ only for $0 \le m \le n$, though in what follows we will define $C(n, m)$ for $m < 0$ and $m \ge n$. Notice that $C(0, 0) = 1$ is meaningful, in that there is precisely one zero-element subset of the empty set.

Example 3.4.15. Your club of 50 members is going to choose a steering committee of 5 people. Each such committee can be thought of as a 5-element subset of a 50-element set. The number of these subsets is

$$\binom{50}{5} = \frac{50!}{5!45!} = \frac{50 \cdot 49 \cdot 48 \cdot 47 \cdot 46}{5 \cdot 4 \cdot 3 \cdot 2 \cdot 1} = 2,118,760. \tag{3.28}$$

To prove rigorously that an n-element set has $C(n, m)$ subsets of cardinality m, you need the following:

Theorem 3.4.16. *Suppose $n \ge 0$ and $1 \le k \le n$. Then*

$$\binom{n}{k-1} + \binom{n}{k} = \binom{n+1}{k}. \tag{3.29}$$

Proving Theorem 3.4.16 does not require induction, only some straightforward algebraic manipulation (Exercise 6). You'll need it in the inductive step when you prove the result we're after (Exercise 7).

Theorem 3.4.17. *Suppose A is a finite set of cardinality n, and let $0 \le m \le n$. Then the number of subsets of A of cardinality m is*

$$C(n, m) = \binom{n}{m} = \frac{n!}{m!(n-m)!}.$$

Example 3.4.18. The state lottery is played by drawing six numbers from $\mathbb{N}_{51}$, and a winner is declared if someone matches all six numbers, without any consideration of order. The number of ways a lottery drawing can turn out is $C(51, 6) = 18{,}009{,}460$.

Choosing an m-element subset of an n-element set is the same as partitioning the set into two subsets, one of size m and one of size $n - m$. We can generalize this process in the following way.

Suppose $|A| = n$ and we want to partition A into $p \geq 1$ distinguishable subsets whose cardinalities are $n_1, n_2, \ldots, n_p$, where $\sum_{k=1}^{p} n_k = n$ and none of the n_k are zero. How many ways can we do it? There are two informal ways to see what the answer is. The first is to imagine choosing n_1 of the n objects as the first subset, then choosing n_2 of the remaining $n - n_1$ to be the second subset, then n_3 of the remaining $n - n_1 - n_2$, etc. It would seem natural that multiplying these combinations would produce the correct answer for this multistep process. If so, then we have

$$\binom{n}{n_1}\binom{n-n_1}{n_2}\binom{n-n_1-n_2}{n_3} \cdots \binom{n_p}{n_p}$$

$$= \frac{n!}{n_1!(n-n_1)!} \cdot \frac{(n-n_1)!}{n_2!(n-n_1-n_2)!} \cdot \frac{(n-n_1-n_2)!}{n_3!(n-n_1-n_2-n_3)!} \cdots \frac{n_p!}{n_p!0!}$$

$$= \frac{n!}{n_1!n_2!n_3!\cdots n_p!}. \tag{3.30}$$

Here's another way to see that the final expression in Eq. (3.30) is correct. Imagine n ordered slots, and group together the first n_1 slots, group the next n_2 slots, group the next n_3, etc. For a given partition of the n objects into (unordered) subsets of size n_1, $n_2, \ldots, n_p$, imagine the n_1 elements of the first subset being placed into the first n_1 slots in any order, the n_2 elements of the second subset placed into the next n_2 slots in any order, etc. Dumping the unordered subsets into the ordered slots in this way produces a permutation of the n objects, so if we can calculate the number of ways of placing the subset elements into the grouped slots, we can know what fraction of the $n!$ permutations of the n objects correspond to this one partition. Now there are $n_1!$ ways to place the elements of the first subset into the first group of slots, $n_2!$ ways to place the elements of the second subset into the second group of slots, etc. Since all steps of the arranging are independent of each other, we conclude that there are $n_1!n_2!\cdots n_p!$ ways to arrange the elements. Thus we have an informal argument for the following.

Theorem 3.4.19. *Suppose $|A| = n$, and let $n_1, n_2, \ldots, n_p \in \mathbb{N}$ be such that $\sum_{k=1}^{p} n_p = n$. Then the number of ways to partition A into subsets of size $n_1, n_2, \ldots, n_p$ is $n!/(n_1!n_2!\cdots n_p!)$.*

Example 3.4.20. Your club of 50 members is going to choose a president, vice-president, secretary/treasurer, and two committees, each with 5 members. No one may serve in more than one capacity. This process can be thought of as partitioning a 50-element set into subsets of size 1, 1, 1, 5, 5, and 37. The number of ways this can be done is

$$\binom{50}{1}\binom{49}{1}\binom{48}{1}\binom{47}{5}\binom{42}{5} = \frac{50!}{1!1!1!5!5!37!} = 153{,}453{,}043{,}779{,}235{,}200. \quad (3.31)$$

3.4.4 Counting examples

Let's put together all the counting techniques we've developed to answer some more complex questions.

Example 3.4.21. Suppose you toss a coin 10 times and observe the sequence of outcomes. From Example 3.4.10, we know that there are 2^{10} possible outcomes. We want to count the number of these outcomes that have exactly three Heads. Imagine having 10 letters, 3 Hs and 7 Ts, and we are going to fill 10 slots with these letters. We want to count the different arrangements of the Hs and Ts. Each such outcome can be thought of as a three-element subset of $\mathbb{N}_{10}$, where the three numbers chosen are the slots where the Hs appear. There are $C(10, 3) = (10 \cdot 9 \cdot 8)/(3 \cdot 2 \cdot 1) = 120$ such choices.

Example 3.4.22. A box of lightbulbs contains 100 bulbs, 3 of which are defective. If we choose 10 bulbs from the box, we want to calculate how many ways there are to get precisely one of the defective bulbs. Such a choice of 10 bulbs consists of nine of the 97 good bulbs chosen without regard to order, and one of the defective bulbs chosen from the three. Multiplying these, we have $C(97, 9) \times C(3, 1)$ ways of choosing the bulbs in which precisely one is defective.

Theorem 3.3.6 says that if A and B are disjoint, then $|A \cup B| = |A| + |B|$. Applied to counting the number of ways of performing a task, we call this theorem the *sum rule*. If counting the number of ways of performing a task must be broken up into disjoint cases, the sum rule says simply that the calculations for the different calculations are added to produce the total number of ways of performing the task.

Example 3.4.23. A license plate consists of up to 7 characters, taken from the 26 letters and the digits 0–9. If there is room for spaces, they are not counted in the arrangement; for example, there is no distinction between `TRU BLU` and `TRUBL U`. We count the total number of plates California can issue. There are different cases based on the number of characters used. The number of one-character license plates is 36; the number of two-character license plates is 36^2, etc. Thus the total number of license plates is $36 + 36^2 + 36^3 + \cdots + 36^7$.

Sometimes it's easier to calculate $|A|$ by calculating $|U|$ and $|A'|$, then using the fact that $|A| + |A'| = |U|$.

Example 3.4.24. Your club of 30 women and 20 men is going to choose a committee of 5. We calculate the number of ways to choose the committee if it must include at least 1 man. The number of ways to choose the committee with no restrictions is $C(50, 5)$, and the number of ways to choose the committee with no men is $C(30, 5)$. Thus the number of ways to choose the committee where there is at least 1 man is $C(50, 5) - C(30, 5)$.

3.4.5 The binomial theorem

Combinations pop up in several places in mathematics, and we want to take some time now to see an important one. The appearance of $C(n, m)$ in a variety of situations reveals some nice ties between seemingly unrelated mathematical ideas. First, there's a nice way to visualize Theorem 3.4.16 in what is called *Pascal's triangle*, though to create it we need to extend the definition of $C(n, m)$. For a fixed $n \geq 0$, we want to allow $m < 0$ and $m > n$, and in either case, we define $C(n, m) = 0$. You can think of this as saying that there are zero ways to choose m elements from an n-element set if $m < 0$ or $m > n$. With these definitions, Theorem 3.4.16 is true for all $n \geq 0$ and all $k \in \mathbb{Z}$ (Exercise 8).

We construct Pascal's triangle row by row, beginning with a row that corresponds to $n = 0$, which we'll therefore call row zero. The next row corresponds to $n = 1$, etc., and for each row, the entries correspond to $k = 1, 2, \ldots, n$. (See Fig. 3.6.) Row zero consists of the single entry 1, which is $C(0, 0)$. To create what we are calling the first row, imagine zeroes to the left and right of $C(0, 0)$ in row zero, which we may think of as $C(0, -1)$ and $C(0, 1)$, respectively. Create the two entries in row one by adding adjacent entries in row zero. By Theorem 3.4.16, the first entry in this first row is $C(0, -1) + C(0, 0) = C(1, 0)$, and the second is $C(0, 0) + C(0, 1) = C(1, 1)$. Thus the entries in row one are all possible $C(1, k)$ for $0 \leq k \leq 1$. Continuing in this fashion to create the second row, imagine $C(1, -1) = 0$ on the left end of the first row and $C(1, 2) = 0$ on the right end. Adding adjacent entries of the first row, we generate $C(2, 0)$, $C(2, 1)$, and $C(2, 2)$ by Theorem 3.4.16. Finishing out as many rows of Pascal's triangle as you like, we have that row n consists of $C(n, k)$ for $0 \leq k \leq n$.

Here's an important place where the entries in Pascal's triangle appear. Suppose you're going to expand the expression $(a + b)^n$ for some $n \in \mathbb{N}$. Rather than multiply out $(a + b)(a + b) \cdots (a + b)$ using the extended distributive property, there is a much easier way to see what the terms are, and it involves $C(n, k)$ in the coefficients.

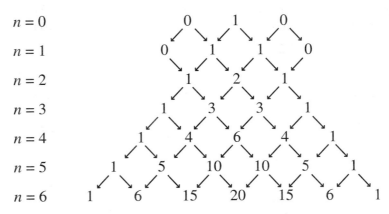

Figure 3.6 Pascal's triangle.

Theorem 3.4.25 (Binomial theorem). *Let $a, b \in \mathbb{R}\backslash\{0\}$, and let $n \in \mathbb{W}$. Then*

$$(a+b)^n = \binom{n}{0}a^n b^0 + \binom{n}{1}a^{n-1}b^1 + \binom{n}{2}a^2 b^2 + \cdots + \binom{n}{n}a^0 b^n$$
$$= \sum_{k=0}^{n}\binom{n}{k}a^{n-k}b^k = \sum_{k=0}^{n}\binom{n}{n-k}a^{n-k}b^k. \tag{3.32}$$

The only reason we do not allow a or b to be zero in Theorem 3.4.25 is that 0^0 is not defined in $\mathbb{R}$, so Eq. (3.32) would produce some undefined terms. But if either a or b is zero, the expansion of $(a+b)^n$ is not particularly interesting. Thus we omit it. Using the entries from Pascal's triangle, we have

$$\begin{aligned}(a+b)^0 &= 1a^0 b^0 = 1, \\ (a+b)^1 &= 1a^1 b^0 + 1a^0 b^1 = a+b, \\ (a+b)^2 &= a^2 + 2ab + b^2, \\ (a+b)^3 &= a^3 + 3a^2 b + 3ab^2 + b^3, \text{ etc.}\end{aligned} \tag{3.33}$$

There are several ways to convince ourselves that the binomial theorem is true, using somewhat open-ended arguments that appeal to some of our previous counting techniques. One way is to note that applying the extended distributive property to $(a+b)(a+b)\cdots(a+b)$ is tantamount to creating a whole bunch of terms of a form like $aaabbaab$, where we go to each $(a+b)$ factor and select either a or b. Since each factor allows for two possible choices, there are 2^n terms generated, and every one is of the form $a^{n-k}b^k$ for some $0 \leq k \leq n$. The question we must answer is, for a given k, how many times does $a^{n-k}b^k$ appear in all the distribution of the multiplication? If k of the factors supply us with b and the remaining factors provide us with a, then the number of times $a^{n-k}b^k$ appears in the grand sum is the same as the number of ways of choosing k of the n terms to provide us with b (or $n-k$ of the terms to provide us with a). That is, the term $a^{n-k}b^k$ is produced $C(n, k)$ times in the distribution.

We're going to provide a more rigorous proof of Theorem 3.4.25 than our informal argument here. The technique we use in making the inductive step contains some manipulation of summations that can really come in handy in some of your other coursework, especially, for example, differential equations and complex analysis, where you must resort to finding what we call a *series solution* to a problem. Watch how we pull off some first and last terms from the summations, then realign the terms by changing the counter variable.

Proof of theorem 3.4.25: Let $a, b \in \mathbb{R}\backslash\{0\}$.

(J1) For $n = 0$, $(a+b)^0 = 1 = C(0,0)a^0 b^0$, so Eq. (3.32) is true for $n = 0$.

(J2) Suppose $n \geq 0$ and that Eq. (3.32) is true for n. Then

$$\begin{aligned}(a+b)^{n+1} = (a+b)^n(a+b) &= \left[\sum_{i=0}^{n}C(n,i)a^{n-i}b^i\right](a+b) \\ &= \sum_{i=0}^{n}C(n,i)a^{n+1-i}b^i + \sum_{i=0}^{n}C(n,i)a^{n-i}b^{i+1} \\ &= \sum_{i=0}^{n}C(n,i)a^{(n+1)-i}b^i + \sum_{i=0}^{n}C(n,i)a^{(n+1)-(i+1)}b^{i+1}.\end{aligned} \tag{3.34}$$

To align the terms in the two summations, let $j = i + 1$ for the second summation (that is, $i = j - 1$), and note that $0 \leq i \leq n$ is the same as $1 \leq j \leq n + 1$. Making this substitution first, then pulling off a few individual terms, we can continue Eq. (3.34) several steps.

$$\begin{aligned}(a+b)^{n+1} &= \sum_{i=0}^{n} C(n,i)a^{(n+1)-i}b^i + \sum_{j=1}^{n+1} C(n, j-1)a^{(n+1)-j}b^j \\ &= \sum_{i=1}^{n} C(n,i)a^{(n+1)-i}b^i + \sum_{j=1}^{n} C(n, j-1)a^{(n+1)-j}b^j \\ &\quad + C(n,0)a^{n+1-0}b^0 + C(n,n)a^0 b^{n+1}.\end{aligned} \quad (3.35)$$

Now $C(n, 0) = C(n + 1, 0)$ and $C(n, n) = C(n + 1, n + 1)$. Also, we may let $k = i = j$ and combine the two summations and apply Theorem 3.4.16 to have

$$\begin{aligned}(a+b)^{n+1} &= \sum_{k=1}^{n}[C(n,k) + C(n, k-1)]a^{(n+1)-k}b^k \\ &\quad + C(n+1, 0)a^{n+1-0}b^0 + C(n+1, n+1)a^0 b^{n+1} \\ &= \sum_{k=1}^{n} C(n+1, k)a^{(n+1)-k}b^k \\ &\quad + C(n+1, 0)a^{n+1-0}b^0 + C(n+1, n+1)a^0 b^{n+1} \\ &= \sum_{k=0}^{n+1} C(n+1, k)a^{(n+1)-k}b^k,\end{aligned} \quad (3.36)$$

which verifies that Eq. (3.32) holds for $n + 1$. ∎

Corollary 3.4.26. *For all $n \in \mathbb{W}$,*

$$\sum_{k=0}^{n} \binom{n}{k} = 2^n. \quad (3.37)$$

Proof: Let $a = b = 1$ in Eq. (3.32). ∎

Suppose A is a finite set with cardinality n. Corollary 3.4.26 implies that A has 2^n subsets, for the left-hand side of Eq. (3.37) is the sum of the number of subsets of an n-element set with all possible numbers of elements. Another way to see the same result is to create a one-to-one correspondence between the family of all subsets of A and the collection of the 2^n functions from A to $\{0, 1\}$. For $A_1 \subseteq A$, we may define

$$f(a) = \begin{cases} 1, & \text{if } a \in A_1; \\ 0, & \text{if } a \notin A_1. \end{cases} \quad (3.38)$$

Every distinct subset of A generates a distinct function, and by Theorem 3.4.9, there are 2^n functions from A to $\{0, 1\}$.

EXERCISES

1. Finish the verification that the equivalence defined on $A \times B$ on page 118 is an equivalence relation by showing it has properties E2–E3.

2. Prove Theorem 3.4.2: If A_1, A_2, and B are sets, then
$$(A_1 \cup A_2) \times B = (A_1 \times B) \cup (A_2 \times B).$$
Furthermore, if $A_1 \cap A_2 = \emptyset$, then $(A_1 \times B) \cap (A_2 \times B) = \emptyset$.

3. Prove Theorem 3.4.4: Suppose A and B are nonempty, finite sets. Then $|A \times B| = |A| \times |B|$.[15]

4. Prove Theorem 3.4.6: Suppose $\{A_k\}_{k=1}^n$ is a finite family of finite sets. Then $\left|\prod_{k=1}^n A_k\right| = \prod_{k=1}^n |A_k|$.

5. Prove Theorem 3.4.12: Let A and B be nonempty sets with $|A| = m \leq n = |B|$. Then there are $P(n,m)$ distinct one-to-one functions from A to B.

6. Prove Theorem 3.4.16: Suppose $n \geq 0$ and $1 \leq k \leq n$. Then $\binom{n}{k-1} + \binom{n}{k} = \binom{n+1}{k}$.

7. Prove Theorem 3.4.17: Suppose A is a finite set of cardinality n, and let $0 \leq m \leq n$. Then the number of subsets of A of cardinality m is
$$C(n,m) = \binom{n}{m} = \frac{n!}{m!(n-m)!}.$$

8. Verify that Eq. (3.29) holds if $k \leq 0$ or $k \geq n+1$.

9. The State Lottery Commission (Example 3.4.18) is considering including the number 52 in its drawing, so that the game is played by drawing six numbers from $\mathbb{N}_{52}$. By what percentage is the number of possible outcomes increased?

10. In California, a standard license plate consists of a digit 1–9, followed by three letters, followed by three more digits 0–9, such as 3AAG045. If all constructions of this form are considered usable, how many standard license plates can California issue?

11. Twenty horses run a race in which prizes are given for win, place, and show (first, second, and third places). How many outcomes are there for the race?

12. Pizza Peddler offers 12 different toppings, and is having a special on their two-topping pizzas. How many distinct ways are there to order one of their special pizzas?

13. You overheard that a TRUE/FALSE test of 10 questions had 7 TRUE and 3 FALSE answers. If you take the test with this knowledge, how many ways are there to fill out the answer sheet?

14. Your club consists of 50 members, 22 men and 28 women. A committee of 5 is to be chosen, which must contain 2 men and 3 women. How many distinct committees can be chosen?

---

[15] Induct on $|A| \geq 1$, thinking of $|B|$ as fixed.

15. Your office is 10 blocks east and 7 blocks north of your apartment, in a section of the city where the streets run uninterruptedly north-south and east-west. Every day you go to work by walking along these streets the 17 blocks, always going either east or north. How many distinct paths to work are there?[16]

16. When you walk home from the office (as in the previous problem), you always like to stop at the ice cream parlor located on the corner two blocks south and five blocks west of the office. How many distinct paths back home are there if you always stop for ice cream?

17. Twelve jurors and 2 alternates are chosen from 30 people summoned for jury duty. How many ways can this be done?

18. Let a *word* be defined as any arrangement of letters from the alphabet, so that, for example, EIEIO is a word. If the vowels are {A, E, I, O, U}, how many four-letter words contain at least one vowel?

19. Your club of 30 women and 20 men is going to choose a committee of 5, which must not consist of 5 people of the same gender. How many ways can it be done?

---

[16] See Example 3.4.21.

PART II

Basic Principles of Analysis

CHAPTER 4

The Real Numbers

Let's look at some defining characteristics of the area of mathematics we call analysis. Given a set S, its elements might be endowed with a measure of size. The size, or *norm*, of an element x is typically denoted $|x|$, like absolute value on $\mathbb{R}$, or perhaps $\|x\|$. The norm of an element will always be a nonnegative real number, so that a norm is really a function $|\cdot| : S \to \mathbb{R}^+ \cup \{0\}$. There are two other characteristics that a norm must have. You've already run into all three in the context of the real numbers in Section 2.3.3. They are properties N1–N3 beginning on page 57.

Measuring the sizes of elements is only one type of structure that can be placed on a set that puts it squarely in the field of analysis, but some notion of measure with nonnegative real numbers is characteristic of structures in analysis. For example, some structures do not have a norm, but they do have a way of measuring some idea of distance between elements. Such a measure of distance is called a *metric*. Whether S is endowed with a norm or a metric, the measure it represents is inextricably tied to $\mathbb{R}$. Naturally, then, $\mathbb{R}$ is at the heart of all analysis, and one could argue that no analysis into any structure except $\mathbb{R}$ should be undertaken until one understands all the axioms and fundamental results of the theory of real numbers.

We said at the outset that this book is not designed to be a rigorous study of the foundations of mathematics. However, most of the assumptions about $\mathbb{R}$ in Chapter 0 are standard axiomatic ones. It was primarily with assumptions A21 and A22 that we glossed over a lot of groundwork that could have been laid. In all fairness, however, it should be said that all the assumed properties of $\mathbb{R}$ in Section 0.2 should be addressed by beginning from scratch with the set $\{0, 1\}$ and progressing through $\mathbb{W}$, $\mathbb{Z}$, $\mathbb{Q}$, to $\mathbb{R}$, and seeing how the assumptions are motivated along the way to developing this smooth continuum we envision as the set of real numbers.

Be that as it may, a norm or metric on S, if one exists, lends itself to much fruitful study: Sequences and their convergence, continuity, and calculus are but a few. In this chapter we address some fundamental properties of $\mathbb{R}$ that arise out of the assumptions from Chapter 0. These properties are important not only because they apply to $\mathbb{R}$, but also because they are typical of properties of many other structures in analysis. In your later coursework in analysis, you'll study concepts that sound peculiarly similar to those here, even though the set you're dealing with might be, say, a set of functions.

4.1 The Least Upper Bound Axiom

The least upper bound (LUB) axiom is a standard axiom of the real numbers, endowing it with some of its familiar features. For example, the way we visualize $\mathbb{R}$ as a numberline with no holes is due to the LUB axiom. There are two other characteristics of $\mathbb{R}$ that are logically equivalent to the LUB axiom. One is called the *nested interval property* (NIP) and has to do with what you get when you intersect a whole bunch of intervals in $\mathbb{R}$ that have certain properties. The other is called *completeness* and has to do with the way certain sequences behave when the terms get close to each other. It is possible to assume any one of the LUB axiom, the NIP, or completeness and derive the other two as theorems. Assuming the LUB property as an axiom is probably most common, so we choose that route. In this chapter, we'll prove the NIP from the LUB axiom and completeness from the NIP, and look into the converses of these theorems to show that all three are logically equivalent. In this section we explore the LUB property of $\mathbb{R}$ in some depth to get a feel for what it means and derive some immediately important implications. First, a definition.

Definition 4.1.1. A set $A \subseteq \mathbb{R}$ is said to be *bounded from above* if there exists $M_1 \in \mathbb{R}$ such that $a \leq M_1$ for all $a \in A$. A is said to be *bounded from below* if there exists $M_2 \in \mathbb{R}$ such that $a \geq M_2$ for all $a \in A$. A is said to be *bounded* if there exists $M > 0$ such that $|a| \leq M$ for all $a \in A$.

The following theorem seems like the most obvious thing in the world, but it must be demonstrated in terms of Definition 4.1.1.

Theorem 4.1.2. *A set $A \subseteq \mathbb{R}$ is bounded if and only if it is bounded from above and below.*

Proving the $\Rightarrow$ direction of Theorem 4.1.2 is easier. For if A is bounded, then the guaranteed $M > 0$ such that $|a| \leq M$ for all $a \in A$ should clearly suggest values of M_1 and M_2 such that $a \leq M_1$ and $a \geq M_2$ for all $a \in A$. However, in proving the $\Leftarrow$ direction, you must use the existence of M_1 and M_2 such that $M_1 \leq a \leq M_2$ for all $a \in A$ to create a single value of $M > 0$ such that $-M \leq a \leq M$ for all $a \in A$ (Exercise 1).

4.1.1 Least upper bounds

Suppose $A \subseteq \mathbb{R}$ is nonempty and bounded from above. Among all possible upper bounds for A, we would call L a *least upper bound* for A if L is an upper bound for A with the additional property that $L \leq N$ for every upper bound N. That is, among all upper bounds for A, L is no bigger than any of them. The LUB axiom says that every nonempty subset of $\mathbb{R}$ that is bounded from above has a least upper bound in $\mathbb{R}$. We state it again here for the sake of reference.

(A20) Least upper bound property of $\mathbb{R}$: If A is a nonempty subset of $\mathbb{R}$ that is bounded from above, then there exists $L \in \mathbb{R}$ with the following properties:

(L1) For every $a \in A$, we have that $a \leq L$, and

(L2) If N is any upper bound for A, it must be that $N \geq L$. ∎

A natural first question to ask is whether there can be more than one LUB for a nonempty set A. Naturally, the answer is no, which you'll prove in Exercise 2.

Theorem 4.1.3. *The least upper bound of* $A \subseteq \mathbb{R}$ *is unique.*

Interval notation is a convenient shorthand for a subset of $\mathbb{R}$ consisting of a single, clean chunk of the numberline. Here are some illustrations of the notation:

$$(a, b) = \{x \in \mathbb{R} : a < x < b\} \tag{4.1}$$

$$[a, b] = \{x \in \mathbb{R} : a \leq x \leq b\} \tag{4.2}$$

$$(a, +\infty) = \{x \in \mathbb{R} : x > a\} \tag{4.3}$$

$$(-\infty, a] = \{x \in \mathbb{R} : x \leq a\}. \tag{4.4}$$

Intervals of the forms (4.1) and (4.3) are called *open* intervals, and those of the form (4.2) and (4.4) are called *closed* intervals. The motivation for these terms will become clear in Section 4.2 where we discuss open and closed sets in general.

The Greek letter ϵ (epsilon) has been used so much in analysis to represent an arbitrary positive real number that it has come to have a personality all its own. Although ϵ is not generally thought to be of any particular size, it is usually present in theorems and proofs because smaller values of ϵ represent the primary obstacle to overcome in concocting the proof. At first, when you read a theorem using the phrase "for all $\epsilon > 0$," you might mentally insert the additional phrase "no matter how small" in there. This might help you catch the spirit of the theorem, but just remember it might not be particularly relevant. With this in mind, we can now present an alternate form of the LUB property that is often easier to work with than A20.

Theorem 4.1.4. *Given* $A \subseteq \mathbb{R}$, L *is the least upper bound for* A *if and only if the following two conditions hold.*

(M1) *For every* $\epsilon > 0$ *(no matter how large)*, $(L, L + \epsilon) \cap A = \emptyset$.

(M2) *For every* $\epsilon > 0$ *(no matter how small), there exists* $a \in A \cap (L - \epsilon, L]$.

Theorem 4.1.4 is a natural way to visualize the LUB property in terms of how intervals to the left and right of L intersect the set A. Every interval of the form $(L, L + \epsilon)$, no matter how large ϵ is, must not contain any elements of A. Also, every interval of the form $(L - \epsilon, L]$, no matter how small ϵ is, must contain some element of A. You'll prove Theorem 4.1.4 in Exercise 3. The logic of the proof deserves a comment, so here's a suggestion about how to proceed. Theorem 4.1.4 says (L1 ∧ L2) ↔ (M1 ∧ M2). Let's look only at the → direction, for the ← direction would be similar. In showing →,

Exercises 3e and 3h from Section 1.2 imply that

$$(L1 \land L2) \to (M1 \land M2) \Leftrightarrow [(L1 \land L2) \to M1] \land [(L1 \land L2) \to M2]$$
$$\Leftrightarrow [(L2 \land \neg M1) \to \neg L1] \land [(L1 \land \neg M2) \to \neg L2]. \quad (4.5)$$

Thus, to prove the $\to$ direction of Theorem 4.1.4, you will want to show L1 $\to$ M1 by contrapositive in the presence of L2, then show L2 $\to$ M2 by contrapositive in the presence of L1. Then you can prove $\leftarrow$ in a similar way.

Example 4.1.5. Show that the set $\{1\}$ has LUB 1.

Solution: Pick $\epsilon > 0$. Because $(1, 1+\epsilon) \cap \{1\} = \emptyset$, property M1 holds. Also, because $1 \in (1-\epsilon, 1] \cap \{1\}$, property M2 holds. Thus 1 is the LUB of $\{1\}$. ∎

With Theorem 4.1.4, it's also easy to see why b is the LUB of intervals of the form (a, b) or $[a, b]$ (Exercise 4).

4.1.2 The Archimedean property of $\mathbb{R}$

The LUB property of $\mathbb{R}$ has an important implication on the natural numbers, saying that they are unbounded in $\mathbb{R}$.

Theorem 4.1.6. *For all $x > 0$, there exists $n \in \mathbb{N}$ such that $n > x$.*

To prove Theorem 4.1.6, suppose $\mathbb{N}$ is bounded in $\mathbb{R}$. Then its LUB can be shown to pit properties M1 and M2 against each other to produce a contradiction (Exercise 5).

Theorem 4.1.6 has a logically equivalent form called the *Archimedean property* of $\mathbb{R}$. We have already pointed out briefly that the ancient Greeks were very sophisticated mathematically. One idea that they used involved what we would call an *infinitesimal* number. Different from zero, an infinitesimal was considered to be smaller than every positive number. One salient property of an infinitesimal is that you can add it to itself any finite number of times and still have an infinitesimal. The following theorem says in effect that there are no real number infinitesimals. It's named after Archimedes, who argued against the use of infinitesimals. You'll prove it in Exercise 6.

Theorem 4.1.7 (Archimedean Property). *For every $\epsilon, r > 0$, there exists $n \in \mathbb{N}$ such that $n\epsilon > r$.*

Thus no matter how small a positive number ϵ is, and no matter how large $r > 0$ is, you can add ϵ to itself some n number of times to produce a sum that exceeds r, so that ϵ is therefore not an infinitesimal. The Archimedean property often proves to be useful in a slightly different form, using the specific case $r = 1$ and writing the inequality as $1/n < \epsilon$. In this form, the Archimedean property says no matter how small $\epsilon > 0$ is, there exists some $n \in \mathbb{Z}$ whose reciprocal is underneath ϵ.

4.1.3 Greatest lower bounds

Now let's turn the LUB property upside down and discuss bounds from below. We don't have to make any new assumptions concerning the existence of what we call the greatest lower bound of a set, for we can derive results from the LUB property by flipping a set upside down, so to speak.

Definition 4.1.8. A number $G \in \mathbb{R}$ is said to be a *greatest lower bound* (GLB) for a set A if G has the following properties:

(G1) For every $a \in A$, we have that $a \geq G$, and

(G2) If N is any lower bound for A, it must be that $N \leq G$.

You will prove the following theorems in Exercises 9 and 10:

Theorem 4.1.9. *If $A \subseteq \mathbb{R}$ is a nonempty set bounded from below, then A has a GLB.*

Theorem 4.1.10. *The greatest lower bound of $A \subseteq \mathbb{R}$ is unique.*

4.1.4 The LUB and GLB properties applied to finite sets

The intervals (a, b) and $[a, b]$ illustrate that the LUB and GLB of a set might or might not actually be in the set. If a set $S = \{a_k\}_{k=1}^n$ is a finite set of real numbers, then we probably expect that the LUB and GLB of S will be elements of S that we might call a *maximum* or *minimum* value of S. In the rest of this section, we prove this fact, which gives us the right to talk about the maximum and minimum values of a finite set. You'll prove all the results in the exercises.

Theorem 4.1.11. *If $S = \{a_k\}_{k=1}^n$ is a finite set of real numbers, then S is bounded.*

Theorem 4.1.12. *Let $S = \{a_k\}_{k=1}^n$ be a finite set of positive real numbers. Then there exists $N \in \mathbb{N}$ such that $1/N < a_k$ for all $1 \leq k \leq n$.*

Theorem 4.1.12 extends the Archimedean property of $\mathbb{R}$ to apply to a finite set. Be careful, though. The Archimedean property applied to each element of S separately guarantees that, for all $1 \leq k \leq n$, there exists N_k such that $1/N_k < a_k$. This is different from saying that there exists a single N satisfying $1/N < a_k$ for all $1 \leq k \leq n$. This distinction occurs quite often in mathematics. It's one thing to say that every question has an answer. It's quite another to say that there is a single answer that works for every question.

The Archimedean property says that numbers of the form $1/n$ are clustered around zero arbitrarily closely. If you zoom in to any interval $(-\epsilon, \epsilon)$, no matter how small ϵ might be, there will be some $1/n$ in this interval. You cannot zoom in close enough to find an interval around zero that is devoid of numbers of the form $1/n$. In general, if it's possible to find some $\epsilon > 0$ such that the interval $(L - \epsilon, L + \epsilon)$ contains no elements of a set A, we say that A can be *bounded away from* L. Theorem 4.1.12 says, in effect, that a finite set of positive numbers can be bounded away from zero. We state this fact as a corollary.

Corollary 4.1.13. *A finite set of positive real numbers can be bounded away from zero. That is, if $S = \{a_k\}_{k=1}^n$ is a finite set of positive real numbers, then there exists $M > 0$ such that $a_k > M$ for all $1 \leq k \leq n$.*

We're ready for the main result. If $S = \{a_k\}_{k=1}^n$ is a finite set, then by Theorem 4.1.11, S is bounded. Thus by Theorem 4.1.2, it is bounded from above and below. By the LUB and GLB properties, S has a LUB and GLB. In fact:

Theorem 4.1.14. *If $S = \{a_k\}_{k=1}^n$ is a finite set of real numbers, then S contains its LUB and GLB.*

You'll prove the LUB part of Theorem 4.1.14 in Exercise 14. The GLB part would be very similar. Theorem 4.1.14 motivates the following definition.

Definition 4.1.15. Let $S = \{a_k\}_{k=1}^n$ be a finite set of real numbers. Then the LUB and GLB of S are called the *maximum* and *minimum* values of S, respectively, and are denoted $\max S$ and $\min S$.

In Section 4.2 the importance of Theorem 4.1.14 will become clear right away. Here are two simple illustrations of how it will be important. First, suppose anyone at least 16 years old can drive, anyone at least 18 can vote, and anyone at least 65 can collect Social Security. How can you guarantee that a chosen person can drive, vote, and collect Social Security? Certainly you would choose a person whose age is at least $\max\{16, 18, 65\}$. For if we write $M = \max\{16, 18, 65\}$, then M satisfies all three inequalities $M \geq 16$, $M \geq 18$ and $M \geq 65$.

As another example, suppose every real number within ϵ_1 distance of zero is in set A, and every real number within ϵ_2 of zero is in set B. That is, if $|x| < \epsilon_1$, then $x \in A$, and if $|x| < \epsilon_2$, then $x \in B$. If we let $\epsilon = \min\{\epsilon_1, \epsilon_2\}$, we can be sure that every real number within ϵ of zero will be in $A \cap B$. For if $|x| < \epsilon$, then both $|x| < \epsilon_1$ and $|x| < \epsilon_2$ are true, so that $x \in A$ and $x \in B$.

The terms $\max S$ and $\min S$ are often used when S is not finite in the event that S contains its LUB and GLB.

EXERCISES

1. Prove Theorem 4.1.2: A set $A \subseteq \mathbb{R}$ is bounded if and only if it is bounded from above and below.[1]

---

[1] To prove $\Leftarrow$, let $M = |M_1| + |M_2|$. Use Exercise 14 from Section 2.3.

4.1 The Least Upper Bound Axiom

2. Prove Theorem 4.1.3: The least upper bound of $A \subseteq \mathbb{R}$ is unique.

3. Prove Theorem 4.1.4: Given $A \subseteq \mathbb{R}$, L is the least upper bound for A if and only if the following two conditions hold:

   (M1) For every $\epsilon > 0$, $(L, L + \epsilon) \cap A = \emptyset$.

   (M2) For every $\epsilon > 0$, there exists $a \in A \cap (L - \epsilon, L]$.

4. Show that b is the LUB of the interval (a, b). (An identical argument will work for $[a, b]$.)

5. Prove Theorem 4.1.6: For all $x > 0$, there exists $n \in \mathbb{N}$ such that $n > x$.

6. Prove Theorem 4.1.7: For every $\epsilon, r > 0$, there exists $n \in \mathbb{N}$ such that $n\epsilon > r$.[2]

7. Prove that between any two real numbers $a < b$, there exists a *nonzero* rational number. Need some hints?[3,4,5,6]

8. Prove that between any two real numbers $a < b$, there exists an irrational number.[7]

9. Prove Theorem 4.1.9 in the following way. Suppose $A \subseteq \mathbb{R}$ is a nonempty set bounded from below. Let M be any lower bound for A, and define $B = \{-a : a \in A\}$.

   (a) Show that B is bounded from above by $-M$.

   (b) Applying the LUB property to B, we conclude the existence of $L \in \mathbb{R}$, the LUB of B. Use the fact that L satisfies properties L1 and L2 with regard to B to show that $-L$ has properties G1 and G2 with regard to A.

10. Prove Theorem 4.1.10: The greatest lower bound of $A \subseteq \mathbb{R}$ is unique.

11. State a theorem analogous to Theorem 4.1.4 for the GLB of a set.

12. Prove Theorem 4.1.11: If $S = \{a_k\}_{k=1}^n$ is a finite set of real numbers, then S is bounded.[8]

13. Prove Theorem 4.1.12: Let $S = \{a_k\}_{k=1}^n$ be a finite set of positive real numbers. Then there exists $N \in \mathbb{N}$ such that $1/N < a_k$ for all $1 \le k \le n$.[9]

14. Prove part of Theorem 4.1.14: If $S = \{a_k\}_{k=1}^n$ is a finite set of real numbers, then S contains its LUB.[10]

---

[2] Apply Theorem 4.1.6 to $x = r/\epsilon$.
[3] First start by assuming $a \ge 0$. Worry about $a < 0$ later.
[4] Apply Theorem 4.1.7 to $b - a$.
[5] If $1/n < b - a$, you can apply Theorem 4.1.7 to a and $1/n$.
[6] Use the WOP for the right m such that $m/n > a$.
[7] Apply the result of Exercise 7 to $a/\sqrt{2} < b/\sqrt{2}$. Why is the resulting c irrational?
[8] Let $M = \sum_{k=1}^n |a_k|$.
[9] Get an N_k for each a_k that satisfies $1/N_k < a_k$, then let $N = \sum_{k=1}^n N_k$.
[10] If L is the LUB of S and $L \notin S$, then you ought to be able to use the set $T = \{L - a_k\}_{k=1}^n$ and Corollary 4.1.13 to contradict the assumption that L is the LUB of S.

4.2 Sets in $\mathbb{R}$

4.2.1 Open and closed sets

Definition 4.2.1. Given $a \in \mathbb{R}$ and $\epsilon > 0$, we define the ϵ-*neighborhood* of a as

$$N_\epsilon(a) = (a - \epsilon, a + \epsilon) = \{x \in \mathbb{R} : a - \epsilon < x < a + \epsilon\} = \{x \in \mathbb{R} : |x - a| < \epsilon\}, \quad (4.6)$$

and we say that the neighborhood has *radius* ϵ.

Notice if $0 < \epsilon_1 \leq \epsilon_2$, then $N_{\epsilon_1}(a) \subseteq N_{\epsilon_2}(a)$. For if $x \in N_{\epsilon_1}(a)$, then

$$a - \epsilon_2 \leq a - \epsilon_1 < x < a + \epsilon_1 \leq a + \epsilon_2, \quad (4.7)$$

so that $x \in N_{\epsilon_2}(a)$.

Definition 4.2.2. A set $A \subseteq \mathbb{R}$ is called *open* if for every $a \in A$, there exists $\epsilon > 0$ such that $N_\epsilon(a) \subseteq A$.

A set A is open if every element of A is in some interval contained entirely within A; that is, every point in A is, in a sense, insulated from the outside of A by a neighborhood that is a subset of A. In Exercise 1, you'll show that the intervals $(a, +\infty)$ and $(-\infty, b)$ are open.

As usual, one of the first questions you ought to ask yourself after a definition is presented is what its negation is. So what does it mean to say that A is not open? Logically,

$$\begin{aligned}\neg(A \text{ is open}) &\Leftrightarrow \neg(\forall a \in A)(\exists \epsilon > 0)(N_\epsilon(A) \subseteq A) \\ &\Leftrightarrow (\exists a \in A)(\forall \epsilon > 0)(N_\epsilon \not\subseteq A) \\ &\Leftrightarrow (\exists a \in A)(\forall \epsilon > 0)(\exists x \in N_\epsilon(a))(x \notin A).\end{aligned} \quad (4.8)$$

Thus A is not open if there is some point in A, every ϵ-neighborhood of which contains some point not in A. That is, there is at least one point $a \in A$ that cannot, as it were, be insulated from the outside.

Example 4.2.3. Show that the set $(a, b]$ is not open.

Solution: Consider the point b, and pick any $\epsilon > 0$. Then the point $b + \epsilon/2$ is in $N_\epsilon(b)$ but not in $(a, b]$. ∎

Two other sets besides $(a, +\infty)$ and $(-\infty, b)$ immediately present themselves as open sets. The empty set is open simply because otherwise there would have to exist some $x \in \emptyset$ with some property, which is a contradiction. $\mathbb{R}$ is open because for any point $a \in \mathbb{R}$, we may let $\epsilon = 1$ and have $N_\epsilon(a) \subseteq \mathbb{R}$.

Let's address unions and intersections of open sets. If A and B are both open, do you suspect $A \cup B$ is open? $A \cap B$?[11] In both cases the answer is yes, but when an arbitrary family of open sets is combined by union or intersection, the answer can change. In Exercise 4 you'll prove the following:

Theorem 4.2.4. *The union of a family of open sets is open. The intersection of a finite family of open sets is open.*

With Theorem 4.2.4, it follows that if $a < b$, then the interval (a, b) is open, for $(a, b) = (-\infty, b) \cap (a, +\infty)$. The intersection of an infinite family of open sets certainly could be open, but it might not be (Exercise 5).

If we were discussing doors or stores or minds, to say that one is closed would mean that it is not open. When it comes to sets, *closed* does not mean not open.

Definition 4.2.5. A set $A \subseteq \mathbb{R}$ is said to be *closed* if A' is open.

The interval $[a, b]$ is closed because $[a, b]' = (-\infty, a) \cup (b, +\infty)$, the union of two open sets. Similarly, a set with one element, $\{a\}$, called a *singleton*, is closed because $\{a\}' = (-\infty, a) \cup (a, +\infty)$. The proof of the following theorem should be quick (Exercise 6):

Theorem 4.2.6. *The intersection of a family of closed sets is closed. The union across a finite family of closed sets is closed.*

Because of Theorem 4.2.6, it follows that any finite set is closed because it is a finite union of singleton sets. The integers provide a good example of an infinite set that is closed. (What is $\mathbb{Z}'$?) In Exercise 8, you'll demonstrate a union of closed sets that is not closed.

Many sets are neither open nor closed. For example, $[a, b)$ is sometimes called a *half open* or *half closed* interval. Believe it or not, some sets are both open and closed. For example, $\mathbb{R}$ and $\emptyset$ are both open and closed because they are both open and they are complementary. Theorem 4.2.7 uses the LUB axiom to show the important fact that these are the only two subsets of $\mathbb{R}$ that are both open and closed. The proof is a little intricate, so we provide it here. It is important for you to work your way through the details because it illustrates an important type of proof in which we show that certain examples of something are the only ones that exist. Keep a pencil and paper handy to sketch some number lines that will help you visualize the details.

Theorem 4.2.7. *The only subsets of $\mathbb{R}$ that are both open and closed are $\emptyset$ and $\mathbb{R}$.*

Proof: We prove by contradiction. Suppose there exists $A \subset \mathbb{R}$ such that A and A' are both open and nonempty. Pick any $a_1 \in A$ and $a_2 \in A'$. Without any loss of generality, we may assume $a_1 < a_2$. Let $E = \{\epsilon > 0 : N_\epsilon(a_1) \subseteq A\}$; that is, E is the set of all $\epsilon > 0$ such that the ϵ-neighborhood of a_1 is contained entirely within A. Since A is open, E is nonempty. Because $a_2 \in A'$, E is bounded from above by $a_2 - a_1$. Thus we may apply the LUB property to E to conclude the existence of some

---

[11] Think intuitively in terms of open intervals.

ϵ_0, the LUB of E. Thus $N_{\epsilon_0}(a_1)$ is the largest neighborhood of a_1 that is contained entirely within A.

In order to arrive at a contradiction, we look at the endpoints of the interval $(a_1 - \epsilon_0, a_1 + \epsilon_0)$. First we ask if it is possible that $a_1 + \epsilon_0 \in A'$. If so, then the fact that A' is open means there exists $\epsilon_1 > 0$ such that $N_{\epsilon_1}(a_1 + \epsilon_0) \subseteq A'$. Because $a_1 \in A$, we know that $\epsilon_1 < \epsilon_0$. But then $a_1 + \epsilon_0 - \epsilon_1/2 \in A'$. This contradicts the fact that $N_{\epsilon_0}(a_1) \subseteq A$. Thus $a_1 + \epsilon_0 \in A$. By an identical argument applied to $a_1 - \epsilon_0$, we have that $a_1 - \epsilon_0 \in A$, so that $[a_1 - \epsilon_0, a_1 + \epsilon_0] \subseteq A$.

Now let's look again at the endpoints of $[a_1 - \epsilon_0, a_1 + \epsilon_0]$. Since A is open and $a_1 - \epsilon_0, a_1 + \epsilon_0 \in A$, there exist $\epsilon_2, \epsilon_3 > 0$ such that $N_{\epsilon_2}(a_1 - \epsilon_0) \subseteq A$ and $N_{\epsilon_3}(a_1 + \epsilon_0) \subseteq A$. Letting $\epsilon_4 = \min\{\epsilon_2, \epsilon_3\}$, we have that $N_{\epsilon_0 + \epsilon_4}(a_1) \subseteq A$. This contradicts the fact that ϵ_0 is the LUB of E, for we have shown that $\epsilon_0 + \epsilon_4 \in E$. Thus it is impossible that A and A' are both open and nonempty. ∎

Theorem 4.2.7 is important both in its own right and because it motivates the idea of *connectedness* of sets in the area of mathematics called *topology*. By Theorem 4.2.7, if A is a nonempty, proper, open subset of $\mathbb{R}$, then A' is a nonempty, proper subset of $\mathbb{R}$ that is *not* open. Thus Theorem 4.2.7 says that $\mathbb{R}$ cannot be written as the disjoint union of two nonempty open sets. Another way to say this is that if $A, B \subseteq \mathbb{R}$ are both open, and if $\mathbb{R} = A \cup B$, then either $A = \emptyset$ or $B = \emptyset$. This property gives us the imagery of the real number line as one connected piece, and as the proof reveals, follows from the LUB axiom. In the more abstract setting of topology, a set is said to be connected *by definition* if it cannot be written as the disjoint union of two nonempty, open sets.

4.2.2 Interior, exterior, and boundary

Suppose $A \subseteq \mathbb{R}$ and we're given some $x \in \mathbb{R}$. Then precisely one of the following must be true.

1. There exists $\epsilon > 0$ such that $N_\epsilon(x) \subseteq A$, in which case we say that x is in the *interior* of A, written $x \in \text{Int}(A)$.

2. There exists $\epsilon > 0$ such that $N_\epsilon(x) \subseteq A'$, in which case we say that x is in the *exterior* of A, written $x \in \text{Ext}(A)$.

3. For all $\epsilon > 0$, there exists $a_1, a_2 \in N_\epsilon(x)$ such that $a_1 \in A$ and $a_2 \in A'$, in which case we say that x is on the *boundary* of A, written $x \in \text{Bdy}(A)$.

Note that if $x \in \text{Int}(A)$, then $x \in A$. Also, if $x \in \text{Ext}(A)$, then $x \notin A$. However, if $x \in \text{Bdy}(A)$, then x might or might not be in A. The gist of $x \in \text{Bdy}(A)$ is that every ϵ-neighborhood of x contains at least one point from each of A and A', and x is allowed to serve as one of these two points, if necessary. Notice that $\text{Bdy}(A) = \text{Bdy}(A')$, and that $\text{Int}(A) = \text{Ext}(A')$ and vice versa.

Example 4.2.8. Consider $A = (2, 3] \cup \{1/n\}_{n=1}^\infty$. Then $2.5 \in \text{Int}(A)$ by letting $\epsilon = 0.25$, $1.5 \in \text{Ext}(A)$ by letting $\epsilon = 0.5$, and $2, 3, 0, 1/n \in \text{Bdy}(A)$.

Theorem 4.2.9. *$A \subseteq \mathbb{R}$ is open if and only if $\text{Int}(A) = A$.*

We noted in the preceding that Int(A) $\subseteq A$, an immediate consequence of the definition of Int(A). The proof of $\supseteq$ should be quick (Exercise 9). With Theorem 4.2.9, we see that openness is logically equivalent to every point being an interior point.

EXERCISES

1. Show that the intervals $(a, +\infty)$ and $(-\infty, b)$ are open.[12]

2. Let $\mathcal{F} = \{(0, 4), (10, 14), (2, 3), (9, 12), (2, 8)\}$. Write $\cup_{A \in \mathcal{F}} A$ in terms of as few intervals as possible.

3. Let $\mathcal{F} = \{(1, 10), (-2, 8), (0, 20), (2, 15)\}$. Write $\cap_{A \in \mathcal{F}} A$ in terms of as few intervals as possible.

4. Prove Theorem 4.2.4: Let $\mathcal{F} = \{A\}$ be a family of open sets, and let $\{B_k\}_{k=1}^n$ be a finite family of open sets. Then

    (a) $\cup_{\mathcal{F}} A$ is open.
    (b) $\cap_{k=1}^n B_k$ is open.

5. Show that the intersection of infinitely many open sets might not be open by showing $\cap_{n=1}^{\infty} (-1/n, 1/n) = \{0\}$.

6. Prove Theorem 4.2.6: Let $\mathcal{F} = \{A\}$ be a family of closed sets, and let $\{B_k\}_{k=1}^n$ be a finite family of closed sets. Then

    (a) $\cap_{\mathcal{F}} A$ is closed.[13]
    (b) $\cup_{k=1}^n B_k$ is closed.

7. Show that if $a < b$, then $[a, b]$ is closed.

8. Give an example of an infinite family $\mathcal{F}$ of closed sets such that $\cup_{\mathcal{F}} A$ is not closed. Prove.

9. Prove Theorem 4.2.9: $A \subseteq \mathbb{R}$ is open if and only if Int(A) = A.

4.3 Limit Points and Closure of Sets

Given $A \subseteq \mathbb{R}$ and any $x \in \mathbb{R}$, we might want to know if elements of A are densely clustered around x.

Definition 4.3.1. A point $x \in \mathbb{R}$ is said to be a *limit point* of $A \subseteq \mathbb{R}$ if every ϵ-neighborhood of x contains a point in A other than x itself. That is, for all $\epsilon > 0$, there exists $a \in A \cap N_\epsilon(x)$, where $a \neq x$.

---

[12] See Exercise 12 from Section 2.3 if you need to.
[13] See Theorem 2.2.7 from Section 2.2 if you need to.

Nothing about Definition 4.3.1 requires a limit point of A to be in A. For example, the Archimedean property implies immediately that 0 is a limit point of $\{1/n\}_{n=1}^{\infty}$. If, however, $x \in A$, then for x to be a limit point of A, the definition requires that every ϵ-neighborhood contain an element of A different from x. This motivates a new term.

Definition 4.3.2. For $x \in \mathbb{R}$ and $\epsilon > 0$, the set $N_\epsilon(x) \setminus \{x\}$ is called the *deleted ϵ-neighborhood* of x, and is denoted $DN_\epsilon(x)$. Another way to write $DN_\epsilon(x)$ is $\{y \in \mathbb{R} : 0 < |y - x| < \epsilon\}$.

With this new notation, we may say that x is a limit point of A if every deleted ϵ-neighborhood of x contains an element of A; that is, for all $\epsilon > 0$, $A \cap DN_\epsilon(x) \neq \emptyset$. Limit points have other common names, for example, *cluster points* or *accumulation points*.

True to form, we ask what it means for x not to be a limit point of A.

$$\begin{aligned}\neg(x \text{ is a limit point of } A) &\Leftrightarrow \neg(\forall \epsilon > 0)(A \cap DN_\epsilon(x) \neq \emptyset) \\ &\Leftrightarrow (\exists \epsilon > 0)(A \cap DN_\epsilon(x) = \emptyset).\end{aligned} \quad (4.9)$$

That is, x is not a limit point of A if there is some ϵ-neighborhood of x that is devoid of elements of A, other than, perhaps, x itself. We'll prove Theorem 4.3.3 here and leave the proof of Theorem 4.3.4 to you in Exercise 3.

Theorem 4.3.3. *If $A \subseteq \mathbb{R}$ is open, then every $x \in A$ is a limit point of A.*

Proof: Suppose A is open, pick any $x \in A$ and any $\epsilon > 0$. Since A is open, there exists $\epsilon_1 > 0$ such that $N_{\epsilon_1}(x) \subseteq A$. Let $\epsilon_2 = \min\{\epsilon, \epsilon_1\}$. Then the point $x + \epsilon_2/2 \in A \cap DN_\epsilon(x)$. ∎

Notice that the converse of Theorem 4.3.3 is not true. For example, every point in $[a, b]$ is a limit point, but $[a, b]$ is not open.

Theorem 4.3.4. *A set is closed if and only if it contains all its limit points.*

4.3.1 Closure of sets

If $A \subseteq \mathbb{R}$ is not closed, we might want to close it off, so to speak, by finding the smallest closed superset of A, if there is one. First we define a term for this smallest closed superset of A and then we address whether it exists, and if so, whether it's unique.

Definition 4.3.5. Suppose $A \subseteq \mathbb{R}$ is given, and suppose $C \subseteq \mathbb{R}$ is a set with the following properties:

(C1) $A \subseteq C$.

(C2) C is closed.

(C3) If $D \subseteq \mathbb{R}$ is closed and $A \subseteq D$, then $C \subseteq D$.

Then C is called a *closure* of A, and is denoted $\overline{A}$.

Notice how Definition 4.3.5 lays down the characteristics that a set C must have in order for it to be a smallest closed superset of A. Property C1 guarantees that any set we would call $\overline{A}$ does, in fact, contain all elements of A. Property C2 guarantees that it is actually closed. Property C3 says that any other closed set containing all elements of A will not be any smaller. That $\overline{A}$ exists uniquely is guaranteed by the following theorem, which gives us one way to visualize its construction.

Theorem 4.3.6. *Suppose $A \subseteq \mathbb{R}$. Then $\overline{A}$ exists uniquely, and can be constructed as*

$$\overline{A} = \bigcap_{\substack{S \supseteq A \\ S \text{ closed}}} S. \tag{4.10}$$

Thus if $\mathcal{F}$ is the family of all closed supersets of A, then the intersection across $\mathcal{F}$ satisfies properties C1–C3, and therefore has earned the right to be called $\overline{A}$. An important question to ask about the construction in Eq. (4.10) is whether there even exist any sets in $\mathcal{F}$. If there are no closed supersets of A, then Eq. (4.10) is a vacuous construction. But since $\mathbb{R} \supseteq A$ and $\mathbb{R}$ is closed, then this is not the case. You'll prove that the construction in Eq. (4.10) satisfies properties C1–C3 and that $\overline{A}$ is unique in Exercise 5.

The construction of $\overline{A}$ in Theorem 4.3.6 is a way of closing off a set A by intersecting down from above as it were, through closed sets containing A as if we were shrink wrapping it. Another way to arrive at the same thing is to build the closure up from below, as the following theorem asserts:

Theorem 4.3.7. $\overline{A} = \text{Int}(A) \cup \text{Bdy}(A)$.

The proof of Theorem 4.3.7 is left to you in Exercise 7. It says, in effect, that the closure of a set A can be thought of as merely A, some points of which might be interior points and some boundary points, with any missing boundary points thrown in. The easiest way to prove Theorem 4.3.7 is a bit surprising. There will be a hint in the exercises if you need it. Here are our last two theorems, whose proofs are left to you in the exercises.

Theorem 4.3.8. *If $A \subseteq B$, then $\overline{A} \subseteq \overline{B}$.*

Theorem 4.3.9. *A is closed if and only if $\overline{A} = A$.*

EXERCISES

1. Prove that if x is a limit point of A, then every ϵ-neighborhood of x contains infinitely many elements of A.

2. Suppose L is the LUB of a set A and $L \notin A$. Show L is a limit point of A.

3. Prove Theorem 4.3.4: A set is closed if and only if it contains all its limit points.

4. Show that $\{1/n : n \in \mathbb{N}\}$ is not closed.

5. Prove Theorem 4.3.6 by showing that the construction in Eq. (4.10) satisfies properties C1–C3, and that $\overline{A}$ is unique.[14]

6. What does it mean to say $x \notin \overline{A}$?

7. Prove Theorem 4.3.7: $\overline{A} = \text{Int}(A) \cup \text{Bdy}(A)$.[15]

8. Determine the closure of the following sets.

    (a) $\emptyset$

    (b) $\mathbb{R}$

    (c) (a, b)

    (d) $(a, b) \cup (b, c)$

9. Prove Theorem 4.3.8: If $A \subseteq B$, then $\overline{A} \subseteq \overline{B}$.[16]

10. Prove Theorem 4.3.9: A is closed if and only if $\overline{A} = A$.

11. One of the following statements is true, and for the other, one direction of subset inclusion is true. Prove the three true subset inclusion statements, and provide a counterexample to illustrate the falsity of the fourth.

    (a) $\overline{A \cup B} = \overline{A} \cup \overline{B}$[17]

    (b) $\overline{A \cap B} = \overline{A} \cap \overline{B}$

4.4 Compactness

There is a special kind of subset of $\mathbb{R}$ that deserves our attention now. Sets that are *compact* have a fundamental feature that makes them very important if they serve as the domain of a function whose domain and codomain are subsets of $\mathbb{R}$. Before we can define compactness, we need another term first.

Definition 4.4.1. Suppose $A \subseteq \mathbb{R}$ and $\mathcal{C}$ is a family of subsets of $\mathbb{R}$. If

$$A \subseteq \bigcup_{S \in \mathcal{C}} S, \tag{4.11}$$

we say that $\mathcal{C}$ *covers* A, or is a *cover* of A. Expression (4.11) simply means that every $x \in A$ is in some $S \in \mathcal{C}$. If every $S \in \mathcal{C}$ is open, we call $\mathcal{C}$ an *open cover*. If $\mathcal{F} \subseteq \mathcal{C}$ also covers A, we call $\mathcal{F}$ a *subcover* of $\mathcal{C}$.

---

[14] See Exercises 3 and 6 from Section 2.2, and Theorem 4.2.6.
[15] Show $(\overline{A})' = \text{Ext}(A)$.
[16] Apply Exercise 4c from Section 2.1 and show $(\overline{B})' \subseteq (\overline{A})'$.
[17] Take a hint from the hint for Exercise 7.

Example 4.4.2. Let $\mathcal{C} = \{N_1(x) : x \in \mathbb{Q}\}$, the set of neighborhoods of radius 1 of all rational numbers. Then $\mathcal{C}$ is an open cover of $\mathbb{R}$ because every real number is within 1 of some rational. Furthermore, $\{N_1(n) : n \in \mathbb{Z}\}$ is a subcover of $\mathcal{C}$ because every real number is within 1 of some integer.

Covers and subcovers are of most interest to us when the cover $\mathcal{C}$ is an open cover with infinitely many sets in it, and $\mathcal{F}$ is a subcover that has only finitely many sets from $\mathcal{C}$ in it.

Example 4.4.3. Let $a \in \mathbb{R}$. Then the set

$$\mathcal{C} = \{(-\infty, a - 1/n) \cup (a + 1/n, +\infty) : n \in \mathbb{N}\} \tag{4.12}$$

covers $\mathbb{R}\setminus\{a\}$, and $\mathcal{C}$ has no finite subcover (Exercise 3).

Example 4.4.4. The set $\{(-n, n) : n \in \mathbb{N}\}$ is a cover of $\mathbb{R}$ with no finite subcover (Exercise 4).

Now we're ready for the definition of compactness.

Definition 4.4.5. A set $A \subseteq \mathbb{R}$ is said to be *compact* if every open cover of A has a finite subcover.

To demonstrate that a given $A \subseteq \mathbb{R}$ is compact from Definition 4.4.5 would appear to be quite a task, for something rather powerful must be shown about every conceivable open cover of A. Showing A is not compact probably would not be particularly difficult, for all we need to do is demonstrate some open cover of A that cannot be reduced to a finite subcover. Therefore, it might be fairly easy to eliminate entire classes of sets from the possibility of being compact, but showing certain entire classes of sets are compact is another story. Thankfully, someone has been down this road before and supplied us with the answer to the question of precisely which sets are compact.

Theorem 4.4.6 (Heine-Borel). *A set $A \subseteq \mathbb{R}$ is compact if and only if A is closed and bounded.*

We'll piece together the proof of Theorem 4.4.6 in several stages. First, you'll prove the $\Rightarrow$ direction in Exercise 5. To prove $\Leftarrow$, we use the following two theorems. We'll prove the first one here, then you'll prove the second one in Exercise 6.

Theorem 4.4.7. *The interval $[a, b]$ is compact.*

Proof: Suppose $\mathcal{C}$ is an open cover of $[a, b]$. Construct $E \subseteq (a, b]$ in the following way. For $x \in (a, b]$, let x be in E if and only if some finite subfamily of $\mathcal{C}$ covers the interval $[a, x]$. First we show $E \neq \emptyset$. Since $\mathcal{C}$ covers $[a, b]$, then there is some $S_1 \in \mathcal{C}$ such that $a \in S_1$. Since S_1 is open, there exists $\epsilon_1 > 0$ such that $N_{\epsilon_1}(a) \subseteq S_1$. Thus the subfamily $\{S_1\}$ covers $[a, a + \epsilon_1/2]$, so that $a + \epsilon/2 \in E$ and $E \neq \emptyset$. Also, E is bounded above by b. Therefore, E has LUB $L \leq b$. First we show that $L \in E$, then we show that $L = b$ in order to have that the entire interval $[a, b]$ can be covered by a finite subfamily of $\mathcal{C}$.

Suppose $L \notin E$. Since $\mathcal{C}$ is an open cover of $[a, b]$ and $L \leq b$, there exists $S_2 \in \mathcal{C}$ such that $L \in S_2$. Since S_2 is open, there exists $\epsilon_2 > 0$ such that $N_{\epsilon_2}(L) \subseteq S_2$. Since L is the LUB of E, $L - \epsilon_2/2 \in E$, so that $[a, L - \epsilon_2/2]$ can be covered by a finite subfamily $\mathcal{F}_2 \subseteq \mathcal{C}$. But then $\mathcal{F}_2 \cup \{S_2\}$ is a finite subfamily of $\mathcal{C}$ that covers $[a, L + \epsilon_2/2]$. This is a contradiction, so $L \in E$.

Now suppose $L < b$. Since some finite $\mathcal{F}_3 \subseteq \mathcal{C}$ covers $[a, L]$, there exists some $S_3 \in \mathcal{F}_3$ such that $L \in S_3$. Since S_3 is open, there exists $\epsilon_3 > 0$ such that $N_{\epsilon_3}(L) \subseteq S_3$. Let $x = \min\{L + \epsilon_3/2, b\}$. Then $x > L$, $x \in S_3$, and $x \in [a, b]$. Thus, the interval $[a, x]$ can be covered by a finite subcover of $\mathcal{C}$. This contradicts the fact that L is the LUB of E. Thus $L < b$ is impossible, and $L = b$. ∎

Knowing that $[a, b]$ is compact moves us very close to showing that any closed and bounded set is compact. The following theorem, which you'll prove in Exercise 6, will get us within ϵ of being finished.

Theorem 4.4.8. *If $A \subseteq B$, where A is closed and B is compact, then A is also compact.*

Now we can put the pieces together and prove the Heine-Borel theorem.

Proof of Theorem 4.4.6. The $\Rightarrow$ direction is Exercise 5. To prove $\Leftarrow$, suppose A is closed and bounded. Then there exists $M > 0$ such that $A \subseteq [-M, M]$. Since $[-M, M]$ is compact and $A \subseteq [-M, M]$, then by Theorem 4.4.8 A is compact. ∎

You might be a little disappointed that we went to all the trouble of defining a new term (compactness) in complicated terms of open covers and subcovers, when it turns out that compact sets are precisely those that are closed and bounded, two ideas we already have a handle on. Well, you would be right to say that there is no reason to define a new term if the class of things to which the term applies can be easily described in already familiar language. However, the point remains that closed and bounded sets have a very important feature: Every open cover can be reduced to a finite subcover. We need this feature in Section 5.7.

EXERCISES

1. What does it mean for $\mathcal{C}$ not to be a cover of A?

2. What does it mean for a cover $\mathcal{C}$ of a set A not to have a finite subcover?

3. Verify the claim in Example 4.4.3: $\mathcal{C} = \{(-\infty, a - 1/n) \cup (a + 1/n, +\infty) : n \in \mathbb{N}\}$ covers $\mathbb{R} \setminus \{a\}$, and $\mathcal{C}$ has no finite subcover.

4. Verify the claim in Example 4.4.4: The set $\{(-n, n) : n \in \mathbb{N}\}$ is a cover of $\mathbb{R}$ with no finite subcover.

5. Prove the $\Rightarrow$ direction of Theorem 4.4.6 in the following way:

   (a) Suppose A is not closed. Demonstrate an open cover of A that has no finite subcover.[18]

   (b) Suppose A is not bounded. Demonstrate an open cover of A that has no finite subcover.[19]

6. Prove Theorem 4.4.8: If B is compact and $A \subseteq B$ is closed, then A is also compact.[20]

4.5 Sequences in $\mathbb{R}$

There are several ways to approach the subject of sequences. Let's move from an informal to a more formal way. A sequence in $\mathbb{R}$ is an ordered list of numbers

$$\langle a_1, a_2, a_3, \ldots, a_n, \ldots \rangle. \tag{4.13}$$

Listing the elements of a sequence might remind you of showing that a set A is countable, for listing the elements of A exhaustively in the fashion of sequence (4.13) is equivalent to constructing a function $f : \mathbb{N} \xrightarrow{\text{onto}} A$, where $f(1) = a_1$, $f(2) = a_2$, etc. Perhaps we can think of a sequence as a function $f : \mathbb{N} \to \mathbb{R}$. To say that a sequence is a function might conjure up slightly different imagery from saying it is a list of numbers, but a little thought will convince you that there is no difference. Let's create some notation. The expression $\langle a_n \rangle_{n=1}^{\infty}$ is one standard way of denoting a sequence. Beginning with term a_1 is a matter of convenience. There might be times it seems more natural to begin with a_0.

If there is a formula for a_n, say, for example, $a_n = 1/n$, we have

$$\left\langle \frac{1}{n} \right\rangle_{n=1}^{\infty} = \left\langle 1, \frac{1}{2}, \frac{1}{3}, \ldots \right\rangle. \tag{4.14}$$

We must distinguish Eq. (4.14) from $\{1/n\}_{n=1}^{\infty}$, which is merely the set of elements of the sequence in Eq. (4.14), and does not have the ordering of the elements as one of its defining characteristics. Saying $a_n = 1/n$ makes the idea of a sequence as a function more natural, for we are talking about nothing other than $f : \mathbb{N} \to \mathbb{R}$ defined by $f(n) = 1/n$, and the set $\{1/n\}$ can then be thought of as the range of the sequence. Then, if we want to visualize the sequence graphically, we can, as in Fig. 4.1. Such a graph will help in visualizing limits in Section 4.6. Example 4.5.1 presents some examples of sequences we'll refer to later.

Example 4.5.1. 1. The sequence $\langle a_n \rangle_{n=1}^{\infty}$ defined by $a_n = 1/2 + (-1)^n \cdot 1/2$ is the sequence $\langle 0, 1, 0, 1, 0, 1, \ldots \rangle$.

2. Letting $a_n = n + (-1)^n$ for $n \geq 0$ generates $\langle 1, 0, 3, 2, 5, 4, 7, 6, \ldots \rangle$.

---

[18] Use the characterization of closed sets from Theorem 4.3.4 and Exercise 3.
[19] Use Exercise 4.
[20] If $\mathcal{C}$ is an open cover of A, then the fact that A' is open should suggest a way to expand $\mathcal{C}$ into an open cover of B.

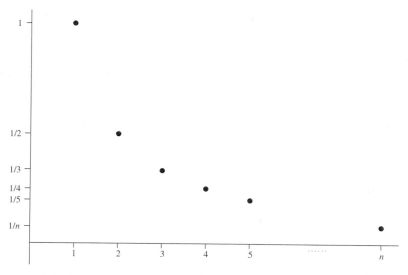

Figure 4.1 The sequence $a_n = f(n) = 1/n$.

3. If $a_n = \sin n\pi$, then we generate the sequence $\langle 0 \rangle$.

4. We can define a_n in cases. For $n \geq 0$, let

$$a_n = \begin{cases} 0, & \text{if } n \text{ is even,} \\ n^2 - 1 & \text{if } n \text{ is odd.} \end{cases} \quad (4.15)$$

to generate the sequence $\langle 0, 0, 0, 8, 0, 24, 0, 48, 0, 80, 0, \ldots \rangle$.

5. If $a_n = n^2/2^n$ for $n \geq 1$, then we generate the sequence

$$\left\langle \frac{1}{2}, 1, \frac{9}{8}, 1, \frac{25}{32}, \frac{36}{64}, \frac{49}{128}, \frac{64}{256}, \ldots \right\rangle.$$

4.5.1 Monotone sequences

An important class of sequences derives from the following definition:

Definition 4.5.2. A sequence $\langle a_n \rangle_{n=1}^{\infty}$ is said to be *increasing* if $a_m \leq a_n$ whenever $m < n$. It is said to be *decreasing* if $a_m \geq a_n$ whenever $m < n$. If $\langle a_n \rangle$ is either increasing or decreasing, then it is said to be *monotone*. If $a_m < a_n$ whenever $m < n$, then $\langle a_n \rangle$ is said to be *strictly increasing*. If $a_m > a_n$ whenever $m < n$, then $\langle a_n \rangle$ is said to be *strictly decreasing*.

Because of the ordering properties of real numbers discussed in Section 2.3, it follows that if $\langle a_n \rangle$ is an increasing (or decreasing) sequence of positive real numbers, then $\langle 1/a_n \rangle$ is a decreasing (or increasing) sequence of all positive numbers. The result is similar for sequences of negative real numbers. Knowing this makes the next example a little easier.

Example 4.5.3. Show that the sequence defined by $a_n = n/(n+1)$ for $n \geq 1$ is strictly increasing.

Solution: We show that the sequence $\langle 1/a_n \rangle$ is strictly decreasing. Suppose $m < n$. Then $1/m > 1/n$, so that $1/a_m = 1 + 1/m > 1 + 1/n = 1/a_n$. Thus the sequence $\langle 1/a_n \rangle$ is strictly decreasing. ∎

Given a fixed $x > 0$, the sequence $\langle x^n \rangle$ proves to be important. In Exercise 1, you'll show that $\langle x^n \rangle$ is increasing if $x > 1$ and decreasing if $0 < x < 1$.

4.5.2 Bounded sequences

In a fashion identical to boundedness of subsets of $\mathbb{R}$, we can define boundedness of sequences.

Definition 4.5.4. Suppose $\langle a_n \rangle$ is a sequence of real numbers. If there exists some $M_1 \in \mathbb{R}$ such that $a_n \leq M_1$ for all n, then the sequence is said to be *bounded from above*. Similarly, if there exists $M_2 \in \mathbb{R}$ such that $a_n \geq M_2$ for all n, then $\langle a_n \rangle$ is said to be *bounded from below*. If there exists $M > 0$ such that $|a_n| \leq M$ for all n, then the sequence is said to be *bounded*. If a sequence is not bounded, it is said to be *unbounded*.

The boundedness terms in Definition 4.5.4 read very much like those in Definition 4.1.1. In fact, to say that a sequence is bounded from above, bounded from below, or bounded is precisely the same as saying that the range of the sequence is bounded from above, bounded from below, or bounded. Furthermore, we may apply Theorem 4.1.2 to the range of a sequence to have the following:

Theorem 4.5.5. *A sequence $\langle a_n \rangle$ is bounded if and only if it is bounded from above and below.*

Example 4.5.6. From Example 4.5.1, we note without details that for

1. $\langle 1/2 + (-1)^n \cdot 1/2 \rangle$ is bounded above by $M_1 = 8$ and below by $M_2 = 0$. It is therefore bounded;

2. $\langle n + (-1)^n \rangle$ is not bounded from above, therefore it is not bounded. It is, however, bounded from below by $M_2 = 0$; and

4. The sequence defined in Eq. (4.15) is not bounded from above, thus not bounded. It is bounded from below by $M_2 = -84$.

Example 4.5.7. We show that the sequence from Example 4.5.1 defined by $a_n = n^2/2^n$ is bounded. In Exercise 12 from Section 2.4, we showed that $2^n > n^2$ for all $n \geq 5$. Thus $0 < n^2/2^n < 1$ for all $n \geq 5$. Let $M = \max\{a_1, a_2, a_3, a_4, 1\}$. We show that $0 \leq a_n \leq M$ for all $n \in \mathbb{N}$.

Pick $n \in \mathbb{N}$. If $1 \leq n \leq 4$, then clearly $0 \leq a_n \leq \max\{a_1, a_2, a_3, a_4\} \leq M$. If $n \geq 5$, then $0 \leq a_n \leq 1 \leq M$. In either case, $0 \leq a_n \leq M$, so that the sequence is bounded.

What does it mean for a sequence to be unbounded? The answer: $\langle a_n \rangle$ is unbounded if for all $M > 0$ there exists $n \in \mathbb{N}$ such that $|a_n| > M$.

Example 4.5.8. Show that the sequence defined by $a_n = \sqrt{n}$ is unbounded.

Solution: Pick $M > 0$, and let n be any integer satisfying $n > M^2$. Then by Exercise 8 from Section 2.8, we have that $a_n = \sqrt{n} > \sqrt{M^2} = M > 0$, so that $|a_n| > M$. ∎

Since the sequence $\langle x^n \rangle$ is decreasing for $0 < x < 1$, it is bounded above by 1. Since every x^n is positive, $\langle x^n \rangle$ is bounded below by zero. Thus if $0 < x < 1$, the sequence $\langle x^n \rangle$ is bounded. However, the next example shows that $\langle x^n \rangle$ is unbounded if $x > 1$.

Example 4.5.9. Suppose $x > 1$. Then $\langle x^n \rangle$ is unbounded.

Solution: Suppose $\langle x^n \rangle$ is bounded, and let L be the LUB of the sequence. Let $\epsilon = x - 1$, which is positive. Then by property M2, there exists $n \in \mathbb{N}$ such that $L - \epsilon < x^n \leq L$, so that $L < x^n + x - 1$. But since $x^n > 1$, we have that

$$x^{n+1} = x^{n+1} - x^n + x^n = x^n(x-1) + x^n > x - 1 + x^n > L. \quad (4.16)$$

This is a contradiction, so that $\langle x^n \rangle$ is unbounded. ∎

To close out this introductory section on sequences, here's an example that illustrates a technique we'll use in Section 4.6. Exercises 6–8 are similar.

Example 4.5.10. Consider the sequence defined by $a_n = (n+3)/(n-1)$ for $n \geq 2$. Find a value of N such that

$$|a_N - 1| < 0.001. \quad (4.17)$$

Does inequality (4.17) hold if $n \geq N$?

Solution: Expanding inequality (4.17) according to Theorem 2.3.19 and then solving for N yields the following logically equivalent inequality statements:

$$-0.001 < \frac{N+3}{N-1} - 1 < 0.001$$

$$-0.001 < \frac{4}{N-1} < 0.001$$

$$-0.001(N-1) < 4 < 0.001(N-1)$$

$$N > -3999 \quad \text{and} \quad N > 4001.$$

The inequality $N > 4001$ is stronger than $N > -3999$, so we may let $N = 4002$. Furthermore, if $n \geq N$, then

$$-0.001 < 0 < \frac{4}{n-1} \leq \frac{4}{N-1} < 0.001, \quad (4.18)$$

so that $|a_n - 1| < 0.001$ for all $n \geq N$. ∎

EXERCISES

1. Let $x > 0$. Show that the sequence $\langle x^n \rangle_{n=1}^{\infty}$ is strictly increasing if $x > 1$ and strictly decreasing if $0 < x < 1$.[21]

2. Give an example of a sequence that is both increasing and decreasing.

3. What does it mean to say that a sequence is not increasing?

4. Determine whether the following sequences are increasing, decreasing, or neither. Prove your claims.

    (a) $\langle 1 - 1/n \rangle$
    (b) $\langle n^2 \rangle$
    (c) $\langle 1 + (-1)^n/n^2 \rangle$

5. Consider the sequence $\langle a_n \rangle$ defined by

$$a_n = \begin{cases} 10, & \text{if } n \text{ is odd}, \\ 1 - n^2, & \text{if } n \text{ is even}. \end{cases}$$

    Prove that $\langle a_n \rangle$ is bounded from above but not from below.

6. Let $\langle a_n \rangle$ be defined as in Example 4.5.10, and let $\epsilon > 0$ be given. Find a value of N (which will be in terms of ϵ) such that $n \geq N$ guarantees $|a_n - 1| < \epsilon$.

7. Let $\langle a_n \rangle$ be defined by $a_n = (n+3)/(2n-1)$, and suppose $\epsilon > 0$ is given. Find a value of N such that $|a_n - 1/2| < \epsilon$ for all $n \geq N$.

8. Let $\langle a_n \rangle$ be defined by $a_n = n/(n+1)$, and suppose $0 < \epsilon < 1$ is given. Find a value of N such that $|a_n| > \epsilon$ for all $n \geq N$.

9. Let $\langle a_n \rangle$ be defined by $a_n = 1/(n^2 + 1)$, and suppose $\epsilon > 0$ is given. Find a value of N such that $|a_n| < \epsilon$ for all $n \geq N$.[22]

10. Suppose $\epsilon > 0$, and $\langle a_n \rangle$ and $\langle b_n \rangle$ are two sequences with the following properties. If $n \geq N_1$, then $|a_n - L_1| < \epsilon/2$, and if $n \geq N_2$, then $|b_n - L_2| < \epsilon/2$. Given this information, you should be able to make a true statement by filling in the blanks:

$$\text{If } n \geq \underline{\quad?\quad}, \text{ then } |(a_n + b_n) - (L_1 + L_2)| < \underline{\quad?\quad} \quad (4.19)$$

    Fill in the blanks with the appropriate numbers and then prove that the resulting statement is true.

4.6 Convergence of Sequences

One of the most important questions we can ask about a sequence of real numbers concerns what we call its *convergence*. The question has to do with whether the terms are "homing in on" or "approaching" some fixed real number L as n takes on larger

---

[21] See Exercise 9b from Section 2.4.
[22] You'll have to treat the cases $\epsilon > 1$ and $\epsilon \leq 1$ separately.

and larger values, or perhaps increasing or decreasing without bound. In this section, we discuss precisely what we mean by the notion of convergence and prove some important basic results about convergent sequences.

It would be misleading not to say up front that the idea of convergence created no small amount of controversy and distress in the historical development of mathematics. It took quite some time for a satisfactory definition to be devised. In developing the calculus, Sir Isaac Newton assumed some pretty sweeping results concerning convergence in his work, only to have its details placed on firm footing later. Because the idea is a bit complicated, we'll work our way into it a little at a time.

4.6.1 Convergence to a real number

Figure 4.2 provides a first glimpse into the idea of convergence to some $L \in \mathbb{R}$. Given some $\epsilon > 0$, the ϵ-neighborhood of L contains all the terms in the sequence beyond some threshold term a_N. In order to visualize how you might prove that the terms of a sequence $\langle a_n \rangle$ are converging to L, imagine the following task. Someone gives you an $\epsilon > 0$. You then take that ϵ value and do a quick calculation to determine some $N \in \mathbb{N}$ (that will be a function of ϵ) with the property that the Nth term and all those after it fall in the ϵ-neighborhood of L. If someone else then gives you an even smaller $\epsilon > 0$, you can still find a threshold value N with the same property but it will probably be a higher threshold. If you are given any $\epsilon > 0$ and are able to find some $N \in \mathbb{N}$ with the property that all terms from the Nth one onward fall in the ϵ-neighborhood of L, then you will have shown that the sequence converges to L.

Before we give the definition of convergence, let's point out that you have already done some of this sort of work in the Exercises of Section 4.5. In Exercise 6, you showed

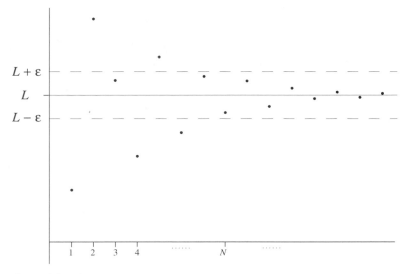

Figure 4.2 A convergent sequence.

that

$$\text{If } n > 1 + \frac{4}{\epsilon}, \quad \text{then} \quad \left|\frac{n+3}{n-1} - 1\right| < \epsilon.$$

That is, all terms with index larger than $1 + 4/\epsilon$ fall in the ϵ-neighborhood of 1. In Exercise 7, you showed that

$$\text{If } n > \frac{1}{2} + \frac{7}{4\epsilon}, \quad \text{then} \quad \left|\frac{n+3}{2n-1} - \frac{1}{2}\right| < \epsilon.$$

That is, all terms with index larger than $1/2 + 7/4\epsilon$ fall in the ϵ-neighborhood of $1/2$. Having worked through this, we're ready for the definition of convergence.

Definition 4.6.1. Suppose $\langle a_n \rangle_{n=1}^{\infty}$ is a sequence of real numbers. We say that the sequence *converges* if there exists $L \in \mathbb{R}$ such that, for every $\epsilon > 0$, there exists $N \in \mathbb{N}$ with the property that $n \geq N$ implies $|a_n - L| < \epsilon$. That is, the sequence converges iff

$$(\exists L \in \mathbb{R})(\forall \epsilon > 0)(\exists N \in \mathbb{N})(\forall n \in \mathbb{N})(n \geq N \to |a_n - L| < \epsilon). \quad (4.20)$$

If $\langle a_n \rangle_{n=1}^{\infty}$ converges to L, we say that L is the *limit* of the sequence as n approaches infinity, and we write this as

$$\lim_{n \to \infty} a_n = L. \quad (4.21)$$

We also say that a_n approaches L as n approaches infinity, and write this as

$$a_n \to L \quad \text{as } n \to \infty. \quad (4.22)$$

If a sequence does not converge, we say it *diverges*.

To construct a proof of convergence is, as always, to follow the logical flow of the definition. A general form for such a proof looks something like this:

Theorem 4.6.2 (Sample). *The sequence $\langle a_n \rangle_{n=1}^{\infty}$ converges to L.*

Proof: Pick $\epsilon > 0$. Let N be any integer satisfying $N > f(\epsilon)$ (where you have already done the scratchwork to find what $f(\epsilon)$ should be). Then if $n \geq N$, we have that

$$|a_n - L| = f^{-1}(n) \leq f^{-1}(N) < \epsilon. \quad (4.23)$$

Thus $a_n \to L$ as $n \to \infty$. ■

Here's a cleaned up example.

Example 4.6.3. Show that the sequence $\langle (n+1)/(4n+3) \rangle$ converges to $1/4$.

Solution: Pick $\epsilon > 0$. Let N be any integer satisfying $N > 1/16\epsilon - 3/4$. Then if $n \geq N$, we have

$$\left|\frac{n+1}{4n+3} - \frac{1}{4}\right| = \left|\frac{1}{16n+12}\right| < \left|\frac{1}{16N+12}\right| < \epsilon. \tag{4.24}$$

Thus $(n+1)/(4n+3) \to 1/4$ as $n \to \infty$. ∎

If you wonder what inspired the statement $N > 1/16\epsilon - 3/4$, it's the result of working backward from inequality (4.24) and solving for N to produce a logically equivalent inequality. This scratchwork is what makes the last step of Eq. (4.24) work.

Specific examples of convergence are great, but we need generalized theorems addressing convergence in order to develop a broader theory of convergence. The following theorems can go a long way by giving us some basic building blocks and ways of combining them. You'll prove the first one in Exercise 1.

Theorem 4.6.4. *The limit of a sequence of real numbers is unique. That is, if $a_n \to L_1$ and $a_n \to L_2$, then $L_1 = L_2$.*

Theorem 4.6.5. *The constant sequence defined by $a_n = c$ for $c \in \mathbb{R}$ has limit c. That is, $\lim_{n\to\infty} c = c$.*

Proof: Pick $\epsilon > 0$. Let $N = 1$. Then if $n \geq N$, we have that

$$|a_n - L| = |c - c| = 0 < \epsilon.$$

∎

Theorem 4.6.6. *The sequence $1/n \to 0$ as $n \to \infty$.*

You'll prove Theorem 4.6.6 in Exercise 2 by using the Archimedean property of $\mathbb{R}$. You prove the next important theorem in Exercise 4.

Theorem 4.6.7. *A convergent sequence is bounded.*

The heart of the proof of Theorem 4.6.7 can be understood in the following way. If $\langle a_n \rangle$ converges, its terms eventually settle down close to L, so that all of its terms beyond a certain point are caught in $(L-1, L+1)$. Thus no matter how wildly $\langle a_n \rangle$ might jump around before this point, there are only finitely many terms to contain. This should suggest a value of M to serve as a bound.

Now for the theorems that allow us to manipulate and combine sequences and their limits.

Theorem 4.6.8. *Suppose $\lim_{n\to\infty} a_n = L_1$ and $\lim_{n\to\infty} b_n = L_2$. Then*

1. $\lim_{n\to\infty}(a_n + b_n) = L_1 + L_2$,
2. $\lim_{n\to\infty}(a_n b_n) = L_1 L_2$.

The proofs of the components of Theorem 4.6.8 are classic, and it would be a shame for you not to discover them for yourself (with some hints, of course). You'll do that in Exercise 6.

Corollary 4.6.9. *If* $\lim_{n\to\infty} a_n = L$ *and* $c \in \mathbb{R}$, *then* $\lim_{n\to\infty}(ca_n) = cL$.

Proof: Let $b_n = c$ for all n in Theorem 4.6.8 and apply Theorem 4.6.5. ∎

Corollary 4.6.10. *If* $a_n \to L_1$ *and* $b_n \to L_2$ *as* $n \to \infty$, *then* $(a_n - b_n) \to (L_1 - L_2)$ *as* $n \to \infty$.

Proof:
$$\lim_{n\to\infty}(a_n - b_n) = \lim_{n\to\infty}[a_n + (-1)b_n] = \lim_{n\to\infty} a_n + \lim_{n\to\infty}(-1)b_n$$
$$= L_1 + (-1)L_2 = L_1 - L_2. \tag{4.25}$$
∎

With an induction argument calling on Theorem 4.6.6 and part 2 of Theorem 4.6.8, you can show the following (Exercise 7):

Corollary 4.6.11. *If* $p \in \mathbb{N}$, *then* $1/n^p \to 0$ *as* $n \to \infty$.

To combine two sequences $\langle a_n \rangle$ and $\langle b_n \rangle$ by division, we first need to do a little work on the sequence $\langle 1/b_n \rangle$ by itself. We would like to be able to say something to the effect that the limit of a quotient is the quotient of the limits. Well, we can, sort of, with some preliminary work.

The sequence $\langle 1/n \rangle$ contains only positive terms. However, $1/n \to 0$, so every ϵ-neighborhood of zero contains some $1/n$. Thus, the terms of the sequence $\langle 1/n \rangle$ cannot be bounded away from zero. However, if the terms of a sequence are all nonzero and converge to a nonzero limit, then we can bound its terms away from zero.

Theorem 4.6.12. *Suppose* $\langle b_n \rangle$ *is a sequence of nonzero real numbers such that* $b_n \to L \neq 0$. *Then there exists* $M > 0$ *such that* $|b_n| \geq M$ *for all* $n \in \mathbb{N}$.

You'll prove Theorem 4.6.12 in Exercise 8. To keep things simple, you'll prove it only for the case $L > 0$. For the case $L < 0$, we can then make the following observation. If $L < 0$, then $-b_n \to -L$. Since $-L > 0$, we can apply the positive case of the theorem to $\langle -b_n \rangle$ to conclude there exists $M > 0$ such that $|-b_n| \geq M$ for all $n \in \mathbb{N}$. But $|-b_n| = |b_n|$, so $|b_n| \geq M$ for all n.

Theorem 4.6.12 helps us derive the following theorem, which you'll prove in Exercise 9.

Theorem 4.6.13. *Suppose* $\langle b_n \rangle$ *is a sequence of nonzero real numbers such that* $b_n \to L_2 \neq 0$. *Then* $\lim_{n\to\infty} 1/b_n = 1/L_2$.

Corollary 4.6.14. *Suppose* $\langle a_n \rangle$ *is a sequence converging to* L_1, *and* $\langle b_n \rangle$ *is a sequence of nonzero real numbers converging to* $L_2 \neq 0$. *Then* $a_n/b_n \to L_1/L_2$ *as* $n \to \infty$.

Proof: Apply part 2 of Theorem 4.6.8 to the sequences $\langle a_n \rangle$ and $\langle 1/b_n \rangle$. ∎

Theorems 4.6.5 through Corollary 4.6.14 can take us a long way in dealing with certain classes of sequences. Here's a pretty general result, one case of which you'll prove in Exercise 10.

Theorem 4.6.15. *Suppose $\langle a_n \rangle$ is a sequence defined by*

$$a_n = \frac{b_r n^r + b_{r-1} n^{r-1} + \cdots + b_1 n + b_0}{c_s n^s + c_{s-1} n^{s-1} + \cdots + c_1 n + c_0}, \tag{4.26}$$

where all $b_k, c_k \in \mathbb{R}$, b_r and c_s are nonzero, and the denominator is nonzero for all n. Then $a_n \to 0$ if $r < s$, and $a_n \to b_r/c_s$ if $r = s$.

Proof: We prove for the case $r < s$ and leave the case $r = s$ to you. We may take the expression for a_n in Eq. (4.26) and multiply by n^{-s}/n^{-s} to have

$$a_n = \frac{b_r n^{r-s} + b_{r-1} n^{r-1-s} + \cdots + b_1 n^{1-s} + b_0 n^{-s}}{c_s + c_{s-1} n^{-1} + \cdots + c_1 n^{1-s} + c_0 n^{-s}}. \tag{4.27}$$

Since $r < s$, every exponent of n in Eq. (4.27) is negative, and we may apply results of this section to obtain

$$\begin{aligned}
\lim_{n \to \infty} a_n &= \frac{\lim(b_r n^{r-s} + \cdots + b_0 n^{-s})}{\lim(c_s + \cdots + c_0 n^{-s})} \\
&= \frac{\lim(b_r n^{r-s}) + \cdots + \lim(b_0 n^{-s})}{\lim c_s + \cdots + \lim(c_0 n^{-s})} \\
&= \frac{(\lim b_r)(\lim n^{r-s}) + \cdots + (\lim b_0)(\lim n^{-s})}{\lim c_s + \cdots + \lim(c_0)(\lim n^{-s})} \\
&= \frac{b_r \cdot 0 + \cdots + b_0 \cdot 0}{c_s + \cdots + c_0 \cdot 0} = \frac{0}{c_s} = 0.
\end{aligned} \tag{4.28}$$

∎

4.6.2 Convergence to $\pm\infty$

Sometimes a sequence does not converge to a real number, but it does exhibit a nice behavior in that it increases or decreases without bound. The term we use for such behavior is that $a_n \to +\infty$ (or $-\infty$). We have already used the somewhat loose expression $n \to \infty$ in Definition 4.6.1, even though ∞ is not a real number. It is not difficult to define the notion of $\lim_{n \to \infty} a_n = \infty$, but there is no way to do it in terms of neighborhoods, unless you can concoct some sort of definition of a neighborhood of infinity. We'll do exactly that in Section 5.4 in the context of functions from $\mathbb{R}$ to $\mathbb{R}$, where we'll spend some quality time with infinity. For now, we'll be content to work with a few basic ideas.

If $a_n \to L$ as $n \to \infty$, the terms a_n might hop all around L, just as long as that hopping stays within an ϵ-radius of L beyond some threshold term. That this bouncing around can be eventually contained for all $\epsilon > 0$ is why the bouncing apparently fades out. To say $a_n \to +\infty$ as $n \to \infty$ means loosely that the terms a_n are hopping increasingly higher and higher, even though there is no reason the terms cannot do at least a little hopping back down. Not too much hopping down, though. Instead of setting an arbitrary ϵ-radius around some L, we set an arbitrary hurdle bar at some $M > 0$ and insist that all terms beyond some threshold N are above the hurdle. If every $M > 0$ is associated with some threshold beyond which $a_n \geq M$, then we say $a_n \to +\infty$ as $n \to \infty$. Although the terms might not be increasing to $+\infty$ in the sense that the sequence eventually becomes monotone, it is in some sense making its way to $+\infty$. Here's the definition.

Definition 4.6.16. Suppose $\langle a_n \rangle$ is a sequence of real numbers. We say that $\lim_{n \to \infty} a_n = +\infty$ if, for all $M > 0$, there exists $N \in \mathbb{N}$ with the property that $n \geq N$ implies $a_n > M$. Logically we may write this as

$$(\forall M > 0)(\exists N \in \mathbb{N})(\forall n \in \mathbb{N})(n \geq N \to a_n > M). \tag{4.29}$$

We also write $a_n \to +\infty$ as $n \to \infty$, and we sometimes say that the sequence *increases without bound*.

Example 4.6.17. Show that $\lim_{n \to \infty} \sqrt{n} = +\infty$.

Solution: Pick $M > 0$ and let N be any natural number satisfying $N > M^2$. Then if $n \geq N$, we have that $\sqrt{n} \geq \sqrt{N} > \sqrt{M^2} = M$. ∎

Theorem 4.6.18. *If $\langle a_n \rangle$ is increasing and unbounded, then $a_n \to +\infty$ as $n \to \infty$.*

You'll prove Theorem 4.6.18 in Exercise 14. An immediate consequence of Theorem 4.6.18 is that if $x > 1$, then $x^n \to +\infty$ as $n \to \infty$. In Exercise 1 (Section 4.5) you showed that $\langle x^n \rangle$ is increasing for $x > 1$, and Example 4.5.9 showed that it is unbounded. With Theorem 4.6.19, it follows that if $0 < x < 1$, then $x^n \to 0$ as $n \to \infty$. You'll prove it in Exercise 15.

Theorem 4.6.19. *Suppose $a_n \to +\infty$ as $n \to \infty$. Then $1/a_n \to 0$ as $n \to \infty$.*

EXERCISES

1. Prove Theorem 4.6.4: The limit of a sequence of real numbers is unique. That is, if $a_n \to L_1$ and $a_n \to L_2$, then $L_1 = L_2$.[23]

2. Prove Theorem 4.6.6: $\lim_{n \to \infty} 1/n = 0$.

3. Show that $1/2^n \to 0$ as $n \to \infty$.[24]

4. Prove Theorem 4.6.7: A convergent sequence is bounded.[25,26]

5. Suppose $\langle a_n \rangle$ converges to zero and $\langle b_n \rangle$ is bounded. Show that $a_n b_n \to 0$.[27]

6. Prove Theorem 4.6.8: Suppose $\lim_{n \to \infty} a_n = L_1$ and $\lim_{n \to \infty} b_n = L_2$. Then

---

[23] If $L_1 < L_2$, then letting $\epsilon = (L_2 - L_1)/2$ should yield a contradiction.
[24] For $n \geq 5$, $1/2^n < 1/n^2 < 1/n$.
[25] Look at the reasoning we used in Example 4.5.7, where we took care of finitely many terms individually and then handled all the rest with a single number. A good start for this proof would be to say: "Let $\epsilon = 1$. Then there exists . . . ."
[26] Let $M = \max\{|a_1|, |a_2|, \ldots, |a_{N-1}|, |L| + 1\}$.
[27] If $\epsilon > 0$ and $|b_n| \leq M$ for $M > 0$, then $\epsilon/M > 0$, also.

(a) $\lim_{n\to\infty}(a_n + b_n) = L_1 + L_2$.[28,29]
(b) $\lim_{n\to\infty}(a_n b_n) = L_1 L_2$.[30,31]

7. Prove Corollary 4.6.11: If $p \in \mathbb{N}$, then $1/n^p \to 0$ as $n \to \infty$.

8. Prove one case of Theorem 4.6.12: Suppose $\langle b_n \rangle$ is a sequence of *nonzero* real numbers such that $b_n \to L > 0$. Then there exists $M > 0$ such that $|b_n| \geq M$ for all $n \in \mathbb{N}$.[32]

9. Prove Theorem 4.6.13: Suppose $\langle b_n \rangle$ is a sequence of nonzero real numbers such that $b_n \to L_2 \neq 0$. Then $\lim_{n\to\infty} 1/b_n = 1/L_2$.

10. Prove the remaining case of Theorem 4.6.15: Suppose $\langle a_n \rangle$ is a sequence defined by
$$a_n = \frac{b_r n^r + b_{r-1} n^{r-1} + \cdots + b_1 n + b_0}{c_s n^s + c_{s-1} n^{s-1} + \cdots + c_1 n + c_0}$$
where all $b_k, c_k \in \mathbb{R}$, b_r, and c_s are nonzero, and the denominator is nonzero for all n. Then $a_n \to b_r/c_s$ if $r = s$.

11. Suppose $a_n \to L_1$ and $b_n \to L_2$ as $n \to \infty$. Suppose also that $a_n \leq b_n$ for all $n \in \mathbb{N}$. Show that $L_1 \leq L_2$.

12. Show that if $a_n \to L$ as $n \to \infty$, then $|a_n| \to |L|$ as $n \to \infty$.[33]

13. From Corollary 4.6.10, if $a_n \to L$ and $b_n \to L$, then $(a_n - b_n) \to 0$. Thus, for any $\epsilon > 0$, there exists $N \in \mathbb{N}$ such that $n \geq N$ implies $|a_n - b_n| < \epsilon/2$. Use this fact to prove the Sandwich theorem: If $a_n \leq c_n \leq b_n$ for all n, and if $a_n \to L$ and $b_n \to L$, then $c_n \to L$.

14. Prove Theorem 4.6.18: If $\langle a_n \rangle$ is monotone increasing and unbounded, then $a_n \to +\infty$ as $n \to \infty$.

15. Prove Theorem 4.6.19: Suppose $a_n \to +\infty$ as $n \to \infty$. Then $1/a_n \to 0$ as $n \to \infty$.

16. Show that if $-1 < x < 0$, then $x^n \to 0$ as $n \to \infty$.

17. In the spirit of Definition 4.6.16, create a definition for the statement $\lim_{n\to\infty} a_n = -\infty$.

4.7 The Nested Interval Property

Imagine a sequence of closed intervals $\langle [a_n, b_n] \rangle_{n=1}^{\infty}$, where $[a_n, b_n] \subset [a_m, b_m]$ whenever $m < n$. Such a sequence of intervals is said to be *nested* because any particular interval contains all those after it. In this section, we derive the *nested interval property* (NIP) of

---
[28] Remember, if $\epsilon > 0$, then so is $\epsilon/2$. See Exercise 10 from Section 4.5.
[29] Given $\epsilon > 0$, there exists $N_1 \in \mathbb{N}$ such that $|a_n - L_1| < \epsilon/2$. Similarly for b_n.
[30] If $L_1 = 0$, the result follows from Exercise 5. Otherwise, do some preliminary playing around with the expression $|a_n b_n - L_1 L_2|$. The sneaky trick here is adding and subtracting the same quantity inside $|a_n b_n - L_1 L_2|$. If you want to know what that quantity is, check the next hint.
[31] Add and subtract $b_n L_1$ and use the triangle inequality. Also, apply Theorem 4.6.7 to b_n.
[32] Let $\epsilon = L/2$. Then there exists $N \in \mathbb{N}$ such that . . . .
[33] See Theorem 2.3.24.

$\mathbb{R}$, which says that a nested sequence of closed intervals whose lengths $(b_n - a_n)$ tend to zero as $n \to \infty$ will intersect down to a single real number. We then investigate an important implication of the NIP as it relates to bounded sequences. We also demonstrate that the NIP can be used axiomatically to prove the LUB property, thus making the two statements logically equivalent.

4.7.1 From LUB axiom to NIP

If $\langle [a_n, b_n] \rangle$ is a nested sequence of closed intervals, then clearly $a_m \leq a_n < b_n \leq b_m$ whenever $m < n$. In particular, $a_1 \leq a_n < b_n \leq b_1$ for all $n \in \mathbb{N}$. Thus, the sequence of left endpoints $\langle a_n \rangle$ is monotone increasing and bounded above by b_1, and the sequence of right endpoints $\langle b_n \rangle$ is monotone decreasing and bounded below by a_1. As we begin to make our way to the NIP, we need the following theorem. You'll prove it in Exercise 1.

Theorem 4.7.1. *Suppose $\langle a_n \rangle$ is an increasing sequence of real numbers that is bounded from above. Then $\langle a_n \rangle$ is convergent. Furthermore, if $\lim a_n = L$, then $a_n \leq L$ for all $n \in \mathbb{N}$.*

Theorem 4.7.1 is the step in the process of proving the NIP where you will use the LUB axiom. The next theorem should follow quickly from Theorem 4.7.1 by the strategic insertion of some negative signs (Exercise 2).

Theorem 4.7.2. *Suppose $\langle b_n \rangle$ is a decreasing sequence of real numbers that is bounded from below. Then $\langle b_n \rangle$ is convergent. Furthermore, if $\lim b_n = L$, then $b_n \geq L$ for all $n \in \mathbb{N}$.*

If the hypothesis conditions for Theorem 4.7.1 are satisfied, we say that $\langle a_n \rangle$ converges to L *from below*, and we write $a_n \nearrow L$. Similarly for Theorem 4.7.2, we say $\langle b_n \rangle$ converges to L *from above*, and we write $b_n \searrow L$.

Therefore we have that a nested sequence of closed intervals $\langle [a_n, b_n] \rangle$ has the property that $a_n \to L_1$ and $b_n \to L_2$ for some $L_1, L_2 \in \mathbb{R}$. If $\langle [a_n, b_n] \rangle$ also has the property that the lengths of the intervals approach zero as $n \to \infty$, then we have $\lim_{n \to \infty}(b_n - a_n) = 0$. Because $\langle a_n \rangle$ and $\langle b_n \rangle$ converge individually, then by Corollary 4.6.10,

$$0 = \lim_{n \to \infty}(b_n - a_n) = \lim_{n \to \infty} b_n - \lim_{n \to \infty} a_n = L_2 - L_1. \qquad (4.30)$$

Thus $L_1 = L_2$. The last piece of the NIP puzzle asserts that this limit is the single element in the intersection of all the nested intervals in the sequence. Here is a statement of the theorem and its proof, with its last detail left to you in Exercise 3.

Theorem 4.7.3 (Nested Interval Property). *Let $\langle [a_n, b_n] \rangle$ be a sequence of nested intervals with the property that $\lim_{n \to \infty}(b_n - a_n) = 0$. Then there exists $x_0 \in \mathbb{R}$ such that*

$$\bigcap_{n=1}^{\infty}[a_n, b_n] = \{x_0\}. \qquad (4.31)$$

Proof: By Theorems 4.7.1 and 4.7.2, and by Corollary 4.6.10, $\langle a_n \rangle$ and $\langle b_n \rangle$ converge to some $L \in \mathbb{R}$. By Exercise 3, $\cap_{n=1}^{\infty}[a_n, b_n] = \{L\}$. ∎

Don't think that the hypothesis condition $(b_n - a_n) \to 0$ by itself allows you to conclude that $\lim a_n = \lim b_n$. Convergence of $\langle a_n \rangle$ and $\langle b_n \rangle$ separately must be demonstrated apart from this in order to be able to apply Corollary 4.6.10. Just because $(b_n - a_n) \to 0$, no conclusion can be drawn about the convergence of either $\langle a_n \rangle$ or $\langle b_n \rangle$. For example, let $a_n = n^2$ and $b_n = n^2 + 1/n$. The lengths of the intervals $[a_n, b_n]$ tend to zero. However, the intervals are not nested, and neither $\langle a_n \rangle$ nor $\langle b_n \rangle$ converge individually.

4.7.2 The NIP applied to subsequences

In Section 4.8, we'll need some results based on taking a given sequence $\langle a_n \rangle$ and creating from it a new sequence by perhaps deleting some of its terms and preserving the order of the kept terms. The new sequence created in this way is called a *subsequence* of $\langle a_n \rangle$. We introduce the term here and note what the NIP has to say about certain sequences and subsequences.

Creating a subsequence of $\langle a_n \rangle$ can make for a slight notational mess when it comes to indexing the subsequence, so let's look at an example. Given a sequence $\langle a_n \rangle$, suppose we want to consider the following subsequence:

$$\langle a_2, a_8, a_{10}, a_{40}, a_{44}, a_{66}, \ldots \rangle. \tag{4.32}$$

The notational dilemma can be resolved by noting that the indices

$$\langle 2, 8, 10, 40, 44, 66, \ldots \rangle \tag{4.33}$$

form a strictly increasing sequence $\langle n_k \rangle_{k=1}^{\infty}$ of natural numbers. If we denote $n_1 = 2$, $n_2 = 8$, $n_3 = 10$, etc., then the sequence $\langle n_1, n_2, n_3, \ldots \rangle$ is a way of indexing our subsequence. The subsequence can then be denoted by

$$\langle a_{n_1}, a_{n_2}, a_{n_3}, a_{n_4}, \ldots \rangle = \langle a_{n_k} \rangle_{k=1}^{\infty}, \tag{4.34}$$

where k is now the counter variable. This brings us to a definition:

Definition 4.7.4. Suppose $\langle a_n \rangle_{n=1}^{\infty}$ is a sequence of real numbers and suppose $\langle n_k \rangle_{k=1}^{\infty}$ is a strictly increasing sequence of natural numbers. Then the sequence $\langle a_{n_k} \rangle_{k=1}^{\infty}$ is called a *subsequence* of the sequence $\langle a_n \rangle$.

Suppose you love to play darts. You play every day from now to eternity, and you're pretty good at it. At least you're good enough so that you always hit somewhere on the target. If every dart you throw leaves a tiny hole behind, what must eventually happen to the set of pinholes on the target as time goes on? Well, it is certainly possible that your darts could always land on one of a finite number of spots on the target, in which case you would hit at least one spot infinitely many times. But if you don't hit any one spot infinitely many times, then your target will have infinitely many holes in it. If so, then the finite size of the target will imply that the pinholes are going to cluster in one or more places. Of course, if we think of two direct hits on the same exact spot as leaving two distinguishable holes (why not?), then it can indisputably be said of your target that the holes are *somewhere* clustered, and around at least one spot.

This little analogy is going somewhere: Suppose $\langle a_n \rangle$ is a sequence of real numbers that is bounded by $M > 0$. Then the interval $[-M, M]$ is the dart board target, and the terms a_n are the spots you hit. The next theorem effectively says that there are infinitely many terms of the sequence clustered around some point in $[-M, M]$, but it does so in the language of subsequences. It will come in handy in Exercise 4, where you will prove the famous Bolzano-Weierstrass theorem.

Theorem 4.7.5. *Every bounded sequence of real numbers contains a convergent subsequence.*

Proof: Suppose $\langle a_n \rangle$ is bounded by M and denote $[c_0, d_0] = [-M, M]$. (See Fig. 4.3.) Let m_0 be the midpoint of $[c_0, d_0]$, and note that either $[c_0, m_0]$ or $[m_0, d_0]$ must contain infinitely many of the a_n. (Perhaps both do.) Let $[c_1, d_1]$ be one of the subintervals of $[c_0, d_0]$ that contains infinitely many of the a_n, pick any term of the sequence from $[c_1, d_1]$, and denote it a_{n_1}. Iterating this process, and letting m_1 be the midpoint of $[c_1, d_1]$, we may let $[c_2, d_2]$ be whichever of $[c_1, m_1]$ or $[m_1, d_1]$ contains infinitely many of the a_n, then pick any term of the sequence from $[c_2, d_2]$ whose index exceeds n_1, and call this term a_{n_2}. Continuing in this fashion, we generate two things: a nested sequence of intervals $[c_k, d_k]$ whose lengths are $2M/2^k$ and therefore tend to zero (see Exercise 3 from Section 4.6), and $\langle a_{n_k} \rangle_{k=1}^{\infty}$, a subsequence of $\langle a_n \rangle_{n=1}^{\infty}$, each term of which falls in $[c_k, d_k]$.

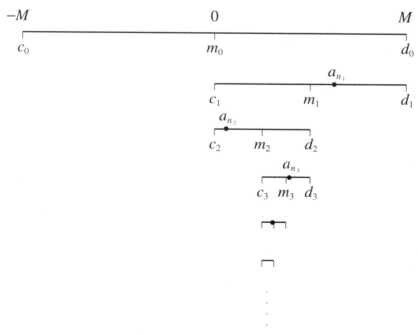

Figure 4.3 Nested intervals generated in the proof of Theorem 4.7.5.

By the NIP, $\bigcap_{k=1}^{\infty}[c_k, d_k] = \{x_0\}$ for some $x_0 \in \mathbb{R}$. Since $c_k \to x_0$ and $d_k \to x_0$ as $k \to \infty$, and since $c_k \leq a_{n_k} \leq d_k$ for all $k \in \mathbb{N}$, the Sandwich theorem (Exercise 13 from Section 4.6) implies that $a_{n_k} \to x_0$. ∎

4.7.3 From NIP to LUB axiom

Let's go back and note how the LUB axiom has contributed to the results developed thus far. We said in Section 2.8 that assumption A22 can be derived from the LUB axiom, but we have not proved this is true. All the results of Section 2.8 are based on A22, hence they depend on the LUB axiom as well. Certainly a lot of our work so far in this chapter has relied on the LUB axiom. Section 4.1 was all about the LUB axiom. In Section 4.2, Theorem 4.2.7 used the LUB axiom to show that ∅ and $\mathbb{R}$ are the only subsets of $\mathbb{R}$ that are both open and closed. In Section 4.5, the only places we used any results from the LUB axiom were in choosing the maximum value of a finite set in proving some results about boundedness. Because of this, the theorems from Section 4.6 involving boundedness stem from the LUB axiom. In what follows, we want to delete the LUB axiom from our list of assumptions and replace it with the NIP. We can then show as a theorem that $\mathbb{R}$ has the LUB property. To do this, we retain our terminology of sequences and convergence. Assuming the NIP, here is the theorem we want to prove.

Theorem 4.7.6. *Suppose $A \subseteq \mathbb{R}$ is nonempty and bounded from above by M. Then there exists $L \in \mathbb{R}$ such that L is the least upper bound of S.*

If you're inclined to tackle the proof of Theorem 4.7.6 without any guidance, go right ahead (Exercise 5). Just make sure you understand what the task is. You have to use the NIP to prove the LUB property of $\mathbb{R}$ as a theorem. That is, somehow you're going to use what you know about A to generate a sequence of nested intervals whose lengths tend to zero, so that you can apply the NIP to it. If you want more guidance, read on for a thumbnail sketch of the proof.

First, because $A \neq \emptyset$, we can choose some $x \in A$. We then split the interval $[x, M]$ in half repeatedly. At each stage, we keep the right half of the split interval if it contains any elements of A; otherwise, we keep the left half. (See Fig. 4.4.) The successive splits will produce a sequence of nested intervals whose lengths tend to zero. The NIP then gives us some x_0 in the intersection of all the intervals, which you can show has properties L1 and L2 from page 135 (or M1 and M2 if you prefer).

EXERCISES

1. Prove Theorem 4.7.1: If $\langle a_n \rangle$ is an increasing sequence of real numbers that is bounded from above, then $\langle a_n \rangle$ is convergent. Furthermore, if $\lim a_n = L$, then $a_n \leq L$ for all $n \in \mathbb{N}$.[34]

---

[34] The fact that $a_n \leq L$ for all n should be a mere observation based on where you obtain L.

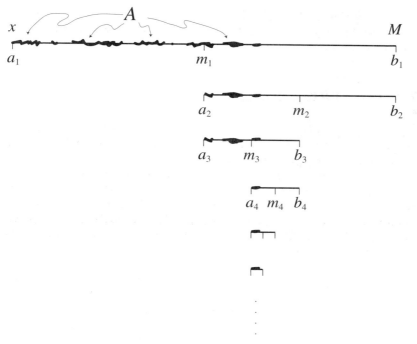

Figure 4.4 Nested intervals generated in the proof of Theorem 4.7.6.

2. Prove Theorem 4.7.2: If $\langle b_n \rangle$ is a decreasing sequence of real numbers that is bounded from below, then $\langle b_n \rangle$ is convergent. Furthermore, if $\lim b_n = L$, then $b_n \geq L$ for all $n \in \mathbb{N}$. [35]

3. Suppose $\langle [a_n, b_n] \rangle$ is a nested sequence of closed intervals with the property that $\langle a_n \rangle$ and $\langle b_n \rangle$ both converge to L. Show that $\cap_{n=1}^{\infty} [a_n, b_n] = \{L\}$.

4. Prove the Bolzano-Weierstrass theorem: Every bounded, infinite set of real numbers has a limit point in $\mathbb{R}$.[36]

5. Assuming the NIP, prove Theorem 4.7.6: Suppose $A \subset \mathbb{R}$ is nonempty and bounded from above by M. Then there exists $L \in \mathbb{R}$ such that L is the least upper bound of S.[37]

4.8 Cauchy Sequences

Now we turn our attention to a certain class of sequences called *Cauchy* sequences. If we can say that convergent sequences are characterized by the fact that the terms all eventually bunch up around some real number L, then we would say that Cauchy sequences are

---

[35] Apply Theorem 4.7.1 to $\langle -b_n \rangle$.
[36] You can apply Theorem 4.7.5 by creating just the right sequence from the set, or you can mimic the proof of Theorem 4.7.5 in its use of the NIP.
[37] Prove that $x_0 \in \cap [a_n, b_n]$ has properties L1 and L2 by contradiction.

characterized by the fact that the terms all eventually bunch up around each other, without any specific reference to any possible limit in $\mathbb{R}$. It turns out that this bunching of terms is logically equivalent convergence to some $L \in \mathbb{R}$ (with the help of the NIP), and this is the main result we want to prove in this section. Once we have worked through these results, we discuss why Cauchy sequences are important, and we relate their convergence to the LUB axiom and the NIP.

4.8.1 Convergence of Cauchy sequences

First let's define precisely what we mean for a sequence to be Cauchy. Then we'll show that a sequence is convergent if and only if it is Cauchy.

Definition 4.8.1. Suppose $\langle a_n \rangle$ is a sequence of real numbers. Then $\langle a_n \rangle$ is said to be *Cauchy* if, for all $\epsilon > 0$, there exists $N \in \mathbb{N}$ such that, for all $m, n \geq N$, we have that $|a_m - a_n| < \epsilon$.

One way to envision a Cauchy sequence in terms of some given $\epsilon > 0$ and the corresponding threshold N is that all terms which have an index of at least N fall within an ϵ distance, not only of a_N, but of each other as well. (See Fig. 4.5.) There is no apparent limit around which the neighborhood is anchored, so the question of whether a Cauchy sequence must converge is not immediately answerable. In showing that convergence and Cauchy are equivalent, let's do the easy part first.

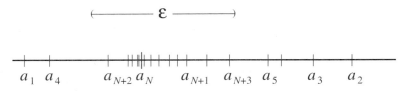

Figure 4.5 A Cauchy sequence.

Theorem 4.8.2. *A convergent sequence is Cauchy.*

Theorem 4.8.2 says that, if the terms of a sequence bunch up around some $L \in \mathbb{R}$, then they must bunch up around each other. You'll prove this in Exercise 1.

What does it mean for a sequence not to be Cauchy (Exercise 2)? If a sequence is not Cauchy, then by Theorem 4.8.2 it does not converge. Sometimes a natural way to show a particular sequence diverges is to show it is not Cauchy. In Exercise 3 you'll show that the sequence $\langle (-1)^n \rangle$ is not Cauchy and hence it diverges.

To prove that a Cauchy sequence is convergent, we need the following theorem:

Theorem 4.8.3. *A Cauchy sequence is bounded.*

You can prove Theorem 4.8.3 in a way almost identical to the proof of Theorem 4.6.7 where you bounded all but a finite number of terms within some neighborhood of the limit. Instead of fencing in the tail of the sequence around a known limit, a Cauchy sequence can be bounded by working around the threshold term a_N for $\epsilon = 1$ (Exercise 4).

At this point, you have all the machinery you need to show that a Cauchy sequence is convergent. Here's a statement of the theorem.

Theorem 4.8.4. *A Cauchy sequence of real numbers is convergent.*

You'll prove Theorem 4.8.4 in Exercise 5. The next paragraph explains the approach, which you're welcome to skip over if you don't want the hint.

If a sequence is Cauchy, then by Theorem 4.8.3, it is bounded. Since it is bounded, then it contains a convergent subsequence by Theorem 4.7.5. This gives you some $L \in \mathbb{R}$ to work with, and you can show the entire sequence converges to this L by using convergence of the subsequence and Cauchiness of the sequence.

The real numbers are only one example of a set where we address the convergence of Cauchy sequences. In any set endowed with a *metric* (a measure of distance between elements), Cauchy sequences can be defined in the same way as in Definition 4.8.1 by interpreting $|a_m - a_n|$ as the distance between elements in that context. Any such *metric space*, as we call it, where Cauchy sequences always converge to an element in the space is said to be *complete*.

In some situations, completeness might be taken as an axiom of the metric space. What we have done here is to show completeness of $\mathbb{R}$ as a theorem, based on the NIP, hence on the LUB axiom. Some authors of texts in analysis prefer to go the other way by assuming completeness as an axiom of $\mathbb{R}$ and demonstrating the NIP and LUB property as theorems. Of course, this means that completeness of $\mathbb{R}$ is logically equivalent to both the LUB property and the NIP, and we will show that completeness of $\mathbb{R}$ implies the NIP. By Theorem 4.7.6, completeness also implies the LUB property.

One good reason an author might want to take completeness as an axiom is that it makes for an interesting way to fill in the spaces between the rational numbers with the irrationals and make the real number line into the smooth continuum we envision. In a nutshell, this is how it goes.

Build up the rational numbers through $\{0, 1\}$, $\mathbb{N}$, $\mathbb{W}$, $\mathbb{Z}$, and $\mathbb{Q}$ in the way we discussed in Section 2.6. Envision $\mathbb{Q}$ as the set of all rational points on the number line with all the irrationals missing. Now imagine the set of all possible sequences of rational numbers *only*. Many of the Cauchy sequences in $\mathbb{Q}$ will not converge in $\mathbb{Q}$. There are several unproved assumptions in the next example, but it still illustrates this point.

Example 4.8.5. The terms

$$a_0 = 1$$
$$a_1 = 1.4$$
$$a_2 = 1.41$$
$$a_3 = 1.414$$
$$\vdots$$
$$a_{11} = 1.41421356237$$
$$\vdots$$

represent the start of a sequence that converges to $\sqrt{2}$. Every term in the sequence is rational because it is a terminating decimal. The sequence is Cauchy, for, given $\epsilon > 0$, let $N \in \mathbb{N}$ satisfy $10^{-N} < \epsilon$. Then if $m, n \geq N$, $|a_m - a_n| < \epsilon$ because a_m and a_n will agree through the Nth decimal place.

Therefore $\mathbb{Q}$ by itself is not complete. What are you going to do? If you want to assume completeness as a characteristic of a set containing $\mathbb{Q}$ as a subset, then you will have to throw in all possible limits of Cauchy sequences of rational numbers. With a little bit of work, you can fill in the holes in the number line with these irrational limits.

Practically speaking, we apply this principle quite often. Anytime you use your calculator to give you $\sqrt{2}$, you are using a terminating decimal approximation, which, of course, is rational. If you need a better approximation, you may use a computer to give you more decimal places of accuracy. You're still working with a rational number, but you're counting on the internal workings of the technology and the theory of analysis to give you a rational number that is actually closer to the exact value. In computation, we always work with rational numbers, except perhaps when we manipulate symbols such as π or $\sqrt{2}$ algebraically.

4.8.2 From completeness to the NIP

If we can show that completeness of $\mathbb{R}$ implies the NIP, then we will have demonstrated the logical equivalence of the LUB property, the NIP, and completeness. You'll supply the details of what remains in Exercise 6. Here are some reminders and a suggestion about how to proceed.

In proving the NIP from the LUB property in Section 4.7, we supposed $\langle [a_n, b_n] \rangle$ is a nested sequence of closed intervals whose lengths tend to zero as $n \to \infty$. One thing we noted is that $(b_n - a_n) \to 0$ by itself does not say anything about convergence of $\langle a_n \rangle$ or $\langle b_n \rangle$ separately. Intervals $[a_n, b_n]$ might become arbitrarily short, but if they are not nested, then nothing prevents them from waltzing forever up the real number line, so that neither $\langle a_n \rangle$ nor $\langle b_n \rangle$ converges. However, the hypothesis condition that the $[a_n, b_n]$ are nested means that $\langle a_n \rangle$ and $\langle b_n \rangle$ are bounded. Then monotonicity and the LUB property applied to the a_n and b_n put a number into our hands that was just the thing we needed to arrive at the NIP.

This time, in proving the NIP from completeness, we still begin with the assumption that the intervals are nested and $(b_n - a_n) \to 0$. The difference this time is that the axiom we need to apply is completeness of $\mathbb{R}$. If we can show that $\langle a_n \rangle$ and $\langle b_n \rangle$ are Cauchy, then the assumption of completeness will give us a real number limit for each. This turns out to be fruitful because $(b_n - a_n) \to 0$ makes $|b_n - a_n|$ very small eventually. To conclude that $\langle a_n \rangle$ is Cauchy, b_n in the expression $|b_n - a_n|$ can be replaced by a_m for $m \geq n$ to have $|a_m - a_n|$, which is even smaller than $|b_n - a_n|$ because a_m and a_n are at least as close together as a_n and b_n. The rest of the work in proving the NIP would be identical to what you did in Section 4.7, where we observed $\lim a_n = \lim b_n$, and where you showed in Exercise 3 that $\cap [a_n, b_n] = \{L\}$, where L is the common limit of $\langle a_n \rangle$ and $\langle b_n \rangle$. Here is the only theorem we need to bridge the gap from completeness to the NIP, which you'll prove in Exercise 6.

Theorem 4.8.6. *Suppose $\langle [a_n, b_n] \rangle$ is a sequence of nested intervals such that $(b_n - a_n) \to 0$ as $n \to \infty$. Then $\langle a_n \rangle$ and $\langle b_n \rangle$ are Cauchy.*

To make the proof of Theorem 4.8.6 easier, all you need to do is show that $\langle a_n \rangle$ is Cauchy, then observe that a similar argument will work for $\langle b_n \rangle$.

EXERCISES

1. Prove Theorem 4.8.2: A convergent sequence is Cauchy.

2. What does it mean for a sequence $\langle a_n \rangle$ not to be Cauchy?

3. Show that the sequence $\langle (-1)^n \rangle$ is not Cauchy, hence diverges.

4. Prove Theorem 4.8.3: A Cauchy sequence is bounded.

5. Prove Theorem 4.8.4: A Cauchy sequence of real numbers is convergent.

6. Prove part of Theorem 4.8.6: Suppose $\langle [a_n, b_n] \rangle$ is a sequence of nested intervals such that $(b_n - a_n) \to 0$ as $n \to \infty$. Then $\langle a_n \rangle$ is Cauchy.

CHAPTER 5

Functions of a Real Variable

Now we turn our attention to functions whose domain and codomain are some subset of $\mathbb{R}$. Functions whose domain is a subset of $\mathbb{R}$ are called *functions of a real variable*, and functions whose range is a subset of $\mathbb{R}$ are called *real-valued*. Although the domain of a function of a real variable might be any subset of $\mathbb{R}$, some especially powerful and interesting results can be deduced if S is compact. Because this chapter deals exclusively with real-valued functions of a real variable, we'll not state every time that the domain and codomain are subsets of $\mathbb{R}$. When we write $f : S \to \mathbb{R}$, it's understood that $S \subseteq \mathbb{R}$.

5.1 Bounded and Monotone Functions

First we discuss boundedness and monotonicity of a function $f : S \to \mathbb{R}$. With these and other ideas from Sections 5.2 and 5.3, we'll point out many links to the properties of sequences we discussed in Chapter 4.

5.1.1 Bounded functions

The definitions of the boundedness terms for functions resemble those for sequences.

Definition 5.1.1. Suppose $f : S \to \mathbb{R}$ is a function and $A \subseteq S$. We say f is *bounded from above* on A if there exists $M_1 \in \mathbb{R}$ such that $f(x) \leq M_1$ for all $x \in A$. Similarly, we say f is *bounded from below* on A if there exists $M_2 \in \mathbb{R}$ such that $f(x) \geq M_2$ for all $x \in A$. If there exists $M > 0$ such that $|f(x)| \leq M$ for all $x \in A$, we say f is *bounded* on A. For each of these characteristics, if it applies on all of S, we say simply that f is bounded from above, bounded from below, or bounded.

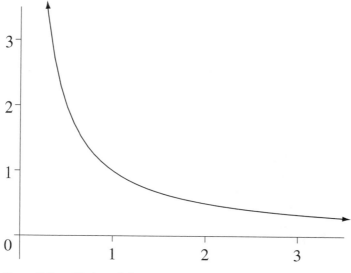

Figure 5.1 $f(x) = 1/x$.

In Exercises 1 and 2, you'll negate these definitions and apply all these terms to $f : (0, \infty) \to \mathbb{R}$ defined by $f(x) = 1/x$. (See Fig. 5.1.)

5.1.2 Monotone functions

When we defined a sequence to be monotone increasing, the fact that the terms of the sequence are lined up in an order made the definition easy to come by—if we pick two indices $m < n$, it must be that $a_m \leq a_n$. A similar definition works for functions.

Definition 5.1.2. Suppose $f : S \to \mathbb{R}$ is a function, $A \subseteq S$, and $a, b \in A$. Then f is said to be *increasing* on A if $f(a) \leq f(b)$ whenever $a < b$. Similarly, f is said to be *decreasing* on A if $f(a) \geq f(b)$ whenever $a < b$. If $f(a) < f(b)$ whenever $a < b$, f is said to be *strictly increasing* on A. If $f(a) > f(b)$ whenever $a < b$, f is said to be *strictly decreasing* on A. For each of these characteristics, if it applies on all of S, we say simply that f is increasing, decreasing, strictly increasing, or strictly decreasing, and we group functions of these types into the class we call *monotone* functions.

Go back and take a look at Exercise 9 from Section 2.4. Though we did not have the language of monotonicity at the time, this exercise proved the following.

Theorem 5.1.3. *Let* $n \in \mathbb{N}$. *Then:*

1. $f(x) = x^{2n-1}$ *is strictly increasing.*
2. $f(x) = x^{2n}$ *is strictly decreasing on* $(-\infty, 0]$ *and strictly increasing on* $[0, \infty)$.

In your precalculus work, you probably learned about inverse functions and whether a given function f even has an inverse by asking whether it passes the *horizontal line test*. The horizontal line test is an intuitive way of determining whether f is one-to-one, for if every horizontal line in the xy-plane crosses the graph of f no more than once, then f is one-to-one. Whether f is onto $\mathbb{R}$ does not matter, for you can crop the codomain of f down to Rng f so that $f : S \to$ Rng f is onto. Thus if $f : S \to \mathbb{R}$ is one-to-one, then $f^{-1} :$ Rng $f \to S$ will exist. Given that f^{-1} exists, then you were probably told that its graph could be sketched by reflecting the graph of f about the diagonal line $y = x$. You were taught to find the formula for f^{-1} by switching the roles of x and y, then solving for y, and that reflecting the graph about $y = x$ is the visual result of swapping x and y. Think intuitively for a moment. If f is a strictly increasing function, then does it pass the horizontal line test? If so, then f^{-1} exists on Rng f. Given this, can you say anything about monotonicity of f^{-1}? If you have the answers to these questions, then you have Theorems 5.1.4 and 5.1.6. You'll prove them in Exercises 4 and 5.

Theorem 5.1.4. *If f is strictly increasing (or strictly decreasing) on A, then it is one-to-one on A.*

As an immediate consequence of Theorems 5.1.4 and 3.2.5, we have the following:

Corollary 5.1.5. *If $f : S \to \mathbb{R}$ is strictly monotone on $A \subseteq S$, then there exists $f^{-1} : f(A) \to A$.*

Theorem 5.1.6. *If $f : S \to \mathbb{R}$ is strictly increasing (or strictly decreasing), then $f^{-1} :$ Rng$(f) \to S$ is strictly increasing (or strictly decreasing).*

Theorems 5.1.3 and 5.1.6 imply that $\sqrt[2n]{x}$ is strictly increasing on $[0, \infty)$ and $\sqrt[2n-1]{x}$ is strictly increasing on $\mathbb{R}$. In addition, with Theorem 5.1.7, which you'll prove in Exercise 6, the function $f(x) = x^r$ is strictly increasing on $[0, \infty)$ for any $r \in \mathbb{Q}^+$. We may write $r = m/n$ for $m, n \in \mathbb{N}$, so that $f(x) = \sqrt[n]{x^m}$.

Theorem 5.1.7. *Suppose f and g are real-valued functions such that $g \circ f$ is defined on some interval (a, b). If f and g are both increasing, then so is $g \circ f$.*

EXERCISES

1. Use the logic of Definition 5.1.1 to state what the following terms mean:

    (a) f is not bounded from above on A

    (b) f is not bounded from below on A

    (c) f is not bounded on A

2. For $f(x) = 1/x$ on the interval $(0, \infty)$ show the following:

    (a) f is bounded on $(1, \infty)$

    (b) f is bounded from below on $(0, 1)$

    (c) f is not bounded from above on $(0, 1)$

3. Show that $f(x) = 1/(x^2 + 1)$ is strictly decreasing on $[0, \infty)$.

4. Prove part of Theorem 5.1.4: If f is strictly increasing on A, then it is one-to-one on A.

5. Prove part of Theorem 5.1.6: If $f : S \to \mathbb{R}$ is strictly increasing, then $f^{-1} : \text{Rng}(f) \to S$ is strictly increasing.

6. Prove Theorem 5.1.7: Suppose f and g are real-valued functions such that $g \circ f$ is defined on some interval (a, b). If f and g are both increasing, then so is $g \circ f$.

7. State and prove other theorems analogous to Theorem 5.1.7 by varying the hypothesis conditions on f and g to include all combinations of increasing and decreasing.

5.2 Limits and Their Basic Properties

Suppose $f : S \to \mathbb{R}$ is a function that is defined on some deleted neighborhood of $a \in \mathbb{R}$. Whether $f(a)$ exists is beside the point in the definition of limit, so we insist on no more than a deleted neighborhood. In a way somewhat similar to our definition of convergence of sequences as $n \to \infty$, we discuss $\lim_{x \to a} f(x)$, the limit as x approaches a of $f(x)$. In this section, we'll work our way into the definition of the limit of a function, then look at some basic theorems involving limits that bear striking parallels to those theorems involving sequences.

5.2.1 Definition of limit

Let's work our way into a definition of $\lim_{x \to a} f(x)$ gradually. Let $f(x)$ be defined in the following way:

$$f(x) = \begin{cases} 4x - 1, & \text{if } x \neq 2 \\ 50, & \text{if } x = 2. \end{cases} \tag{5.1}$$

Defining f in this way produces a very familiar straight line, but with a hole punched in the graph at $x = 2$ and the value of $f(2)$ artificially (but purposefully) defined to be a number way out of line with what you probably expect it to be. (See Fig. 5.2.) Although $f(2) = 50$, the function does not take on values anywhere near 50 if x is close to but different from 2. To the contrary, if x is in some small *deleted* neighborhood of 2, $f(x)$ appears to take on values close to 7. In fact, we can ask the following question:

Example 5.2.1. Let f be defined as in Eq. (5.1) and suppose $\epsilon > 0$ is given. For the ϵ-neighborhood of 7 on the y-axis in Fig. 5.2, we want to find a radius $\delta > 0$ of a deleted neighborhood of 2 on the x-axis so that every $x \in DN_\delta(2)$ maps into $N_\epsilon(7)$. That is, we want to find $\delta > 0$ so that

$$7 - \epsilon < f(x) < 7 + \epsilon \tag{5.2}$$

for every $x \in DN_\delta(2)$.

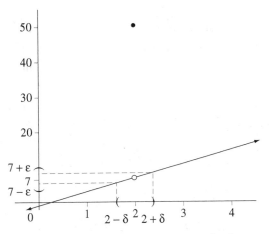

Figure 5.2 Shows f defined from Eq. (5.1).

Solution: Since the only values of x we're interested in are different from 2, we may use $f(x) = 4x - 1$ in inequality (5.2). Thus we have

$$7 - \epsilon < 4x - 1 < 7 + \epsilon \tag{5.3}$$

$$2 - \frac{\epsilon}{4} < x < 2 + \frac{\epsilon}{4}. \tag{5.4}$$

Inequalities (5.3) and (5.4) are logically equivalent, so $x \in DN_{\epsilon/4}(2)$ implies $7 - \epsilon < f(x) < 7 + \epsilon$. Letting $\delta = \epsilon/4$ guarantees that if $x \in DN_\delta(2)$, then $|f(x) - 7| < \epsilon$. ∎

Example 5.2.1 is effectively the determination of a deleted δ-neighborhood of 2 that maps into a given ϵ-neighborhood of $y = 7$. The value of δ depends on the given value of ϵ. It illustrates the following point, which is the heart of the definition of $\lim_{x \to a} f(x)$: Given a "target" of radius ϵ around $y = 7$, we can find a deleted neighborhood of $x = 2$ such that all values of the function in the deleted neighborhood hit somewhere in the target. If you imagine a sequence of smaller and smaller ϵ-values, it is always possible to find sufficiently small deleted neighborhoods of $x = 2$ that map entirely within the smaller and smaller targets. If such a deleted δ-neighborhood can be found for all ϵ-neighborhoods of $y = 7$, then we say that $\lim_{x \to 2} f(x) = 7$. Here's the definition.

Definition 5.2.2. Suppose $f : S \to \mathbb{R}$ is defined on some deleted neighborhood of $a \in \mathbb{R}$, and $L \in \mathbb{R}$. Then we say $\lim_{x \to a} f(x) = L$ if for all $\epsilon > 0$, there exists $\delta > 0$ such that for all $x \in S$, $0 < |x - a| < \delta$ implies $|f(x) - L| < \epsilon$. If $\lim_{x \to a} f(x) = L$, we say f converges to L as $x \to a$. Another way to write this is $f(x) \to L$ as $x \to a$.

To write Definition 5.2.2 symbolically, we would have

$$\lim_{x \to a} f(x) = L \Leftrightarrow (\forall \epsilon > 0)(\exists \delta > 0)(\forall x \in S)(0 < |x - a| < \delta \to |f(x) - L| < \epsilon). \tag{5.5}$$

Here's an example of a proof based on Definition 5.2.2. Some scratch work has been omitted that leads to the value of δ. See if you can figure out what δ should be on your own.

Example 5.2.3. Let g be defined by

$$g(x) = \frac{3x^2 + 17x + 20}{2x + 8}. \tag{5.6}$$

Show that $\lim_{x \to -4} g(x) = -7/2$.

Solution: Pick $\epsilon > 0$. Let $\delta = 2\epsilon/3$. Now if $x \neq -4$, a factor of $(x + 4)$ may be canceled from the numerator and denominator of g. Thus if $0 < |x + 4| < \delta$, we have

$$\begin{aligned}
\left| g(x) + \frac{7}{2} \right| &= \left| \frac{3x^2 + 17x + 20}{2x + 8} + \frac{7}{2} \right| \\
&= \left| \frac{(3x + 5)(x + 4)}{2(x + 4)} + \frac{7}{2} \right| \\
&= \left| \frac{3x + 5}{2} + \frac{7}{2} \right| = \left| \frac{3x + 12}{2} \right| \\
&= \left| \frac{3(x + 4)}{2} \right| = \frac{3}{2} |x + 4| < \frac{3\delta}{2} = \epsilon.
\end{aligned} \tag{5.7}$$

Thus $\lim_{x \to -4} g(x) = -7/2$. ∎

In the same way that f from Eq. (5.1) seems a little contrived because of the special effort we took to define $f(2) = 50$, g from Example 5.2.3 might seem contrived because nothing would seem to prevent us from canceling the $(x + 4)$ factor from Eq. (5.6). Granted, g is merely the straight line $y = 3x + 5$ with a hole punched in the domain at $x = -4$ by the introduction of an $(x + 4)$ factor in the numerator and denominator. But f and g are good first examples to illustrate that the limit of a function as $x \to a$ has nothing to do with the value or even the existence of $f(a)$. Whether $f(a)$ exists and is the same as the limit as $x \to a$ is called *continuity*, and we'll look at that in Section 5.5. There are plenty of examples where a hole in the domain occurs naturally, and we'll look at one later in this section.

5.2.2 Basic theorems of limits

There is a real similarity between the definition of convergence of a sequence $\lim_{n \to \infty} a_n$ and convergence of a function $\lim_{x \to a} f(x)$. In convergence of sequences, a given $\epsilon > 0$ must be associated with a threshold term beyond which all terms of the sequence fall in the ϵ-neighborhood of the limit. In the convergence of a function, a given $\epsilon > 0$ must be associated with a δ-radius of a deleted neighborhood of a, within which all values of the function, except perhaps $f(a)$, must fall in the ϵ-neighborhood of the limit. The good news is that the similarity of these definitions makes for plenty of similar theorems involving $\lim_{x \to a} f(x)$, with similar proofs to match. Here is a barrage of theorems involving convergence of a function. As we present them, we'll often link them back to sequence theorems and properties. We'll prove some of them here to illustrate the similarities, but you'll prove most of them in the exercises by mimicking your work from Section 4.6.

Theorem 5.2.4. *For any $a \in \mathbb{R}$, the constant function $f(x) = c$ satisfies $\lim_{x \to a} f(x) = c$.*

Proof: Pick any $a \in \mathbb{R}$ and $\epsilon > 0$. Let $\delta = 1$. Then if $0 < |x - a| < \delta$, $|f(x) - c| = |c - c| = 0 < \epsilon$. Thus $\lim_{x \to a} f(x) = c$. ∎

Theorem 5.2.5. *For any $a \in \mathbb{R}$, $\lim_{x \to a} x = a$.*

Theorem 5.2.6. *If $f(x) \to L$ as $x \to a$, then there exists a deleted neighborhood of a where f is bounded.*

Theorem 5.2.7. *If $f(x) \to 0$ as $x \to a$, and if g is bounded on a deleted neighborhood of a, then $f(x)g(x) \to 0$ as $x \to a$.*

Theorem 5.2.8. *If $f(x) \to L_1$ and $g(x) \to L_2$ as $x \to a$, then as $x \to a$:*

1. $f(x) + g(x) \to L_1 + L_2$,
2. $f(x)g(x) \to L_1 L_2$.

Proof: We prove part 1 and leave part 2 to you. Pick $\epsilon > 0$. Since $f(x) \to L_1$ as $x \to a$, there exists $\delta_1 > 0$ such that $0 < |x - a| < \delta_1$ implies $|f(x) - L_1| < \epsilon/2$. Similarly, since $g(x) \to L_2$, there exists $\delta_2 > 0$ such that $0 < |x - a| < \delta_2$ implies $|g(x) - L_2| < \epsilon/2$. Let $\delta = \min\{\delta_1, \delta_2\}$. Then $\delta > 0$, and if $0 < |x - a| < \delta$, both $0 < |x - a| < \delta_1$ and $0 < |x - a| < \delta_2$ are satisfied. Thus we have

$$|[f(x) + g(x)] - (L_1 + L_2)]| \leq |f(x) - L_1| + |g(x) - L_2| \leq \frac{\epsilon}{2} + \frac{\epsilon}{2} = \epsilon. \quad (5.8)$$

∎

Three results follow as immediate corollaries of Theorems 5.2.4 and 5.2.8.

Corollary 5.2.9. *If $f(x) \to L$ as $x \to a$, and if $c \in \mathbb{R}$, then $cf(x) \to cL$ as $x \to a$.*

Corollary 5.2.10. *If $f(x) \to L_1$ and $g(x) \to L_2$ as $x \to a$, then $f(x) - g(x) \to L_1 - L_2$ as $x \to a$.*

Corollary 5.2.11. *If $P(x) = a_n x^n + a_{n-1} x^{n-1} + \cdots + a_1 x + a_0$, then for all $a \in \mathbb{R}$, $P(x) \to P(a)$ as $x \to a$.*

Theorem 5.2.12. *Suppose $g(x) \to L \neq 0$ as $x \to a$. Then there exists a deleted neighborhood of a where g can be bounded away from zero. In particular, if $L > 0$, there exists $M > 0$ and $\delta > 0$ such that $0 < |x - a| < \delta$ implies $g(x) > M$. If $L < 0$, there exists $M < 0$ and $\delta > 0$ such that $0 < |x - a| < \delta$ implies $g(x) < M$.*

Theorem 5.2.13. *Suppose $g(x) \to L \neq 0$ as $x \to a$. Then $1/g(x) \to 1/L$ as $x \to a$.*

Corollary 5.2.14. *Suppose $f(x) \to L_1$ and $g(x) \to L_2 \neq 0$ as $x \to a$. Then $f(x)/g(x) \to L_1/L_2$ as $x \to a$.*

Theorem 5.2.15. *Suppose f is defined by*

$$f(x) = \frac{P_1(x)}{P_2(x)} = \frac{a_r x^r + a_{r-1} x^{r-1} + \cdots + a_1 x + a_0}{b_s x^s + b_{s-1} x^{s-1} + \cdots + b_1 x + b_0}.$$

Then, provided $P_2(a) \neq 0$, $f(x) \to P_1(a)/P_2(a)$ as $x \to a$.

Proof: Apply Corollaries 5.2.11 and 5.2.14 and the result follows. ∎

Similar to the comments on sequences in Exercise 13 from Section 4.6, if $f(x) \to L$ and $g(x) \to L$ as $x \to a$, then $f(x) - g(x) \to 0$. Thus if $\epsilon > 0$, there exists $\delta > 0$ such that $0 < |x - a| < \delta$ implies $|f(x) - g(x)| < \epsilon/2$. This fact will come in handy in proving the following:

Theorem 5.2.16 (Sandwich Theorem). *Suppose f, g, and h are functions with the property that $g(x) \leq h(x) \leq f(x)$ for all x in some deleted neighborhood of a. Suppose also that $f(x) \to L$ and $g(x) \to L$ as $x \to a$. Then $h(x) \to L$ as $x \to a$.*

Earlier we promised an example of a function with a natural hole in the domain where we want to address the limit. The function is $f(x) = \sin x / x$ at $a = 0$. Except for a passing glance at sine in Section 4.5, we have never mentioned any trigonometric, exponential, or logarithmic functions. There is a reason for this; definitions of these functions that are rooted in the axioms of $\mathbb{R}$ are not possible to come by at this stage of our game. Only the algebraic functions arise from the theory we've developed thus far, and they can make for some pretty sticky proofs all by themselves. Strict definitions of these nonalgebraic functions, called *transcendental functions,* come later. However, if we kick back for a while and give ourselves the freedom to talk about sine, cosine, and tangent in the familiar language of the unit circle, then we can apply Theorem 5.2.16 to show

$$\lim_{x \to 0} \frac{\sin x}{x} = 1. \tag{5.9}$$

In trigonometry, we take a real x-numberline and wrap it, as it were, around the unit circle in the uv-plane with $x = 0$ placed at $(u, v) = (1, 0)$ and the positive half of the x-axis wrapped counterclockwise. For $x \in \mathbb{R}$, $\cos x$ and $\sin x$ are defined to be the u and v coordinates, respectively, of the point where x falls in the uv-plane. If $0 < x < \pi/2$, Fig. 5.3 shows how we may view x as an arc length, and $\sin x$, $\cos x$, and $\tan x$ as the lengths of segments in Quadrant I. Similar triangles in the figure will convince you that the length labeled $\tan x$ is correct. If $0 < x < \pi/2$ as is suggested in the sketch, the geometry of the unit circle reveals

$$\sin x \leq x \leq \tan x. \tag{5.10}$$

If $-\pi/2 < x < 0$ so that x is in Quadrant IV, then $\sin x$ and $\tan x$ are also negative, and we have

$$\tan x \leq x \leq \sin x. \tag{5.11}$$

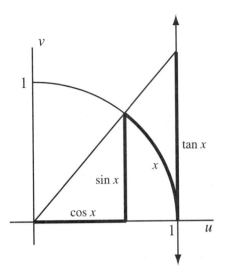

Figure 5.3 Basic trigonometric functions.

If we take both parts of inequality (5.10) separately and solve each for $\sin x/x$, we can put them back together into the single inequality

$$\cos x \le \frac{\sin x}{x} \le 1. \tag{5.12}$$

If we do the same thing for (5.11) remembering that $x < 0$ and $\sin x < 0$, we also arrive at (5.12), so that (5.12) holds for all $x \in DN_{\pi/2}(0)$. Let's accept from the geometry of the unit circle and the definition of $\cos x$ that $\cos x \to 1$ as $x \to 0$. Then, by Theorem 5.2.16, Eq. (5.9) holds.

Let's remind ourselves of some of the theory of Cauchy sequences in order to motivate our last theorem from this section. An immediate result of convergence of a sequence is that it is Cauchy (Theorem 4.8.2). If the terms of a sequence cluster around some $L \in \mathbb{R}$, then they must cluster around each other. Similarly, we can prove the following theorem, which states if $f(x) \to L$ as $x \to a$, then the values of f must not vary much from each other in small deleted neighborhoods of a. You'll prove it in Exercise 9.

Theorem 5.2.17. *Suppose $f(x) \to L$ as $x \to a$. Then for every $\epsilon > 0$, there exists $\delta > 0$ such that $x_1, x_2 \in DN_\delta(a)$ implies $|f(x_1) - f(x_2)| < \epsilon$.*

If a sequence is not Cauchy, it is not convergent. In Exercise 6, Section 4.8, you stated what it means for a sequence not to be Cauchy: There exists $\epsilon > 0$ such that, for all $N \in \mathbb{N}$, there exist $m, n \ge N$ where $|a_m - a_n| \ge \epsilon$. That is, there is some fixed $\epsilon > 0$ so that no matter where in the sequence you put your finger, somewhere farther down the line there will be two terms that are at least ϵ apart. What does it mean for the property in Theorem 5.2.17 *not* to hold? The answer: There exists $\epsilon > 0$ such that for all $\delta > 0$, there exist $x_1, x_2 \in DN_\delta(a)$ with $|f(x_1) - f(x_2)| \ge \epsilon$. That is, there is some fixed $\epsilon > 0$ such that every deleted δ-neighborhood of a, no matter how small, contains two points whose functional values are at least ϵ apart. If a function f has this property, then $\lim_{x \to a} f(x)$

does not exist.

Example 5.2.18. Let f be defined in the following way:

$$f(x) = \begin{cases} 1, & \text{if } x \text{ is rational,} \\ 0, & \text{if } x \text{ is irrational.} \end{cases} \quad (5.13)$$

If $a \in \mathbb{R}$, then $\lim_{x \to a} f(x)$ does not exist. For we may let $\epsilon = 1/2$ and choose any $\delta > 0$. Then by Exercises 7 and 8, Section 4.1, the deleted δ-neighborhood of a contains a rational number x_1 and an irrational number x_2, and $|f(x_1) - f(x_2)| = 1 > \epsilon$. This function is sometimes called the *salt and pepper* function because its values of 0 and 1 are sprinkled up and down the domain like grains of salt and pepper. No one really knows which is the salt and which is the pepper.

EXERCISES

1. Prove Theorem 5.2.5: For any $a \in \mathbb{R}$, $\lim_{x \to a} x = a$.

2. Prove Theorem 5.2.6: If $f(x) \to L$ as $x \to a$, then there exists a deleted neighborhood of a where f is bounded.

3. Prove Theorem 5.2.7: If $f(x) \to 0$ as $x \to a$, and if g is bounded on a deleted neighborhood of a, then $f(x)g(x) \to 0$ as $x \to a$.

4. Prove part 2 of Theorem 5.2.8: If $f(x) \to L_1$ and $g(x) \to L_2$ as $x \to a$, then $f(x)g(x) \to L_1 L_2$.

5. Prove Theorem 5.2.12: Suppose $g(x) \to L \neq 0$ as $x \to a$. Then there exists a deleted neighborhood of a where g can be bounded away from zero. In particular, if $L > 0$, there exists $M > 0$ and $\delta > 0$ such that $0 < |x - a| < \delta$ implies $g(x) > M$. If $L < 0$, there exists $M < 0$ and $\delta > 0$ such that $0 < |x - a| < \delta$ implies $g(x) < M$.

6. Prove Theorem 5.2.13: Suppose $g(x) \to L \neq 0$ as $x \to a$. Then $1/g(x) \to 1/L$ as $x \to a$.

7. Prove Theorem 5.2.16: Suppose f, g, and h are functions with the property that $g(x) \leq h(x) \leq f(x)$ for all x in some deleted neighborhood of a. Suppose also that $f(x) \to L$ and $g(x) \to L$ as $x \to a$. Then $h(x) \to L$ as $x \to a$.

8. Assuming $-1 \leq \sin x \leq 1$ for all $x \in \mathbb{R}$, show that $\lim_{x \to 0} x \sin x = 0$.

9. Prove Theorem 5.2.17: Suppose $f(x) \to L$ as $x \to a$. Then for every $\epsilon > 0$, there exists $\delta > 0$ such that $x_1, x_2 \in DN_\delta(a)$ implies $|f(x_1) - f(x_2)| < \epsilon$.

10. The *signum* function, $\operatorname{sgn} x : \mathbb{R} \setminus \{0\} \to \mathbb{R}$ is defined in the following way:

$$\operatorname{sgn} x = \begin{cases} -1, & \text{if } x < 0, \\ 1, & \text{if } x > 0. \end{cases} \quad (5.14)$$

Show that $\lim_{x \to 0} \operatorname{sgn} x$ does not exist.

11. Use familiar values of $\sin x$ to show that $\lim_{x \to 0} \sin(1/x)$ does not exist.

5.3 More on Limits

In this section, we consider two topics relating to the limit of a function. First, we discuss one-sided limits, where we consider x approaching a either from the right or from the left separately. Then, instead of merely commenting on parallels between convergence of sequences and convergence of a function, we'll actually tie the two ideas together with an important theorem.

5.3.1 One-sided limits

In defining $\lim_{x \to a} f(x) = L$, we insisted that f be defined on both sides of a and all points to the nearby left and right of a map into $N_\epsilon(L)$. If f is defined only on one side of a, or perhaps if values of f to the immediate left of a behave differently from those to the immediate right (as in sgn x), we can discuss the limit of $f(x)$ as x approaches a from the left or from the right separately. Instead of using entire deleted neighborhoods of a, we use only the left- or right-half of them.

Definition 5.3.1. Suppose f is defined on some interval (a, b). Then we say that $\lim_{x \to a^+} f(x) = L$ (read "... as x approaches a from the right") if for all $\epsilon > 0$, there exists $\delta > 0$ such that for all $x \in (a, b)$, $a < x < a + \delta$ implies $|f(x) - L| < \epsilon$; L is called the *right-hand limit* of f at a. Similarly, if f is defined on some interval (c, a), we say that $\lim_{x \to a^-} f(x) = L$ (x approaches a from the left) if for all $\epsilon > 0$, there exists $\delta > 0$ such that $x \in (a, b)$, $a - \delta < x < a$ implies $|f(x) - L| < \epsilon$; L is called the *left-hand limit* of f at a.

Theorem 5.3.2. *For a function* f, $\lim_{x \to a} f(x) = L$ *if and only if*

$$\lim_{x \to a^+} f(x) = \lim_{x \to a^-} f(x) = L. \tag{5.15}$$

Theorem 5.3.2 can be argued in an interesting way from the logical structure of the definitions of the limits involved, thus we'll provide a rather conversational argument. In arguing the $\Rightarrow$ direction, we would pick an $\epsilon > 0$ and note that there exists $\delta > 0$ such that $x \in DN_\delta(a)$ implies $|f(x) - L| < \epsilon$. However, the inequality $x \in DN_\delta(a)$ is weaker than either $a < x < a + \delta$ or $a - \delta < x < a$. Thus the statement

$$x \in DN_\delta(a) \Rightarrow |f(x) - L| < \epsilon$$

is stronger than either

$$a < x < a + \delta \Rightarrow |f(x) - L| < \epsilon$$

or

$$a - \delta < x < a \Rightarrow |f(x) - L| < \epsilon.$$

Thus the existence of $\lim_{x \to a} f(x)$ implies the existence of $\lim_{x \to a^+} f(x)$ and $\lim_{x \to a^-} f(x)$.

In arguing $\Leftarrow$, for a given $\epsilon > 0$, the right-hand limit provides us with δ_1, and the left-hand limit provides us with δ_2, each having the requisite properties. If we let $\delta = \min\{\delta_1, \delta_2\}$, then $x \in DN_\delta(a)$ implies either

$$a < x < a + \delta \leq a + \delta_1 \quad \text{or} \quad a - \delta_2 \leq a - \delta < x < a,$$

so that $|f(x) - L| < \epsilon$ in either case.

Example 5.3.3. Let f be defined in the following way:

$$f(x) = \begin{cases} x^2 - 1, & \text{if } x < 2, \\ \dfrac{1}{x-1}, & \text{if } x > 2. \end{cases} \tag{5.16}$$

By Theorem 5.2.15, $\lim_{x \to 2}(x^2 - 1) = 3$ and $\lim_{x \to 2} 1/(x-1) = 1$ (in the two-sided sense). Thus, by the $\Rightarrow$ direction of Theorem 5.3.2, $\lim_{x \to 2^-}(x^2 - 1) = 3$ and $\lim_{x \to 2^+} 1/(x-1) = 1$. However, $\lim_{x \to 2^-} f(x) = 3$ and $\lim_{x \to 2^+} f(x) = 1$, so that $\lim_{x \to 2} f(x)$ fails to exist by the $\Leftarrow$ direction of Theorem 5.3.2.

5.3.2 Sequential limit of f

The following theorem states that our ϵ-δ definition of $\lim_{x \to a} f(x) = L$ from Definition (5.2.2) is logically equivalent to a characteristic involving sequences.

Theorem 5.3.4. *Suppose $f : S \to \mathbb{R}$ is a function. Then $\lim_{x \to a} f(x) = L$ if and only if every sequence $\langle a_n \rangle$ with the properties that $a_n \to a$ and $a_n \neq a$ for all $n \in \mathbb{N}$ also has the property that $f(a_n) \to L$.*

You'll prove Theorem 5.3.4 in Exercise 3. The $\Rightarrow$ direction is the easier one, for if entire δ-neighborhoods of a map within ϵ of L, then the fact that $a_n \to a$ will certainly guarantee that all terms of $\langle a_n \rangle$ beyond some threshold map within ϵ of L also. Proving $\Leftarrow$ is probably best done by contrapositive. If $\lim_{x \to a} f(x) \neq L$ (What does this mean?), then you should be able to use a sequence of progressively smaller and smaller δ-values to create a sequence $\langle a_n \rangle$ such that $a_n \neq a$ for all n, $a_n \to a$, and $f(a_n) \not\to L$.

The logical equivalence of our ϵ-δ definition of limit and the sequential limit characteristic in Theorem 5.3.4 means that it is possible to define the statement $\lim_{x \to a} f(x) = L$ either in terms of ϵ and δ or in terms of sequences. Some authors prefer to define function limits sequentially by making Theorem 5.3.4 their definition of the limit of a function. Our ϵ-δ definition would then be a theorem. They then construct proofs of theorems such as those from Section 5.2 from the theory of sequences we discussed in Section 4.6. If you have the theory of sequences already under your belt, function limit proofs become very easy. For example, to show that if $f(x) \to L_1$ and $g(x) \to L_2$ as $x \to a$, then $f(x) + g(x) \to L_1 + L_2$, the proof would go something like this.

Proof of Theorem 5.27: We show that every sequence $\langle a_n \rangle$ such that $a_n \neq a$ for all n and $a_n \to a$ also satisfies $f(a_n) + g(a_n) \to L_1 + L_2$. Suppose $a_n \to a$, where $a_n \neq a$ for all n. Then since $f(x) \to L_1$ as $x \to a$, we have that $f(a_n) \to L_1$ as $n \to \infty$. Similarly, since $g(x) \to L_2$ as $x \to a$, then $g(a_n) \to L_2$ as $n \to \infty$. By Theorem 4.6.8, $f(a_n) + g(a_n) \to L_1 + L_2$. Since $\langle a_n \rangle$ was chosen arbitrarily, we have $f(x) + g(x) \to L_1 + L_2$ as $x \to a$. ∎

Perhaps you feel a little cheated at this point. After all, if only we had proved Theorem 5.3.4 at the beginning of Section 5.2, then all our proofs in Section 5.2 would have been one-liners. There is some truth to that. However, ϵ-δ proofs pervade mathematics, and the fact that your first ϵ-δ proofs could be done easily by merely mimicking work from Section 4.6 was probably a humane way for you to be introduced to them.

EXERCISES

1. Let f be defined in the following way:

$$f(x) = \begin{cases} 2 \operatorname{sgn} x, & \text{if } x \leq 1, \\ x^3 + 2, & \text{if } x > 1. \end{cases} \quad (5.17)$$

   Evaluate the following, with reference to applicable theorems:

   (a) $\lim_{x \to 0^-} f(x)$
   (b) $\lim_{x \to 0^+} f(x)$
   (c) $\lim_{x \to 0} f(x)$
   (d) $\lim_{x \to 1^-} f(x)$
   (e) $\lim_{x \to 1^+} f(x)$
   (f) $\lim_{x \to 1} f(x)$

2. Show that $\lim_{x \to 0^+} \sqrt{x} = 0$.

3. Prove theorem 5.3.4: Suppose $f : S \to \mathbb{R}$. Then $\lim_{x \to a} f(x) = L$ if and only if every sequence $\langle a_n \rangle$ such that $a_n \neq a$ for all $n \in \mathbb{N}$ and $a_n \to a$ satisfies $f(a_n) \to L$.[1]

5.4 Limits Involving Infinity

Now we want to take the concepts and language of limits and extend them to include the symbols $\pm \infty$, even though ∞ is not a real number. We have used the symbol ∞ before in our work with sequences, and when we defined $\lim_{n \to \infty} a_n$, we probably raised no eyebrows because we defined precisely what we mean by $\lim_{n \to \infty} a_n$ in terms of $\mathbb{N}$, the domain of the sequence. In this section, we want to take the expression $\lim_{x \to a} f(x) = L$ and replace either a or L (or both) with one of the symbols $\pm \infty$. Up until now, the fact that a and L are real numbers allowed us to discuss ϵ-neighborhoods of L and δ-neighborhoods of a. With our current definition of neighborhood, it makes no sense to talk about a *neighborhood of infinity*, unless of course we decide to give this term meaning. We will do precisely this. In fact, as we go, we'll concoct extended, new meanings for old language and revamp some of our visual imagery to show that there just might be some way to understand ∞ so that we can almost treat it as a number.

---

[1] Prove $\Leftarrow$ contrapositively by using a sequence $\delta_n = 1/n$ to generate a sequence a_n where $a_n \to a$ and $f(a_n) \not\to L$.

5.4 Limits Involving Infinity

The language we'll use in replacing a or L with $\pm\infty$ will go something like this. In discussing $\lim_{x\to\pm\infty} = L$, we'll call these limits *at* positive or negative infinity. When we discuss $\lim_{x\to a} f(x) = \pm\infty$, we'll call these limits *of* positive or negative infinity. Graphically, you can imagine that a limit L at infinity corresponds to a *horizontal asymptote* in the graph of f, and a limit of infinity at a corresponds to a *vertical asymptote*. There are buckets and buckets of ways to combine and specialize the ideas we'll discuss here. With both $+\infty$ and $-\infty$, and with either a or L or both being replaced by $\pm\infty$, with two-sided and one-sided limits, there are several new terms we can define. By hitting a few, you will probably catch on to what the others ought to be, so we'll not address all of them.

5.4.1 Limits at infinity

Let's begin by replacing a with $+\infty$ because it almost exactly replicates the theory of sequences. Since a sequence $\langle a_n \rangle$ is really a real-valued function whose domain is $\mathbb{N}$, then our way of graphing a sequence as we did in Fig. 4.1 makes convergence to L easily visualizable as a horizontal asymptote of discrete points in the plane. If we imagine filling in values of the function to other real numbers so that the domain is some interval $(a, +\infty)$, then the following definition seems to be a natural adaptation of Definition 4.6.1.

Definition 5.4.1. Suppose $f : S \to \mathbb{R}$ is defined on some interval $(a, +\infty)$. Then we say $\lim_{x\to +\infty} f(x) = L$ if for all $\epsilon > 0$, there exists $M > 0$ such that $x > M$ implies $|f(x) - L| < \epsilon$.

Notice the similarity between Definitions 5.4.1 and 4.6.1. Given $\epsilon > 0$, there exists a threshold point in the domain beyond which all values of the function fall within ϵ of L. A little thought will convince you that this definition gives rise to a whole slew of theorems involving limits of functions at $+\infty$.

Theorem 5.4.2. $\lim_{x\to +\infty} c = c$.

Theorem 5.4.3. $\lim_{x\to +\infty} 1/x = 0$.

Theorem 5.4.4. If $f(x) \to L_1$ and $g(x) \to L_2$ as $x \to +\infty$, then $f(x) + g(x) \to L_1 + L_2$ and $f(x)g(x) \to L_1 L_2$ as $x \to +\infty$. If $L_2 \neq 0$, then $f(x)/g(x) \to L_1/L_2$ as $x \to +\infty$.

On and on the theorems go that exactly parallel our previous work. The limit of polynomial over polynomial and even a sandwich theorem seem strangely translucent. The logic of the proofs is almost identical to that from before, so we will not ask you to supply all the proofs in full detail. However, you should always make sure you're able to provide them when necessary. Do we need to bother defining $\lim_{x\to -\infty} f(x) = L$? Perhaps an adaptation of Definition 5.4.1 is so clear that we do not need to state it formally here. And, of course, all the theorems follow.

Now let's extend our language of neighborhood to include infinity. When we say $f(x) \to L$ as $x \to a$, we mean that any neighborhood of L has a corresponding deleted neighborhood of a that maps into it. Is there a way to use the same language for the

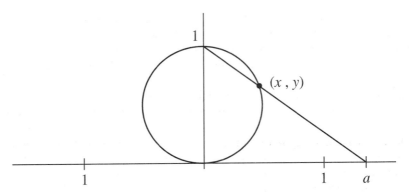

Figure 5.4 $\mathbb{R}$ with the point at infinity.

statement $f(x) \to L$ as $x \to +\infty$? Could we say that any neighborhood of L has a corresponding neighborhood of $+\infty$ that maps into it? Of course we could, if we were to define a *neighborhood of* $+\infty$ to be an interval of the form $(M, +\infty)$. Similarly, a neighborhood of $-\infty$ could be defined as an interval of the form $(-\infty, M)$.

We're catching a glimpse of the *extended real numbers*, and developing an imagery of two phantom points $\pm\infty$ somewhere way off the left and right ends of the numberline. An apparent difference is that neighborhoods of $+\infty$ or $-\infty$ are not two-sided. If we use the single symbol ∞ instead of both $\pm\infty$, then we can create a nice way of visualizing $\mathbb{R} \cup \{\infty\}$, the extended real numbers, where $+\infty$ and $-\infty$ are merged into the one point. Here's how.

Imagine a real numberline with a circle sitting on top of it as in Fig. 5.4. Imagine each $a \in \mathbb{R}$ being mapped to a corresponding point (x, y) on the circle with the help of the diagonal line illustrated in the figure. This geometric way of mapping all points in $\mathbb{R}$ to points on the circle suggests a one-to-one function from $\mathbb{R}$ onto all points of the circle except the north pole $(0, 1)$. If you want an explicit formula for the coordinates of the point (x, y) in terms of the value of $a \in \mathbb{R}$, you can obtain it easily enough with the help of similar triangles (Exercise 1), but it is not necessary to understanding the principle. The point is that $\mathbb{R}$ is now visualized in a different way. Instead of $\mathbb{R}$ being illustrated as an infinite line, it's now a circle, except that there is one point of the circle that has not been associated with a real number in this mapping. Let's call the point at $(0, 1)$ the *point at infinity*, ∞. The extended real numbers then become $\mathbb{R} \cup \{\infty\}$, where $+\infty$ and $-\infty$ are really the same point, and ∞ is nothing more than a formal symbol thrown in, along with a bunch of rules about how you're supposed to use it.

With this imagery, a natural definition for a *neighborhood of* ∞ would seem to be

$$N_M(\infty) = (-\infty, -M) \cup (M, +\infty) = \{x : |x| > M\}, \tag{5.18}$$

even though such a neighborhood does not have radius M. Limits as $x \to +\infty$ or $x \to -\infty$ then look like one-sided limits:

$$\lim_{x \to +\infty} f(x) = \lim_{x \to \infty^-} f(x) \quad \text{and} \quad \lim_{x \to -\infty} f(x) = \lim_{x \to \infty^+} f(x). \tag{5.19}$$

Thus we have arrived at the point where $\lim_{x \to a} f(x) = L$ has meaning for all $L \in \mathbb{R}$ and all $a \in \mathbb{R} \cup \{\infty\}$. The only thing we have to keep in mind is that the neighborhoods of ∞ are not δ-neighborhoods as they are for real numbers in our standard view of $\mathbb{R}$. However, if we view $\mathbb{R} \cup \{\infty\}$ as the circle, and as long as the idea of a neighborhood of ∞ is properly understood, then all the theorems will look exactly the same. After all, we concocted the definition of a neighborhood of ∞ in order to make that happen. In this rather novel view, the notation $\lim_{x \to \infty} f(x) = L$ would mean f has a horizontal asymptote at $y = L$ both to the left and right.

The extended real numbers are not usually the context for a real analysis course. Also, the point at infinity is usually reserved for a course in complex analysis, where the entire xy-plane is mapped to a sphere sitting on $(0, 0)$ in a way analogous to what we described here in two dimensions. The north pole of the sphere becomes the point at infinity. Keep in mind when you see the expression $x \to \infty$ in the context of the real numbers that it is generally understood to mean $x \to +\infty$.

5.4.2 Limits of infinity

Now let's replace L with $\pm\infty$ in the expression $\lim_{x \to a} f(x) = L$.

Definition 5.4.5. Suppose f is defined on some deleted neighborhood of $a \in \mathbb{R}$. We say $\lim_{x \to a} f(x) = +\infty$ if for all $M > 0$, there exists $\delta > 0$ such that $0 < |x - a| < \delta$ implies $f(x) > M$.

Definition 5.4.5 is a two-sided definition, so the graph of a function f for which $\lim_{x \to a} f(x) = +\infty$ will have a vertical asymptote at a, both sides of which head upward. Definitions for $\lim_{x \to a} f(x) = -\infty$ and corresponding one-sided limits should be clear, so we won't bother to state them here. However, you will in Exercise 6.

Example 5.4.6. Show $\lim_{x \to 0} 1/x^2 = +\infty$.

Solution: Pick $M > 0$. Let $\delta = 1/\sqrt{M}$. Then if $0 < |x| < \delta$, it follows that

$$f(x) = \frac{1}{x^2} = \frac{1}{|x|^2} > \frac{1}{\delta^2} = M. \tag{5.20}$$

∎

From Definition 5.4.5 and an argument such as that for Theorem 5.3.2, we have the following:

Theorem 5.4.7. *For a function f, $\lim_{x \to a} f(x) = +\infty$ if and only if*

$$\lim_{x \to a^-} f(x) = \lim_{x \to a^+} f(x) = +\infty. \tag{5.21}$$

What other kinds of theorems can we expect for limits of infinity? Can we prove something that resembles Theorem 5.4.4? The answer is "yes," but the theorems will look different because the limits are not necessarily real numbers. Consequently, we'll have to begin with Definition 5.4.5 and do some of the work from scratch. The parts of Theorem 5.4.8 in what follows are ordered so that some of the subsequent ones follow quickly from the earlier ones. This should make your work a little more efficient. Theorems 5.4.8 and 5.4.9 could just as easily be stated and proved in terms of one-sided limits.

Theorem 5.4.8. *Suppose $f(x) \to L$, $g(x) \to +\infty$, and $h(x) \to +\infty$ as $x \to a$. Then as $x \to a$,*

1. $-g(x) \to -\infty$,
2. $1/g(x) \to 0$,
3. $f(x) + g(x) \to +\infty$,
4. $f(x)g(x) \to +\infty$ *if $L > 0$,*
5. $f(x)g(x) \to -\infty$ *if $L < 0$,*
6. *No conclusion can be drawn about $f(x)g(x)$ if $L = 0$,*
7. $f(x)/g(x) \to 0$,
8. $g(x) + h(x) \to +\infty$,
9. $g(x)h(x) \to +\infty$,
10. *No conclusion can be drawn about $g(x) - h(x)$ or $g(x)/h(x)$.*

If $f(x) \to 0$ as $x \to a$, then we can say something about $1/f(x)$ under certain circumstances.

Theorem 5.4.9. *Suppose there exists a deleted δ-neighborhood of a such that $f(x) > 0$ for all $x \in DN_\delta(a)$. Suppose also that $f(x) \to 0$ as $x \to a$. Then $1/f(x) \to +\infty$. Similarly, if $f(x) < 0$ for all $x \in DN_\delta(a)$ and $f(x) \to 0$, then $1/f(x) \to -\infty$.*

In the same way that we use the notation $x \to a^+$ and $x \to a^-$ to mean x approaches a from the right and left, respectively, we can create a shorthand notation for functions that behave as in Theorem 5.4.9. We write $f(x) \to 0^+$ to mean that $f(x)$ approaches zero and is positive on a deleted neighborhood of a, and we say $f(x)$ approaches zero through positive values. Similarly, we write $f(x) \to 0^-$ to mean that $f(x)$ approaches zero and is negative on a deleted neighborhood of a, and we say $f(x)$ approaches zero through negative values. Theorem 5.4.9 could be stated in terms of one-sided limits also.

EXERCISES

1. Use similar triangles and the equation of the circle from Fig. 5.4 to determine the xy-coordinates of the image of $a \in \mathbb{R}$ under the bijection that sends $\mathbb{R} \cup \{\infty\}$ to the circle.

2. Prove Theorem 5.4.7: For a function f, $\lim_{x \to a} f(x) = +\infty$ if and only if
$$\lim_{x \to a^-} f(x) = \lim_{x \to a^+} f(x) = +\infty. \tag{5.22}$$

3. Prove Theorem 5.4.8: Suppose $f(x) \to L$, $g(x) \to +\infty$, and $h(x) \to +\infty$ as $x \to a$. Then as $x \to a$,

   (a) $-g(x) \to -\infty$,
   (b) $1/g(x) \to 0$,
   (c) $f(x) + g(x) \to +\infty$,
   (d) $f(x)g(x) \to +\infty$ if $L > 0$,
   (e) $f(x)g(x) \to -\infty$ if $L < 0$,
   (f) No conclusion can be drawn about $f(x)g(x)$ if $L = 0$,
   (g) $f(x)/g(x) \to 0$,
   (h) $g(x) + h(x) \to +\infty$,
   (i) $g(x)h(x) \to +\infty$,
   (j) No conclusion can be drawn about $g(x) - h(x)$ or $g(x)/h(x)$.

4. Prove Theorem 5.4.9: Suppose there exists a deleted δ-neighborhood of a such that $f(x) > 0$ for all $x \in DN_\delta(a)$. Suppose also that $f(x) \to 0$ as $x \to a$. Then $1/f(x) \to +\infty$. Similarly, if $f(x) < 0$ for all $x \in DN_\delta(a)$ and $f(x) \to 0$, then $1/f(x) \to -\infty$.

5. Let $f(x) = (x^2 - 4)/(x - 1)$. Find with verification $\lim_{x \to 1^-} f(x)$ and $\lim_{x \to 1^+} f(x)$.

6. In the spirit of Definitions 5.4.1 and 5.4.5, create definitions for the following terms involving limits:

   (a) $\lim_{x \to +\infty} f(x) = +\infty$;
   (b) $\lim_{x \to +\infty} f(x) = -\infty$;
   (c) $\lim_{x \to -\infty} f(x) = +\infty$;
   (d) $\lim_{x \to -\infty} f(x) = -\infty$.

5.5 Continuity

The word *continuous* is possibly already a part of your vocabulary of functions. It might be that your calculus class delved into continuity enough to provide an ϵ-δ definition. More than likely your notions of continuity are probably best summarized as a belief that a continuous function can be sketched without lifting your pencil off the paper. The graph is one clean, easily drawable piece. Even though such a view can be helpful in your understanding of some characteristics of continuity, it is far from true that all continuous functions are so easily drawable. The bizarre examples of undrawable continuous functions will come later in your study of mathematics. For now, we define the terms and study the basic results. We begin with continuity at a single point in the domain of f, then consider continuity on a subset of the domain. Then, in the same way that we talk about left-hand and right-hand limits, we'll talk about left continuity and right continuity.

5.5.1 Continuity at a point

If $\lim_{x \to a} f(x)$ exists, it describes the behavior of f near a but not at a. It's possible that $\lim_{x \to a} f(x) = L$, while $f(a)$ might either fail to exist or exist and be different from L. If $\lim_{x \to a} f(x)$ exists and is the value of $f(a)$, we give that phenomenon a name.

Definition 5.5.1. Suppose $f : S \to \mathbb{R}$ and $a \in \text{Int}(S)$. We say that f is *continuous at a* if

$$\lim_{x \to a} f(x) = f(a). \tag{5.23}$$

If f is not continuous at a, we say f is *discontinuous* there, or that f has a *discontinuity* at a.

Let's reword Definition 5.5.1 in the language of ϵ, δ, and neighborhoods. Continuity is a small step logically from limit, for all we do is delete the word *deleted* when we discuss neighborhoods of a. Definition 5.5.1 can be reworded to state that f is continuous at a if

$$(\forall \epsilon > 0)(\exists \delta > 0)(\forall x \in S)(|x - a| < \delta \to |f(x) - f(a)| < \epsilon). \tag{5.24}$$

Compare the logical statement (5.24) to the symbolic form of the definition of limit in statement (5.5). The only significant difference between these logical statements in (5.24) is that $|x - a| < \delta$ replaces $0 < |x - a| < \delta$ in (5.5). And since $|x - a| < \delta$ is *weaker* than $0 < |x - a| < \delta$, continuity is therefore *stronger* than the existence of a limit.

Because of the short step from limit to continuity, our work with limits in Section 5.2 takes us a long way in the theory of continuous functions. If we go back to the theorems from Section 5.2, replacing limits L with $f(a)$ and deleting the word deleted, we arrive immediately at the following theorems:

Theorem 5.5.2. *The constant function $f(x) = c$ is continuous at every $a \in \mathbb{R}$.*

Theorem 5.5.3. *The identity function $i(x) = x$ is continuous at every $a \in \mathbb{R}$.*

Theorem 5.5.4. *If f is continuous at a, then there exists a neighborhood of a where f is bounded.*

Theorem 5.5.5. *If f and g are both continuous at a, then so are $f + g$, fg, and $f - g$.*

Theorem 5.5.6. *Suppose g is continuous at a and $g(a) \neq 0$, then there exists a neighborhood of a where g can be bounded away from zero. In particular, if $g(a) > 0$, there exists $M > 0$ and $\delta > 0$ such that $|x - a| < \delta$ implies $g(x) > M$. If $g(a) < 0$, there exists $M < 0$ and $\delta > 0$ such that $|x - a| < \delta$ implies $g(x) < M$.*

Theorem 5.5.7. *If f and g are both continuous at a, and if $g(a) \neq 0$, then f/g is continuous at a.*

Corollary 5.5.8. *A polynomial function is continuous at every $a \in \mathbb{R}$. Every rational function $f(x) = P_1(x)/P_2(x)$, where P_1 and P_2 are polynomials, is continuous at every $a \in \mathbb{R}$ for which $P_2(a) \neq 0$.*

5.5 Continuity

Theorem 5.5.9. *A function $f : S \to \mathbb{R}$ is continuous at $a \in S$ if and only if every sequence $\langle a_n \rangle$ such that $a_n \to a$ satisfies $f(a_n) \to f(a)$.*

Notice that the hypothesis condition of Theorem 5.5.9 does not require $a_n \neq a$ as does Theorem 5.3.4. For if any of the $a_n = a$, then $f(a_n) = f(a)$, so that the sequence $\langle f(a_n) \rangle$ is defined for all n.

Here's an important theorem we could not address with limits alone. You'll prove it in Exercise 1.

Theorem 5.5.10. *Suppose f is continuous at a and g is continuous at $f(a)$. Then $g \circ f$ is continuous at a.*

To prove Theorem 5.5.10, an arbitrarily chosen ϵ-neighborhood of $g[f(a)]$ produces a δ_1-neighborhood of $f(a)$, and this δ_1-neighborhood of $f(a)$ produces a δ_2-neighborhood of a. Another way to write what Theorem 5.5.10 states is

$$\lim_{x \to a} g[f(x)] = g\left[\lim_{x \to a} f(x)\right] = g\left[f(\lim_{x \to a} x)\right] = g[f(a)]. \tag{5.25}$$

With a little more mathematical machinery than we have up to now, Theorem 5.5.10 comes in handy in certain manipulations of limits. If we assume for the moment that all the functions in the following are continuous at the points involved, Theorem 5.5.10 would allow us to write something like

$$\begin{aligned}
\lim_{x \to 3} \sqrt{1 + e^{-x^2}} &= \sqrt{\lim_{x \to 3}(1 + e^{-x^2})} \\
&= \sqrt{1 + \lim_{x \to 3} e^{-x^2}} \\
&= \sqrt{1 + e^{\lim_{x \to 3}(-x^2)}} \\
&= \sqrt{1 + e^{-9}}.
\end{aligned} \tag{5.26}$$

The reason we could not have a theorem in Section 5.2 that said something like "If $f(x) \to L_1$ as $x \to a$ and $g(x) \to L_2$ as $x \to L_1$, then $(g \circ f)(x) \to L_2$ as $x \to a$" is that a possible hole in the domain of g at $f(a)$ might cause troublesome gaps in the domain of $g \circ f$ around a. For example, let $f(x) = 0$ and $g(x) = \sin x / x$. Then $g \circ f$ is not defined at any $x \in \mathbb{R}$.

If f is discontinuous at a it could be for one or more of three basic reasons. If Eq. (5.23) is not satisfied it might be that:

(D1) $\lim_{x \to a} f(x)$ does not exist;

(D2) $f(a)$ does not exist; or

(D3) $\lim_{x \to a} f(x)$ and $f(a)$ both exist, but are not equal.

Figure 5.5 illustrates some of these possibilities at $a = 1, 2, 3, 4, 5$. At $a = 1$, f behaves much like $\sin(1/x)$ near $x = 0$. (See Exercise 11 from Section 5.2.) In Exercise 2, you'll be asked to state which of the possibilities D1–D3 explains the discontinuities of f. The discontinuities at 2 and 4 are called *removable* discontinuities because it is possible to define or redefine the value of f there to make it continuous there. The discontinuity at 5 is called a *jump discontinuity*.

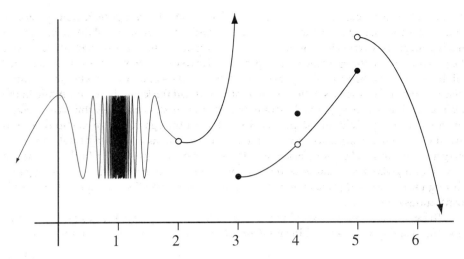

Figure 5.5 Some examples of discontinuities.

Example 5.5.11. The function $f(x) = \sin x / x$ has a removable discontinuity at zero. Thus the function

$$g(x) = \begin{cases} \dfrac{\sin x}{x}, & \text{if } x \neq 0; \\ 1, & \text{if } x = 0; \end{cases} \tag{5.27}$$

is continuous at $a = 0$ because $\lim_{x \to 0} f(x) = 1 = f(0)$.

5.5.2 Continuity on a set

The definition of continuity of $f : S \to \mathbb{R}$ on a subset $A \subseteq S$ is fairly straightforward if A is open, for then we are guaranteed that every $a \in A$ is contained in a neighborhood entirely within the domain of f.

Definition 5.5.12. If $f : S \to \mathbb{R}$ is a function and $A \subseteq S$ is open, we say f is *continuous on A* if it is continuous at every $a \in A$. Logically, we may write this as

$$(\forall a \in A)(\forall \epsilon > 0)(\exists \delta > 0)(\forall x \in A)(|x - a| < \delta \to |f(x) - f(a)| < \epsilon). \tag{5.28}$$

Notice the only difference between the logical statement (5.24) and (5.28) is that the latter begins with $(\forall a \in A)$. To construct a proof that f is continuous on a set A by going back to ϵ and δ, you'll begin by picking both $a \in A$ and $\epsilon > 0$. We want to do precisely this in an extended example now. We want to show that $f(x) = 1/x$ is continuous on $(0, 1)$. Yes, it can be said that continuity of $f(x) = 1/x$ on $(0, 1)$ follows from Theorem 5.5.8. However, we go back to an ϵ-δ proof as a valuable learning experience. There are plenty of functions whose continuity we have not addressed. As you work your way through the details of this demonstration, you're going to find it to be full of algebraic manipulations and observations, none of which is particularly difficult to understand. However, you're likely going to be wondering how anyone would ever know these are the right steps to take. Be patient—these things come with experience. However, do make sure you understand all the steps along the way. We'll present the details in the same order in which you would discover them if you were starting from scratch and had the know-how to stumble in the right direction. We're trying to illustrate three points. First, sometimes you have to stand on your head, especially in ϵ-δ proofs. Second, even though you have to choose $\epsilon > 0$ arbitrarily, there are times when you can make convenient assumptions about ϵ that lose no generality and make your work easier. Third, and most importantly, finding a suitable $\delta > 0$ for a given $a \in A$ and $\epsilon > 0$ will depend on the values of both a and ϵ. Digging through this work will prepare you for Exercise 3, where you'll show that $f(x) = \sqrt{x}$ is continuous on $(0, \infty)$.

To begin the construction of a proof that $f(x) = 1/x$ is continuous on $(0, 1)$, pick $a \in (0, 1)$ and $\epsilon > 0$. As usual in an ϵ-δ proof, we must find $\delta > 0$ so that the inequality

$$|x - a| < \delta \tag{5.29}$$

will be at least as strong an inequality as

$$\left| \frac{1}{x} - \frac{1}{a} \right| < \epsilon. \tag{5.30}$$

Thus we work backwards from Eq. (5.30), trying to transform it into something of the form of inequality (5.29), making sure that the steps we take do not produce inequalities that become any weaker. Unfolding Eq. (5.30) and trying to convert it to something of the form $a - \delta < x < a + \delta$, we have

$$-\epsilon < \frac{1}{x} - \frac{1}{a} < \epsilon$$

$$\frac{1}{a} - \epsilon < \frac{1}{x} < \frac{1}{a} + \epsilon. \tag{5.31}$$

At this point, we're tempted simply to take the reciprocal of everything in Eq. (5.31) to have an x in the middle, but we have to be careful about the signs. In Exercise 11 of Section 2.3, you showed the following:

$$0 < \alpha < \beta \rightarrow \frac{1}{\alpha} > \frac{1}{\beta}, \tag{5.32}$$

$$\alpha < 0 < \beta \rightarrow \frac{1}{\alpha} < \frac{1}{\beta},$$

$$\alpha < \beta < 0 \rightarrow \frac{1}{\alpha} > \frac{1}{\beta}.$$

Looking at all the terms in Eq. (5.31), we're sure that $1/a + \epsilon > 0$. Also, the unknown x will be positive because $a > 0$, and we'll make sure that our δ-neighborhood of a is small enough to include only positive numbers. The problem is $1/a - \epsilon$, which could be positive or negative, depending on the size of ϵ. In fact, $1/a - \epsilon > 0$ if and only if $\epsilon < 1/a$. This might seem troublesome at first, but in actuality we can ignore the problem and assume $1/a - \epsilon > 0$. Here's why.

If someone presents us with an $\epsilon > 0$ to serve as the radius of a target on the y-axis that we must hit from the x-axis, then we are certainly free to aim for a target of smaller radius. If we make the statement "Pick $\epsilon > 0$," and discover that values of ϵ in excess of $1/a$ make things inconvenient, then we could freely replace ϵ with some ϵ_1 that satisfies $\epsilon_1 = \min\{\epsilon, 1/a\}$. Proceeding from there to find δ, we would be safe because

$$\left|\frac{1}{x} - \frac{1}{a}\right| < \epsilon_1 \rightarrow \left|\frac{1}{x} - \frac{1}{a}\right| < \epsilon. \tag{5.33}$$

Thus, finding a $\delta > 0$ that guarantees $|1/x - 1/a| < \epsilon_1$ will also guarantee that Eq. (5.30) is true. This all boils down to the fact that we're free to write: "We may assume without loss of generality that $\epsilon < 1/a$." The rest of the work is then easier.

Continuing from Eq. (5.31) and applying Eq. (5.32), we have

$$\frac{a}{1+\epsilon a} < x < \frac{a}{1-\epsilon a}. \tag{5.34}$$

If we're trying to transform this into something of the form $a - \delta < x < a + \delta$, we still have some work to do because Eq. (5.34) does not suggest a value of δ in its present form. This is where we do some sneaky algebra involving Eq. (2.67) from Section 2.4, Exercise 10. This is just one of many places where Eq. (2.67) demonstrates its usefulness. If we rearrange Eq. (2.67) and write it two ways, once letting $x = y$ and once letting $x = -y$, we have

$$1 + y + y^2 + \cdots + y^n = \frac{1 - y^{n+1}}{1 - y} \tag{5.35}$$

and

$$1 - y + y^2 - \cdots + (-y)^n = \frac{1 - (-y)^{n+1}}{1 + y}. \tag{5.36}$$

If we know that $0 < y < 1$, then letting $n = 1$ in Eq. (5.35) yields

$$1 + y = \frac{1 - y^2}{1 - y} < \frac{1}{1 - y}. \tag{5.37}$$

Letting $n = 2$ in Eq. (5.36), we have

$$1 - y(1 - y) = 1 - y + y^2 = \frac{1 + y^3}{1 + y} > \frac{1}{1 + y}. \tag{5.38}$$

Since $0 < \epsilon < 1/a$, we have $0 < \epsilon a < 1$, so that Eqs. (5.37) and (5.38) both hold for $y = \epsilon a$. Thus

$$\frac{1}{1 + \epsilon a} < 1 - \epsilon a(1 - \epsilon a) \quad \text{and} \quad 1 + \epsilon a < \frac{1}{1 - \epsilon a}. \tag{5.39}$$

Now we're trying to guarantee that Eq. (5.34) is true, for then Eq. (5.30) will be true. If we take both inequalities from Eq. (5.39) and multiply them through by $a > 0$, we have the following:

$$\frac{a}{1 + \epsilon a} < a - \epsilon a^2(1 - \epsilon a) \quad \text{and} \quad a + \epsilon a^2 \leq \frac{a}{1 - \epsilon a}. \tag{5.40}$$

Thus, if
$$a - \epsilon a^2(1 - \epsilon a) < x < a + \epsilon a^2, \tag{5.41}$$
then Eq. (5.34) will also be true. We're almost there. Inequality (5.41) does not quite look like something of the form $a - \delta < x < a + \delta$. It looks more like something of the form $a - \delta_1 < x < a + \delta_2$. However, since $a > 0$, $\epsilon > 0$, and $\epsilon a < 1$, then $0 < 1 - \epsilon a < 1$. Thus, $\epsilon a^2(1 - \epsilon a)$ is a smaller positive number than ϵa^2 so that the left-hand side of inequality (5.41) is closer to x than the right-hand side. Thus, we should let $\delta = \epsilon a^2(1 - \epsilon a)$, and $|x - a| < \delta$ will imply Eq. (5.41), which will imply Eq. (5.34), which will imply Eq. (5.30). This is the heart of all the scratchwork, and we can then present all this in the form of a cleaned-up proof. Because of all the effort we exerted in working through this, let's call it a theorem.

Theorem 5.5.13. *The function defined by $f(x) = 1/x$ is continuous on $(0, 1)$.*

Proof: Pick $a \in (0, 1)$ and $\epsilon > 0$. Without loss of generality, we may assume $\epsilon < 1/a$. Let $\delta = \epsilon a^2(1 - \epsilon a)$, which is clearly positive. Then if $|x - a| < \delta$, we have
$$a - \epsilon a^2(1 - \epsilon a) < a - \delta < x < a + \delta < a + \epsilon a^2(1 - \epsilon a). \tag{5.42}$$

Because
$$\frac{a}{1 + \epsilon a} < a(1 - \epsilon a + \epsilon^2 a^2) = a - \epsilon a^2(1 - \epsilon a) \tag{5.43}$$

and
$$a + \epsilon a^2(1 - \epsilon a) < a + \epsilon a^2 = a(1 + \epsilon a) < \frac{a}{1 - \epsilon a} \tag{5.44}$$

we have
$$\frac{a}{1 + \epsilon a} < x < \frac{a}{1 - \epsilon a}. \tag{5.45}$$

Since every term in Eq. (5.45) is positive, we may reciprocate to have
$$\frac{1}{a} + \epsilon = \frac{1 + \epsilon a}{a} > \frac{1}{x} > \frac{1 - \epsilon a}{a} = \frac{1}{a} - \epsilon, \tag{5.46}$$

or
$$|f(x) - f(a)| < \epsilon. \tag{5.47}$$

Thus, f is continuous at a. ∎

Notice a very important fact. The value of δ we proposed does depend on both ϵ and a. In fact, the closer a is to zero, the smaller must be our value of δ. On the basis of our work, it appears there is no single value of δ that would depend only on ϵ and work for all $a \in (0, 1)$. It might occur to you, however, that someone else might have taken a different approach and stumbled across a value of δ that was not tied to the value of a but depended only on ϵ. The question is an important one and will be answered when we study uniform continuity in Section 5.7. Suffice it to say for now that no such δ as a function of ϵ alone exists for $f(x) = 1/x$ on $(0, 1)$. Intuitively, it might seem plausible when we consider the vertical asymptote $1/x$ has as $x \to 0^+$. Look at Fig. 5.6 and imagine setting an ϵ-tolerance around a point $1/a$ on the y-axis. If a is very close

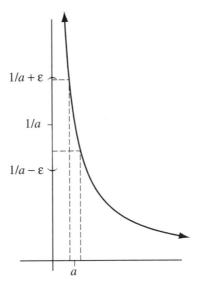

Figure 5.6 $f(x) = 1/x$.

to zero, the graph of f is very steep around $(a, 1/a)$ and varies considerably even in a small neighborhood of a. Granted, for any a there is a δ-neighborhood of a that maps into $N_\epsilon[f(a)]$, but if you imagine smaller and smaller values of a, then the asymptote of the graph necessitates smaller and smaller values of δ. No single δ-value will work for all $a \in (0, 1)$ because of this asymptote.

The scratchwork you'll go through in Exercise 3 to show $\sqrt{x}$ is continuous on $(0, \infty)$ is not nearly as involved as it has been for our example here, but you will learn a bit more about using the inequality $f(a) - \epsilon < f(x) < f(a) + \epsilon$ and working backwards to find an inequality of the form $a - \delta < x < a + \delta$ that is at least as strong. A peculiar thing for $\sqrt{x}$ on $(0, \infty)$, however, is that it actually is possible to find a δ that is independent of the value of a. Although it is true that the graph of $\sqrt{x}$ becomes very steep as $x \to 0^+$, it does not behave asymptotically there. Why such a δ can be found will become clear in Section 5.7.

5.5.3 One-sided continuity

Even if $\lim_{x \to a} f(x)$ does not exist, there might still be a hope that either $\lim_{x \to a^-} f(x)$ or $\lim_{x \to a^+} f(x)$ exists. Similarly, although f might not be continuous at a, we might have continuity from one side or the other, as it were.

Definition 5.5.14. A function f is said to be *left continuous* at a if

$$\lim_{x \to a^-} f(x) = f(a). \tag{5.48}$$

Similarly, f is said to be *right continuous* at a if

$$\lim_{x \to a^+} f(x) = f(a). \tag{5.49}$$

An immediate consequence of Theorem 5.3.2 is the following.

Theorem 5.5.15. *A function f is continuous at a if and only if it is both left and right continuous at a.*

In defining continuity on A, Definition 5.5.12 stipulated that A must be open. This allowed for a definition of continuity on (a, b), but not on $[a, b]$. One-sided continuity now allows us to define continuity on $[a, b]$ in such a way that we do not concern ourselves with how f behaves or whether it even exists for $x < a$ or $x > b$.

Definition 5.5.16. A function $f : S \to \mathbb{R}$ is said to be continuous on $[a, b] \subseteq S$ if the following hold:

1. f is continuous on (a, b);
2. f is right continuous at a; and
3. f is left continuous at b.

Definition 5.5.16 can be naturally adapted to apply to the following example by omitting stipulation 3.

Example 5.5.17. $f(x) = \sqrt{x}$ is continuous on $[0, +\infty)$. From Exercise 3, f is continuous at all $a \in (0, \infty)$. From Exercise 2 in Section 5.3, $\lim_{x \to 0^+} f(x) = f(0)$.

EXERCISES

1. Prove Theorem 5.5.10: Suppose f is continuous at a and g is continuous at $f(a)$. Then $g \circ f$ is continuous at a.

2. For f sketched in Fig. 5.5, state which of the characteristics D1–D3 describes the discontinuity of f at $x = 1, 2, 3, 4, 5$.

3. Use an ϵ-δ proof to show that $f(x) = \sqrt{x}$ is continuous at any $a > 0$.

5.6 Implications of Continuity

Having defined and illustrated continuity in Section 5.5, let's look at some of its implications.

5.6.1 The intermediate value theorem

The imagery of a continuous function being drawable without picking up the pencil makes this first theorem seem plausible. It says, in short, that a continuous function cannot be negative at one point and positive somewhere else without crossing the x-axis somewhere in between the two points.

Theorem 5.6.1. *Suppose $a < b$, $f(a) < 0 < f(b)$, and f is continuous on $[a, b]$. Then there exists $c \in (a, b)$ such that $f(c) = 0$.*

The proof of Theorem 5.6.1 (which you'll supply in Exercise 1) is yet another nice application of the LUB property of $\mathbb{R}$. There will be a hint if you need it. Suffice it to say for now that if you imagine yourself standing on the x-axis at $x = a$ and looking up the x-axis, there's a natural subset of $\mathbb{R}$ that is both nonempty and bounded from above by b to which the LUB property can apply. The LUB of this set is the number you need. If g is a continuous function such that $g(a) > 0 > g(b)$ for some $a < b$, then Theorem 5.6.1 can be applied to $-g$ to produce some $c \in (a, b)$ such that $g(c) = 0$.

Theorem 5.6.1 provides most of the machinery needed to prove a theorem whose name you might remember from calculus—the Intermediate Value Theorem. We'll supply the proof here to illustrate a rather common convenience in mathematics. If you can derive some result about a real-valued function g involving zero, then you might be able to derive a similar, more general result about other functions by altering them to fit the hypothesis conditions of the theorem applied to g.

Theorem 5.6.2 (Intermediate Value Theorem [IVT]). *Suppose f is continuous on $[a, b]$, and suppose $f(a) < f(b)$. Let $y_0 \in \mathbb{R}$ be any number satisfying $f(a) < y_0 < f(b)$. Then there exists $c \in (a, b)$ such that $f(c) = y_0$.*

In the IVT, f does not necessarily change from negative to positive as x runs from a to b. However, if we create a new function g by dropping or raising f to make $g(a) < 0 < g(b)$, then we can apply Theorem 5.6.1 to g and see what it says about f. Here is the proof.

Proof: Define $g(x) = f(x) - y_0$. Since f and the constant function y_0 are both continuous on $[a, b]$, Theorem 5.5.5 and our work with one-sided continuity imply g is continuous on $[a, b]$. Furthermore, $g(a) = f(a) - y_0 < 0$ and $g(b) = f(b) - y_0 > 0$. By Theorem 5.6.1, there exists $c \in (a, b)$ such that $g(c) = 0$. Thus $f(c) = y_0$. ∎

Theorem 5.1.4 with its Corollary 5.1.5 demonstrated that strict monotonicity of a function on a set implies one-to-oneness (hence invertibility) there. The converse of Corollary 5.1.5 is clearly not true, as is illustrated by $f(x) = 1/x$ on $\mathbb{R} \setminus \{0\}$, which is one-to-one but not monotone. However, if f is one-to-one and continuous on A, then it must be strictly monotone. You will prove the following theorem in Exercise 4. The IVT will come in handy.

Theorem 5.6.3. *If $f : S \to \mathbb{R}$ is continuous and one-to-one on $[a, b] \subseteq S$, then f is strictly monotone on A.*

The IVT can also help us prove that a continuous, invertible function has a continuous inverse. Specifically, if f is continuous and invertible on $[a, b]$, then it is continuous and one-to-one. By Theorem 5.6.3, f is strictly monotone on $[a, b]$ also. By picking $d \in f([a, b])$, writing $d = f(c)$, and choosing $\epsilon > 0$, you should be able to find a δ-neighborhood of d such that $d - \delta < x < d + \delta$ implies $c - \epsilon < f^{-1}(x) < c + \epsilon$. (Draw a picture!) You'll do precisely this in Exercise 5. Here's the statement of the theorem.

Theorem 5.6.4. *Suppose $f : S \to \mathbb{R}$ is continuous and invertible on $[a, b] \subseteq S$. Then $f^{-1} : f([a, b]) \to [a, b]$ is continuous on $f([a, b])$.*

5.6.2 Continuity and open sets

The logical equivalence of the ϵ-δ form of continuity and the sequential limit form conveyed by Theorem 5.5.9 is useful not only because it gives us freedom to exchange one property for another, but also because it suggests an alternative way to define continuity that might be preferred in the creation of some mathematical structures. Theorem 5.6.5 provides another statement that is logically equivalent to continuity, this time in terms of the pre-images of open subsets of $\mathbb{R}$. This result does not immediately have the intuitive appeal of the ϵ-δ definition in terms of the graph being drawable without picking up the pencil. This is arguably a good thing, for that feature of drawability is not really valid.

In the theorems that follow, we're going to assume the functions involved are defined on all of $\mathbb{R}$. This will keep the proofs relatively simple and get the point across, though similar theorems can be addressed on restricted domains.

Theorem 5.6.5. *A function $f : \mathbb{R} \to \mathbb{R}$ is continuous if and only if the pre-image of every open set is open.*

When you prove Theorem 5.6.5 in Exercise 6, you can write a very elegant proof if you'll use a slightly different form of the definition of continuity than that in Eq. (5.24). The statement $|x - a| < \delta \to |f(x) - f(a)| < \epsilon$ is equivalent to $x \in N_\delta(a) \to f(x) \in N_\epsilon[f(a)]$, or if you prefer

$$f[N_\delta(a)] \subseteq N_\epsilon[f(a)]. \tag{5.50}$$

If you will use this form of the definition of continuity along with the standard definition of open set, you'll find that Exercises 6 and 7 from Section 3.2 make for some interesting manipulations of the sets involved in the proof. From Theorem 5.6.5, the following should drop right into your lap (Exercise 7).

Corollary 5.6.6. *A function $f : \mathbb{R} \to \mathbb{R}$ is continuous if and only if the pre-image of every closed set is closed.*

Let's digress just for a moment to comment on the significance of Theorem 5.6.5. First, the logical equivalence of the two statements in Theorem 5.6.5 suggests yet another place where one might begin in defining continuity of a function. If one were to begin by defining a function to be continuous provided the pre-image of every open set is open, then our ϵ-δ definition and the sequential limit property of Theorem 5.3.4 would become theorems in the analysis of real numbers. If it seems a little unnatural to begin with such an open set pre-image definition, consider the following, which is a glimpse into topology.

At the beginning of Chapter 4, we said that a defining characteristic of analysis is that elements of a set have either a measure of size (norm) or of distance between them (metric). In $\mathbb{R}$, the measure most commonly used is absolute value, so that $|x|$ is the size of $x \in \mathbb{R}$ and $|a - b|$ is the distance between. If S has a metric, then a neighborhood of a is defined as all points within a certain distance of a. Then a definition of open set as Definition 4.2.2 becomes meaningful because openness is defined in terms of neighborhoods.

In topology there is no such notion of size or distance either defined or assumed on S. Instead, we begin with the idea of open set in what might seem a peculiar way. Given S, we declare some subsets to be open just by definition. If you want to declare every subset of S open, fine; however, that might not prove to be especially interesting. More than likely you'll want some subsets of S to be open and some not to be open. What prevents your freedom to define open sets however you want from degenerating into mathematically useless anarchy is that your collection of open sets needs to have some of the same properties of open sets we derived as theorems for $\mathbb{R}$. Specifically, if we lump all the subsets of S that we declare to be open into the family $\mathcal{O}$, then in forming a topology we insist that:

(T1) both $\emptyset$ and S are members of $\mathcal{O}$;

(T2) if $\mathcal{F} \subseteq \mathcal{O}$ is a family of open sets, then $\bigcup_{A \in \mathcal{F}} A$ is open. That is, $\mathcal{O}$ is closed under union; and

(T3) if $\{A_k : 1 \leq k \leq n\}$ is a finite collection of open sets, then $\bigcap_{k=1}^{n} A_k$ is open. That is, $\mathcal{O}$ is closed under finite intersection.

Closed sets are then defined to be those whose complement is open, and you're on your way to building a structure that will have some parallels to those of $\mathbb{R}$ we've studied, but will be more abstract and austere because the definition of open set does not probe as deeply into some assumed structure of S.

Theorem 5.6.5 will help you prove the following in Exercise 8.

Theorem 5.6.7. *The continuous image of a compact set is compact. That is, if $f : \mathbb{R} \to \mathbb{R}$ is continuous on a compact $S \subseteq \mathbb{R}$, then $f(S)$ is compact.*

In Exercise 2 from Section 4.2, you showed that if L is the LUB of a set S and $L \notin S$, then L is a limit point of S. Since a compact set is bounded, it has an LUB, and since it's also closed, it contains all its limit points. Thus a compact set contains its LUB. Similarly, a compact set also contains its GLB. If $f : \mathbb{R} \to \mathbb{R}$ is continuous and $S \subseteq \mathbb{R}$ is compact, then $f(S)$ is compact by Theorem 5.6.7. Consequently, $f(S)$ contains its LUB and GLB. Write $M = \max[f(S)]$ and $m = \min[f(S)]$. Then there exist $x_1, x_2 \in S$ such that $f(x_1) = m$ and $f(x_2) = M$. We have just proved the following theorem, which you might remember from your calculus class.

Theorem 5.6.8 (Extreme Value Theorem [EVT]). *If $f : \mathbb{R} \to \mathbb{R}$ is continuous on a compact set S, then f attains a maximum and minimum value on S.*

EXERCISES

1. Prove Theorem 5.6.1: If $a < b$, $f(a) < 0 < f(b)$, and f is continuous on $[a, b]$, then there exists $c \in (a, b)$ such that $f(c) = 0$.[2]

---

[2] Let $A = \{x \in [a, b] : f(x) < 0\}$. What can you say about A?

2. Prove the *fixed point theorem*: If $f : [a, b] \to [a, b]$ is continuous, then there exists $c \in [a, b]$ such that $f(c) = c$.[3]

3. The *fundamental theorem of algebra* states that every polynomial $P : \mathbb{R} \to \mathbb{R}$ whose degree is odd has a root in $\mathbb{R}$. That is, for every polynomial $P(x) = a_{2n+1}x^{2n+1} + a_{2n}x^{2n} + \cdots + a_1 x + a_0$, there exists some $c \in \mathbb{R}$ such that $P(c) = 0$. In this exercise you prove the fundamental theorem of algebra, first for the case $a_{2n+1} > 0$, then for the case $a_{2n+1} < 0$.

    (a) Use your definitions from Exercise 6 in Section 5.4 to prove
    $$\lim_{x \to +\infty} x = +\infty \quad \text{and} \quad \lim_{x \to -\infty} x = -\infty. \qquad (5.51)$$

    (b) With part (a) in hand, a theorem analogous to Theorem 5.4.8 would be demonstrable for $x \to \pm\infty$ by paralleling its proof. Assuming this result, show that if $a_{2n+1} > 0$, then $P(x)$ satisfies
    $$\lim_{x \to -\infty} P(x) = -\infty \quad \text{and} \quad \lim_{x \to +\infty} P(x) = +\infty. \qquad (5.52)$$

    (c) Use your result from part (b) to prove the fundamental theorem of algebra for the case $a_{2n+1} > 0$.

    (d) Prove the fundamental theorem of algebra for the case $a_{2n+1} < 0$ by applying part (c) to $-P(x)$.

4. Prove Theorem 5.6.3: Suppose $f : S \to \mathbb{R}$ is continuous and one-to-one on $[a, b] \subseteq S$. Then f is strictly monotone on $[a, b]$.[4,5]

5. Prove Theorem 5.6.4: Suppose $f : S \to \mathbb{R}$ is continuous and invertible on $[a, b] \subseteq S$. Then $f^{-1} : f([a, b]) \to [a, b]$ is continuous on $f([a, b])$.

6. Prove Theorem 5.6.5: A function $f : \mathbb{R} \to \mathbb{R}$ is continuous if and only if the pre-image of every open set is open.

7. Prove Corollary 5.6.6: A function $f : \mathbb{R} \to \mathbb{R}$ is continuous if and only if the pre-image of every closed set is closed.[6]

8. Prove Theorem 5.6.7: The continuous image of a compact set is compact. That is, if $f : \mathbb{R} \to \mathbb{R}$ is continuous on a compact $S \subseteq \mathbb{R}$, then $f(S)$ is compact.

9. Give two examples to illustrate that compactness is necessary for the EVT to apply to a continuous function, one example where S is bounded but not closed, and one where S is closed but not bounded.

---

[3] Consider the function $g(x) = f(x) - x$. If either $g(a)$ or $g(b)$ is zero, you're done. Otherwise, apply the IVT.
[4] See Exercise 3h from Section 1.2.
[5] Follow your nose. For f not monotone and $f(a) < f(b)$, spend some time showing that there exist $c_1 < c_2 < c_3$ such that $f(c_1) < f(c_2) > f(c_3)$ or $f(c_1) > f(c_2) < f(c_3)$.
[6] See Exercise 8 from Section 3.2.

5.7 Uniform Continuity

We've mentioned earlier in this text the distinction between every question having an answer and the existence of a single answer that applies to every question. In Definition 5.5.12 we wrote the definition of continuity on a set A logically as

$$(\forall a \in A)(\forall \epsilon > 0)(\exists \delta > 0)(\forall x \in A)(|x - a| < \delta \to |f(x) - f(a)| < \epsilon). \quad (5.53)$$

In general, the value of δ will depend on both ϵ and a. In this section, we want to define a form of continuity where δ depends only on ϵ. We'll look at some examples to illustrate the point, and then study two important theorems.

5.7.1 Definition and examples

Let's take the definition of continuity in Eq. (5.53) and leapfrog the first component piece $(\forall a \in A)$ two jumps to the right:

$$(\forall \epsilon > 0)(\exists \delta > 0)(\forall a \in A)(\forall x \in A)(|x - a| < \delta \to |f(x) - f(a)| < \epsilon). \quad (5.54)$$

Taking $(\forall a \in A)$ and moving it to the right of $(\forall \epsilon > 0)$ does not change anything logically from statement (5.53). However, moving $(\forall a \in A)$ to the right of $(\exists \delta > 0)$ makes a big difference. If you were writing a proof of continuity from (5.53), you would begin by picking $a \in A$ and $\epsilon > 0$. Then, with both a and ϵ in hand, you would go hunting for $\delta > 0$ with certain required properties. However, with the phrase $(\forall a \in A)$ repositioned as in (5.54), things are different. If you were writing a proof of a theorem where (5.54) was involved, you would begin by picking only $\epsilon > 0$. Then, without knowing any particular value of $a \in A$, you would have to find $\delta > 0$ from ϵ alone, and this δ would have to serve for all $a \in A$. Thus, (5.54) suggests that the value of δ can be found after having specified only ϵ, and this δ-value will work for all $a, x \in A$ satisfying $|x - a| < \delta$. Because (5.53) specifies $a \in A$ first, we think of a as being the center of a δ-neighborhood and x being an arbitrary point in that neighborhood. In (5.55), there is really no reason to call one point a and the other one x, as if one were fixed before the other. The point is that there exists $\delta > 0$ such that if any two points are within δ of each other, then their functional values are within ϵ of each other. For clarity and convenience, we change the symbols slightly in the following definition.

Definition 5.7.1. A function $f : S \to \mathbb{R}$ is said to be *uniformly continuous* on $A \subseteq S$ if for all $\epsilon > 0$ there exists $\delta > 0$ such that, for all $x, y \in A$, $|x - y| < \delta$ implies $|f(x) - f(y)| < \epsilon$. Logically, we may write this as

$$(\forall \epsilon > 0)(\exists \delta > 0)(\forall x, y \in A)(|x - y| < \delta \to |f(x) - f(y)| < \epsilon). \quad (5.55)$$

To show that a function f is uniformly continuous on A, we work backwards from $|f(x) - f(y)| < \epsilon$. Watch what happens in the following example.

Example 5.7.2. Show that $f(x) = x^2/(x + 1)$ is uniformly continuous on $[0, \infty)$.

Solution: Before we write a demonstration, we need to do some scratchwork. If you play with the expression $|f(x) - f(y)| < \epsilon$ and try to factor $|x - y|$ out of it, you can arrive at the following:

$$|f(x) - f(y)| = \left|\frac{x^2}{x+1} - \frac{y^2}{y+1}\right| = |x-y|\left|\frac{xy+x+y}{(x+1)(y+1)}\right|. \quad (5.56)$$

Next, let's split up the fraction in the right-hand side of Eq. (5.56) and apply the triangle inequality. Also, notice that $x, y > 0$ implies $1/(x+1), 1/(y+1), x/(x+1)$, and $y/(y+1)$ are all less than 1. So we have

$$\left|\frac{xy+x+y}{(x+1)(y+1)}\right| \leq \left|\frac{xy}{(x+1)(y+1)}\right| + \left|\frac{x}{(x+1)(y+1)}\right| + \left|\frac{y}{(x+1)(y+1)}\right|$$

$$= \left|\frac{x}{(x+1)}\right|\left|\frac{y}{(y+1)}\right| + \left|\frac{x}{(x+1)}\right|\left|\frac{1}{(y+1)}\right| \quad (5.57)$$

$$+ \left|\frac{y}{(y+1)}\right|\left|\frac{1}{(x+1)}\right| \leq 3.$$

Having arrived at $|f(x) - f(y)| \leq 3|x - y|$, we're ready to write a demonstration. Choose $\epsilon > 0$, and let $\delta = \epsilon/3$. Then if $x, y \geq 0$ and $|x - y| < \delta$, we have

$$|f(x) - f(y)| = \left|\frac{x^2}{x+1} - \frac{y^2}{y+1}\right| = |x-y|\left|\frac{xy+x+y}{(x+1)(y+1)}\right|$$

$$\leq |x-y|\left(\left|\frac{x}{(x+1)}\right|\left|\frac{y}{(y+1)}\right|\right. \quad (5.58)$$

$$\left. + \left|\frac{x}{(x+1)}\right|\left|\frac{1}{(y+1)}\right| + \left|\frac{y}{(y+1)}\right|\left|\frac{1}{(x+1)}\right|\right)$$

$$\leq 3|x-y| < 3\delta = \epsilon. \quad \blacksquare$$

The scratchwork of Example 5.7.2 suggests a theorem that you'll prove in Exercise 1.

Theorem 5.7.3. *If there exists $m > 0$ such that $|f(x) - f(y)| \leq m|x - y|$ for all $x, y \in A$, then f is uniformly continuous on A.*

If $y \neq x$, the hypothesis condition of Theorem 5.7.3 is equivalent to

$$\left|\frac{f(x) - f(y)}{x - y}\right| \leq m, \quad (5.59)$$

which means that f has a bound on the slopes of lines through any two points on its graph. Loosely speaking, if the steepness of f (as measured by slopes of secant lines) is bounded, then f is uniformly continuous. The converse of Theorem 5.7.3 is not true. Shortly, we'll point out a function f that is uniformly continuous on a set A, but for which inequality (5.59) does not hold across A for any $m > 0$.

Naturally, we want to include a demonstration that a continuous function need not be uniformly continuous on a set. What does it mean for f not to be uniformly continuous on A?

$$(\exists \epsilon > 0)(\forall \delta > 0)(\exists x, y \in A)(|x - y| < \delta \text{ and } |f(x) - f(y)| \geq \epsilon). \quad (5.60)$$

That is, there exists some $\epsilon > 0$ so that, no matter how small $\delta > 0$ might be, there will be two points somewhere in the set A that are within δ distance of each other, but whose functional values are at least ϵ apart.

Example 5.7.4. Show $f(x) = 1/x$ is not uniformly continuous on $(0, 1)$.

Solution: First some scratchwork. It's the vertical asymptote at $x = 0$ that provides us with x and y values that are very close together and whose functional values can differ as much as a strategic ϵ we'll choose. We have to play with the inequality

$$\left| \frac{1}{x} - \frac{1}{y} \right| = \frac{|y - x|}{|x||y|} \geq \epsilon, \quad (5.61)$$

and find some $\epsilon > 0$ so that, regardless of $\delta > 0$, we can find two points x and y that are within δ of each other and satisfy inequality (5.61). No obvious ϵ-value jumps out at us, so we'll try letting $\epsilon = 1$ and see if we can proceed.

Inequality (5.61) itself suggests a way to find x and y. Whatever we decide to let x be, we can let $y = x + \delta/2$ so that $|y - x| = \delta/2 < \delta$. Then the trick is to let x be sufficiently close to zero so that $|x||y|$ is small enough to make inequality (5.61) true. Furthermore, since smaller values of δ represent our primary obstacle, we may assume δ is smaller than any convenient positive number, if such an assumption presents itself as helpful. If we let $x = \delta/2$ and $y = \delta$, inequality (5.61) falls right into place as long as $\delta < 1$. Here is our demonstration.

Let $\epsilon = 1$, and pick any $\delta > 0$. We may assume WLOG that $\delta < 1$. Let $x = \delta/2$ and $y = \delta$. Then $|x - y| < \delta$, and

$$|f(x) - f(y)| = \frac{|x - y|}{|x||y|} = \frac{\delta/2}{\delta^2/2} = \frac{1}{\delta} > 1 = \epsilon. \quad (5.62)$$

Thus, f is not uniformly continuous on $(0, 1)$. ∎

In Exercise 2, you'll show that $f(x) = x^2$ is not uniformly continuous on $[0, +\infty)$.

5.7.2 Uniform continuity and compact sets

Perhaps the most beloved theorem dealing with uniform continuity is the following.

Theorem 5.7.5. *If $f : S \to \mathbb{R}$ is continuous and S is compact, then f is uniformly continuous.*

There is something about the fact that every open cover of S is reducible to a finite subcover that allows us to liberate the value of δ from any specific points in the domain and determine it from ϵ alone. We'll supply the proof here because it requires a few sneaky shrinkings of neighborhoods. If you want to try to prove it on your own, here's a thumbnail sketch of how to proceed.

As usual, we pick $\epsilon > 0$. Then since f is continuous at every point in S, we can cover S with a slew of neighborhoods, one centered at each $a \in S$, whose radius is half the δ-value that guarantees $f[N_\delta(a)] \subseteq N_{\epsilon/2}[f(a)]$. Since every $a \in S$ is in its own δ-neighborhood, the set of all these neighborhoods covers S. Compactness of S then allows us to reduce this cover to a finite subcover. These finitely many neighborhoods supply us with a single δ-value—the minimum of the finitely many $\delta/2$-values from the subcover. We can then show that if $|x - y| < \delta$, then $|f(x) - f(y)| < \epsilon$. See if you can fill in the details. If not, here's the whole proof.

Proof: Pick $\epsilon > 0$. Because f is continuous at every $a \in S$, then for any particular $a \in S$, there exists $\delta(a)$ (the notation illustrating that δ is a function of a) such that $|x - a| < \delta(a)$ implies $|f(x) - f(a)| < \epsilon/2$. Cover S with the set $\mathcal{C} = \{N_{\delta(a)/2}(a) : a \in S\}$. Since every $a \in S$ is in its own neighborhood of radius $\delta(a)/2$, $\mathcal{C}$ does in fact cover S. Since S is compact, $\mathcal{C}$ has a finite subcover $\mathcal{C}_1 = \{N_{\delta(a_k)/2}(a_k) : 1 \leq k \leq n\}$. Let $\delta = \min\{\delta(a_k)/2 : 1 \leq k \leq n\}$, and pick $x, y \in S$ such that $|x - y| < \delta$. Now since $\mathcal{C}_1$ covers S, there exists some k such that $x \in N_{\delta(a_k)/2}$. Furthermore, $y \in N_{\delta(a_k)}$ because

$$|y - a_k| \leq |y - x| + |x - a_k| < \delta + \frac{\delta(a_k)}{2} \leq \frac{\delta(a_k)}{2} + \frac{\delta(a_k)}{2} = \delta(a_k). \quad (5.63)$$

Thus

$$|f(x) - f(y)| \leq |f(x) - f(a_k)| + |f(a_k) - f(y)| < \frac{\epsilon}{2} + \frac{\epsilon}{2} = \epsilon, \quad (5.64)$$

so that f is uniformly continuous on S. ∎

In Exercise 3 from Section 5.5, you showed that $f(x) = \sqrt{x}$ is continuous on $(0, +\infty)$. Since $\lim_{x \to 0^+} \sqrt{x} = 0 = \sqrt{0}$, f is continuous on $[0, 1]$, hence it is uniformly continuous there. However, the graph of f becomes very steep as $x \to 0^+$, so that (5.59) is not satisfied. In Exercise 3, you'll demonstrate that this is true.

In Example 5.7.4, the fact that $(0, 1)$ is not closed allows for $1/x$ to have a vertical asymptote at one endpoint. Even though $1/x$ is continuous throughout $(0, 1)$, the asymptote is where we look to disprove uniform continuity. When you show that $f(x) = x^2$ is not uniformly continuous on $[0, +\infty)$ in Exercise 2, it is by looking among sufficiently large numbers that you'll find x and y within δ of each other whose functional values differ by at least ϵ.

EXERCISES

1. Prove Theorem 5.7.3: If there exists $m > 0$ such that $|f(x) - f(y)| \leq m|x - y|$ for all $x, y \in A$, then f is uniformly continuous on A.

2. Show $f(x) = x^2$ is not uniformly continuous on $[0, +\infty)$.

3. Show that inequality (5.59) is not satisfied by $f(x) = \sqrt{x}$ on $[0, 1]$ for any $m > 0$.

PART III

Basic Principles of Algebra

CHAPTER 6

Groups

In its simplest terms, algebra can be thought of as the study of sets on which binary operations provide the defining internal structure for the set. For example, we may construct a set and define a form of addition or multiplication, then look at the structure of S and the relationships between its elements that result from these operations.

One particularly interesting feature of algebra is the study of mappings between sets S_1 and S_2 where the structure of the binary operation on S_1 is preserved among the images of the elements in S_2. More concretely, if S_1 has binary operation $*$, if S_2 has binary operation $\cdot$, and if $f : S_1 \to S_2$, we address the question of whether $f(a*b) = f(a) \cdot f(b)$ for all $a, b \in S_1$. Such is a glimpse into what constitutes algebra. The internal structure of a set is addressed in the way that those elements combine to produce other elements of the set.

In this chapter, we begin our study of some of the most basic concepts of algebraic structures by starting with groups. Some of the theorems are actually restatements of results we have already seen for $\mathbb{R}$. Now, however, the context is broader and more abstract hence the air might seem a little thinner. Instead of proving algebraic theorems where we have the real numbers specifically in mind, we prove theorems based on assumptions that certainly apply to real numbers, but of which the real numbers are only a specific example.

6.1 Introduction to Groups

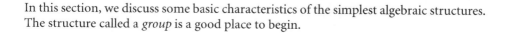

In this section, we discuss some basic characteristics of the simplest algebraic structures. The structure called a *group* is a good place to begin.

6.1.1 Basic characteristics of algebraic structures

An algebraic structure begins with a nonempty set S and builds its internal structure in several stages. First, there must be some notion of equality either defined or understood on S, which naturally must be an equivalence relation. In Chapter 0, we laid out assumptions of $\mathbb{R}$, one of which was that equality in $\mathbb{R}$ satisfies properties E1–E3. We didn't really probe into what's behind equality in $\mathbb{R}$. However, our work in Section 2.6 was a very important example of how equality sometimes arises in a set. Assuming without question that equality in $\mathbb{Z}$ is an equivalence relation, we built the rationals as the set of pairs of integers p/q, with $q \neq 0$. Then as soon as the elements of $\mathbb{Q}$ were constructed, we defined a form of equality in $\mathbb{Q}$ in terms of equality in $\mathbb{Z}$, and showed that this definition is an equivalence relation on $\mathbb{Q}$.

Once the set S is constructed, we define one or more binary operations. If we denote an arbitrary binary operation by $*$, we want the definition of $*$ to have two features. First, $*$ should be well defined, as addition and multiplication are assumed to be on $\mathbb{R}$. That is, given $a, b, c, d \in S$, where $a = b$ and $c = d$ (equality here being the equivalence relation defined on S), we want $a * b = c * d$. Second, $*$ should be closed. Just as $+$ and $\times$ are closed on $\mathbb{R}$, we want assurance that combining two elements $a, b \in S$ by the operation $*$ will produce $a * b \in S$ as well.

Example 6.1.1. Let A be a nonempty set, and let S be the set of all one-to-one, onto functions from A to itself. Then the binary operation of composition is closed on A by Exercise 2 from Section 3.2 (Questions Q2 and Q4).

Example 6.1.2. Consider the set of irrational numbers with the binary operation $\times$. Since $\sqrt{2}$ is irrational, but $\sqrt{2} \times \sqrt{2} = 2 \in \mathbb{Q}$, then $\times$ is not closed as an operation on the irrationals.

If you covered Section 2.9, you can see that a binary operation on S can be defined as a function $f : S \times S \to S$. The formal notation that would be used to represent addition as such a function would be $+ : S \times S \to S$, writing $+(a, b)$ to mean $a + b$. The fact that $+$ as a function has property F1 is precisely what it means for the binary operation $+$ to be closed as we've used the term throughout this text. That $+$ has property F2 means $+$ is well defined as a binary operation, as we've used the term.

All the binary operations we'll address will have the associative property, as do $+$ and $\times$ on $\mathbb{R}$. However, we might have to verify associativity if the context is a new one and we've created a binary operation from scratch. Some binary operations are not associative, but they are indeed rare.

Many, but not all, of the algebraic structures we'll consider will have a binary operation that is commutative. Some very interesting results of algebra derive from binary operations that are not commutative. Be careful in your work! Unless commutativity is explicitly given or proven, you might be tempted to reverse the order of elements without any permission to do so.

The existence of an identity element for $*$ and inverses for elements is rather context specific. We insist in Definition 6.1.3 that an identity element must commute with every element of the set, regardless of whether the binary operation is commutative.

Definition 6.1.3. Suppose S is endowed with binary operation $*$. Then $e \in S$ is called an *identity* for the operation $*$ if $a * e = e * a = a$ for all $a \in S$.

If there is an identity element in S for the binary operation $*$, then it might be that some or all elements have an inverse under $*$.

Definition 6.1.4. Suppose S has binary operation $*$, for which there is an identity element e. If for a given $a \in S$ there exists $b \in S$ such that $a * b = b * a = e$, then we say that b is an *inverse* of a, and we write it as a^{-1}.

For convenience we sometimes list the features of an algebraic structure as an ordered list. For example, if S is a set with binary operation $*$, identity e, and with the feature that every $a \in S$ has an inverse a^{-1}, then we might write such a structure as $(S, *, e, ^{-1})$.

Of the basic algebraic properties we assumed on $\mathbb{R}$ (A2–A14), the only one we have not mentioned here is the distributive property A14. If a set is endowed with two binary operations, it might be that they are linked in their behavior by the distributive property. We'll address algebraic structures with two binary operations in Chapter 7.

If S is a small finite set, it might be convenient to describe the binary operation in what is called a *Cayley table*. Since a binary operation might not be commutative, it's important to read $a * b$ from a Cayley table by going down the left column to find a and across the top to find b.

Example 6.1.5. Consider $S = \{0, 1, 2, 3, 4, 5\}$ and let $\oplus$ be described as in Table 6.1:

$\oplus$	0	1	2	3	4	5
0	0	1	2	3	4	5
1	1	2	3	4	5	0
2	2	3	4	5	0	1
3	3	4	5	0	1	2
4	4	5	0	1	2	3
5	5	0	1	2	3	4

(6.1)

The fact that $\oplus$ is well defined is immediate from the Cayley table, for there is a unique value in each position in the table. Closure is obvious, for every entry is an element of S. Notice $\oplus$ is commutative. How can you tell?[1] Is there an identity element?[2] Does every element have an inverse under $\oplus$?[3] One way to verify associativity would be to work through every possible calculation. In Section 6.3, we'll construct this algebraic structure formally and prove associativity as a theorem.

---

[1] Look for a diagonal symmetry.
[2] Yes, 0.
[3] Yes. $0^{-1} = 0$, $1^{-1} = 5$, $2^{-1} = 4$, and so on.

Example 6.1.6. Let S be as in Example 6.1.5 and define the operation $\otimes$ as in Table 6.2:

$$
\begin{array}{c|cccccc}
\otimes & 0 & 1 & 2 & 3 & 4 & 5 \\
\hline
0 & 0 & 0 & 0 & 0 & 0 & 0 \\
1 & 0 & 1 & 2 & 3 & 4 & 5 \\
2 & 0 & 2 & 4 & 0 & 2 & 4 \\
3 & 0 & 3 & 0 & 3 & 0 & 3 \\
4 & 0 & 4 & 2 & 0 & 4 & 2 \\
5 & 0 & 5 & 4 & 3 & 2 & 1
\end{array}
\tag{6.2}
$$

Is $\otimes$ commutative?[4] Is there an identity element?[5] Does every element have an inverse?[6]

6.1.2 Groups defined

We have different names to refer to algebraic structures with different features. Our first one is the following.

Definition 6.1.7. Suppose G is a set with associative binary operation $*$, identity element e, and with the property that every $g \in G$ has an inverse g^{-1} under $*$. Then the algebraic structure $(G, *, e, ^{-1})$ is called a *group*. If $*$ is a commutative binary operation, then G is called an *abelian* group (after the mathematician Niels Henrik Abel (1802–1829)). If $|G| \in \mathbb{N}$, G is said to be *finite*, and $|G|$ is called the *order* of the group (sometimes denoted $o(G)$ instead of $|G|$).

According to Definition 6.1.7, a group has the following defining features:

(G1) the operation $*$ is well defined;

(G2) the operation $*$ is closed;

(G3) the operation $*$ is associative;

(G4) there is an identity element e; and

(G5) every element of G has an inverse under $*$.

Example 6.1.8. The real numbers with binary operation addition, identity zero, and additive inverses form an abelian group: $(\mathbb{R}, +, 0, -)$. Note that G1 is property A2, G2 is A3, G3 is A4, G4 is A6, and G5 is A7. That $\mathbb{R}$ is abelian is property A5. Also, $\mathbb{Q}$ and $\mathbb{Z}$ are abelian groups under addition.

Example 6.1.9. The nonzero real numbers $\mathbb{R}\backslash\{0\}$ with multiplication form an abelian group: $(\mathbb{R}\backslash\{0\}, \times, 1, ^{-1})$. Also, $\mathbb{Q}\backslash\{0\}$, $\mathbb{R}^+$ and $\mathbb{Q}^+$ are abelian groups under multiplication.

---

[4]Yes.
[5]Yes, 1.
[6]No, only 1 and 5.

Example 6.1.10. The algebraic structure $(S, \oplus, 0, -)$ from Example 6.1.5 is an abelian group, while $(S, \otimes, 1, ^{-1})$ from Example 6.1.6 is not a group, because some elements do not have inverses.

Example 6.1.11. Here is another example of a finite abelian group. In Exercise 9 from Section 2.3, you observed that $x^2 = -1$ has no solution for $x \in \mathbb{R}$. Nothing prevents us from creating a symbol, say, i, declaring $i^2 = -1$, and noting that $i \notin \mathbb{R}$. Now consider the set $S = \{\pm 1, \pm i\}$ with binary operation $\times$ defined according to Table 6.3. Then $(S, \times, 1, ^{-1})$ is an abelian group.

$$\begin{array}{c|cccc} \times & 1 & -1 & i & -i \\ \hline 1 & 1 & -1 & i & -i \\ -1 & -1 & 1 & -i & i \\ i & i & -i & -1 & 1 \\ -i & -i & i & 1 & -1 \end{array} \qquad (6.3)$$

Example 6.1.12. Here's an example that is similar to Example 6.1.11 on the set $Q = \{\pm 1, \pm i, \pm j, \pm k\}$. See Table 6.4:

$$\begin{array}{c|cccccccc} \times & 1 & -1 & i & -i & j & -j & k & -k \\ \hline 1 & 1 & -1 & i & -i & j & -j & k & -k \\ -1 & -1 & 1 & -i & i & -j & j & -k & k \\ i & i & -i & -1 & 1 & k & -k & -j & j \\ -i & -i & i & 1 & -1 & -k & k & j & -j \\ j & j & -j & -k & k & -1 & 1 & i & -i \\ -j & -j & j & k & -k & 1 & -1 & -i & i \\ k & k & -k & j & -j & -i & i & -1 & 1 \\ -k & -k & k & -j & j & i & -i & 1 & -1 \end{array} \qquad (6.4)$$

The gist of this algebraic structure can be understood by noticing that $i^2 = j^2 = k^2 = -1$, but i, j, and k do not commute with each other. If you think of the letters i, j, and k being written on the face of a clock at 12, 4, and 8 o'clock, respectively, then multiplication of two different elements in the clockwise direction produces the third. Multiplication of two elements in the counterclockwise direction produces the negative of the third. For example, $j \times k = i$ and $i \times k = -j$. These three square roots of -1 are called *quaternions*, and they motivate a group on eight elements.

If someone gives you a set with a binary operation and asks you to show that it's a group, you must verify properties G1–G5. To give you an idea of how that might look, we're going to walk through most of the details of a specific example now.

Define the complex numbers by

$$\mathbb{C} = \{a + bi : a, b \in \mathbb{R}\}. \qquad (6.5)$$

We should note immediately that i is a mere symbol with no meaning at this point. We should think of $\mathbb{C}$ merely as a set of ordered real number pairs, the first of which stands alone, and the second of which is tagged with an adjacent i.

Next, we define equality in $\mathbb{C}$, which for clarity we temporarily denote $=_\mathbb{C}$. We define $a_1 + b_1 i =_\mathbb{C} a_2 + b_2 i$ provided $a_1 =_\mathbb{R} a_2$ and $b_1 =_\mathbb{R} b_2$. Notice how $=_\mathbb{C}$ is defined in terms of $=_\mathbb{R}$, which we assume is an equivalence relation. To show that $=_\mathbb{C}$ is an equivalence relation is pretty trivial. For example, to show $=_\mathbb{C}$ has property E2, suppose $a_1 + b_1 i =_\mathbb{C} a_2 + b_2 i$. Then $a_1 =_\mathbb{R} a_2$ and $b_1 =_\mathbb{R} b_2$. Calling on property E2 in $\mathbb{R}$, $a_2 =_\mathbb{R} a_1$ and $b_2 =_\mathbb{R} b_1$. Thus, $a_2 + b_2 i =_\mathbb{C} a_1 + b_1 i$.

Define the binary operation $\oplus$ in the following way:

$$(a + bi) \oplus (c + di) = (a + c) + (b + d)i. \qquad (6.6)$$

We claim that $\mathbb{C}$ is an abelian group under $\oplus$. Here are the steps of the proof in meticulous detail. Working through them will give you a good sense of direction in Exercise 1, where you'll show that $\mathbb{C} \setminus \{0 + 0i\}$ with a form of multiplication is a group.

(G1) Suppose $a_1 + b_1 i =_\mathbb{C} a_2 + b_2 i$ and $c_1 + d_1 i =_\mathbb{C} c_2 + d_2 i$. Then $a_1 =_\mathbb{R} a_2, b_1 =_\mathbb{R} b_2$, $c_1 =_\mathbb{R} c_2$, and $d_1 =_\mathbb{R} d_2$. Since addition in $\mathbb{R}$ is well defined, $a_1 + c_1 =_\mathbb{R} a_2 + c_2$ and $b_1 + d_1 =_\mathbb{R} b_2 + d_2$. Thus

$$\begin{aligned}
(a_1 + b_1 i) \oplus (c_1 + d_1 i) &\stackrel{\text{def}}{=} (a_1 + c_1) + (b_1 + d_1)i \\
&=_\mathbb{C} (a_2 + c_2) + (b_2 + d_2)i \qquad (6.7) \\
&\stackrel{\text{def}}{=} (a_2 + b_2 i) \oplus (c_2 + d_2 i).
\end{aligned}$$

(G2) Pick $a + bi, c + di \in \mathbb{C}$. Since $\mathbb{R}$ is closed under addition, $a + c, b + d \in \mathbb{R}$, so that $(a + bi) \oplus (c + di) = (a + c) + (b + d)i \in \mathbb{C}$.

(G3)

$$\begin{aligned}
[(a + bi) \oplus (c + di)] \oplus (e + fi) &= [(a + c) + (b + d)i] \oplus (e + fi) \\
&= [(a + c) + e] + [(b + d) + f]i \\
&= [a + (c + e)] + [b + (d + f)]i \qquad (6.8) \\
&= (a + bi) \oplus [(c + e) + (d + f)i] \\
&= (a + bi) \oplus [(c + di) \oplus (e + fi)].
\end{aligned}$$

(G4) Pick $a + bi \in \mathbb{C}$. Then $(a + bi) \oplus (0 + 0i) = (a + 0) + (b + 0)i = a + bi$, so $0 + 0i$ functions as an additive identity in $\mathbb{C}$.

(G5) Pick $a + bi$. Then $a, b \in \mathbb{R}$, so that $-a, -b \in \mathbb{R}$ also. Thus $(-a) + (-b)i \in \mathbb{C}$, and $(a + bi) \oplus [(-a) + (-b)i] = (a - a) + (b - b)i = 0 + 0i$.

Finally, $(\mathbb{C}, \oplus, 0 + 0i, -)$ is abelian because

$$\begin{aligned}
(a + bi) \oplus (c + di) &= (a + c) + (b + d)i \\
&= (c + a) + (d + b)i \qquad (6.9) \\
&= (c + di) \oplus (a + bi).
\end{aligned}$$

In Exercises 6 and 7 from Section 2.4, we defined a^n for $a \in \mathbb{R} \setminus \{0\}$ and $n \in \mathbb{Z}$ and proved the rules for exponents

$$a^m \cdot a^n = a^{m+n} \qquad (6.10)$$

$$(a^m)^n = a^{mn} \qquad (6.11)$$

$$(ab)^n = a^n b^n. \qquad (6.12)$$

In the context of an arbitrary group G with binary operation $*$, we can now fix any $a \in G$ and make the same definitions for a^n, where we write

$$a^0 = e, \tag{6.13}$$

$$a^{n+1} = a^n * a \quad \text{for } n \geq 0, \tag{6.14}$$

$$a^{-n} = (a^{-1})^n. \tag{6.15}$$

Furthermore, by mimicking exactly the proofs from these exercises done in the context of $\mathbb{R}\setminus\{0\}$, we arrive at similar exponent rules for $*$ on G, except that one rule depends on G being abelian.

Theorem 6.1.13. *If G is a group, $a, b \in G$, and $n \in \mathbb{Z}$, then*

$$a^m * a^n = a^{m+n} \tag{6.16}$$

$$(a^m)^n = a^{mn}. \tag{6.17}$$

Furthermore, if G is abelian,

$$(a * b)^n = a^n * b^n. \tag{6.18}$$

6.1.3 Subgroups

You might have noticed that S from Example 6.1.11 is a subset of Q from Example 6.1.12. More than that, the binary operation on S is the same as that on Q, in the sense that Table 6.3 is a subtable of Table 6.4. Thus, S is closed under $\times$, contains 1, and is closed under inverses. We give such a subset a name.

Definition 6.1.14. Suppose $(G, *, e, ^{-1})$ is a group, and $H \subseteq G$ is itself a group under the same operation. Then $(H, e, *, ^{-1})$ is called a *subgroup* of G, and we write $H < G$. If $H \subset G$, then H is called a *proper subgroup* of G.

If we are given a group $(G, *, e, ^{-1})$ and a set $H \subseteq G$, and are asked to show that $H < G$, we must show H is itself a group under $*$. Properties G1 and G2 are automatically satisfied on H, for if $*$ is well defined and associative on all of G, then certainly it is well defined and associative when restricted to H. We say that H *inherits* these properties from G. We must demonstrate the following:

(H1) the operation $*$ is closed on H;

(H2) the identity element e is in H; and

(H3) the inverse of every element of H is also in H; that is, H is closed under $^{-1}$.

Example 6.1.15. For any group G, $\{e\}$ and G themselves satisfy properties H1–H3, and are therefore subgroups of G. We call $\{e\}$ the *trivial subgroup*.

Example 6.1.16. The set of even integers is a subgroup of $(\mathbb{Z}, +, 0, -)$ since it is closed under addition, contains 0, and is closed under negation.

Convenience is desirable if it costs nothing in clarity. For this reason, we'll often refer to a group $(G, *, e, ^{-1})$ simply as the group G, and we'll sometimes use juxtaposition of terms ab to indicate the binary operation $a * b$, just like we do with multiplication in $\mathbb{R}$. If we're working with two groups G and H, we'll probably be more careful at first to distinguish between the symbols for the binary operations, and we might denote the identity elements as e_G and e_H, respectively, just to be clear.

Suppose $H < G$ and $H \subseteq H_1 \subseteq G \subseteq G_1$. We can make the following observations about relationships between these sets. First, if H_1 is a group under $*$, then the fact that $H < G$ implies $H < H_1$ also. For clearly, if H exhibits properties H1–H3 as a subset of G, it also does as a subset of H_1. Similarly, if G_1 is a group under $*$, then $H < G_1$. The most efficient way to state this latter relationship is to say that $<$ is transitive: If $H < G$ and $G < G_1$, then $H < G_1$.

EXERCISES

1. Define a form of multiplication $\otimes$ on $\mathbb{C}\setminus\{0 + 0i\}$ by

    $$(a + bi) \otimes (c + di) = (ac - bd) + (ad + bc)i. \qquad (6.19)$$

    Show that $\mathbb{C}\setminus\{0 + 0i\}$ is an abelian group under $\otimes$.[7,8]

2. Define a binary operation $*$ on $\mathbb{Z}$ by $a * b = a + b + 1$. Prove that $\mathbb{Z}$ with $*$ forms an abelian group.

3. If we define a binary operation $*$ by $a * b = a + b - ab$, then there exists $x_0 \in \mathbb{R}$ such that $\mathbb{R}\setminus\{x_0\}$ is a group under $*$. Find x_0, and show that $\mathbb{R}\setminus\{x_0\}$ with $*$ is an abelian group.

4. Suppose $(G, *, e, ^{-1})$ is a group.

    (a) Show that the identity element is unique.

    (b) Prove the left and right cancellation laws:

        i. If $c * a = c * b$, then $a = b$.

        ii. If $a * c = b * c$, then $a = b$.

    (c) Show that the inverse of $a \in G$ is unique.

    (d) Show that $(a * b)^{-1} = b^{-1} * a^{-1}$.

    (e) Show that $(a_1 * a_2 * \cdots * a_n)^{-1} = a_n^{-1} * a_{n-1}^{-1} * \cdots * a_1^{-1}$.

5. Find all subgroups of $(S, \oplus, 0, -)$ from Example 6.1.5.

---

[7]Showing closure of $\otimes$ involves verifying that the product is never $0 + 0i$. Prove contrapositively by showing that if $(a + bi) \otimes (c + di) = 0 + 0i$, then either $a = b = 0$ or $c = d = 0$, so that either $a + bi$ or $c + di$ is not an element of $\mathbb{C}\setminus\{0\}$.
[8]If $(a + bi) \otimes (c + di) = 0 + 0i$, then $ac - bd = 0$ and $ad + bc = 0$. Square these and add.

6. From the multiplicative group $\mathbb{C}\setminus\{0+0i\}$ in Exercise 1, let

$$H = \{a + bi \in \mathbb{C} : a^2 + b^2 = 1\}. \tag{6.20}$$

Show that $H < \mathbb{C}\setminus\{0+0i\}$.

7. Let $G = \{a + b\sqrt{2} : a, b \in \mathbb{Q}, a$ and b not both zero$\}$. Clearly, $G \subseteq \mathbb{R}$, and by Exercise 5 from Section 2.8, $a + b\sqrt{2} = c + d\sqrt{2}$ if and only if $a = c$ and $b = d$. We want to show that G is a subgroup of $(\mathbb{R}\setminus\{0\}, \times, 1, ^{-1})$.

    (a) Show $0 \notin G$.
    (b) Show that G is a subgroup of $(\mathbb{R}\setminus\{0\}, \times, 1, ^{-1})$ by showing it has properties H1–H3.

8. Define the *center* of a group G to be the set of all elements that commute with all elements of G. That is, the center is

$$C = \{a \in G : a * x = x * a \text{ for all } x \in G\}. \tag{6.21}$$

Show that the center of G is a subgroup of G.

9. Suppose $\{H_\alpha\}_{\alpha \in A}$ is a family of subgroups of a group G. Show that

$$\bigcap_{\alpha \in A} H_\alpha < G. \tag{6.22}$$

10. Suppose $\{H_\alpha\}_{\alpha \in A}$ is a family of subgroups of a group G. Is it true that

$$\bigcup_{\alpha \in A} H_\alpha < G? \tag{6.23}$$

Prove or give a counterexample.

11. Suppose G is a group such that $a * a = e$ for all $a \in G$. Show that G is abelian.[9]

12. Let G be a group, and fix some $g \in G$. Define $f : G \to G$ by $f(x) = g * x$ for all $x \in G$. Show that f is a one-to-one function from G onto G.

6.2 Generated and Cyclic Subgroups

From the quaternion group Q in Example 6.1.12, let's take some subset of Q, say, $A = \{i, j\}$. Although $A \subseteq Q$, A is clearly not a subgroup of Q. However, we can talk about the smallest subgroup of Q that contains all elements of A. In this section, we address the existence and possible uniqueness of what we call the *subgroup generated by A*. Then we'll look specifically at the special case of a subgroup generated by a single element. Finally, we'll look at some properties of groups with the special feature that they can be generated in their entirety by some single element.

---

[9]$a * a = e$ is equivalent to $a = a^{-1}$. Use Exercise 4d.

6.2.1 Subgroup generated by $A \subseteq G$

Let's begin by defining the term that conceptually means the smallest subgroup containing all elements of $A \subseteq G$.

Definition 6.2.1. Suppose G is a group and $A \subseteq G$ is nonempty. Suppose also that $H \subseteq G$ has the following properties:

(U1) $A \subseteq H$;

(U2) $H < G$; and

(U3) If $B < G$ and $A \subseteq B$, then $H \subseteq B$.

Then H is called a *subgroup generated by* A, and is denoted (A).

Exactly as in Definition 2.7.12, where we defined $\gcd(a, b)$, we note that Definition 6.2.1 is merely a definition, and does nothing to guarantee the existence of such a subgroup. However, as a definition it does what it's supposed to do—it describes what we would call a smallest subgroup containing all elements of A. Notice how properties U1–U3 do just that. Property U1 guarantees that any H we might be tempted to call (A) does in fact contain all elements of A, and property U2 guarantees that it is indeed a subgroup of G. Property U3 guarantees that no other subgroup of G that contains all elements of A will be any smaller. The fact that (A) exists uniquely is guaranteed by the following theorem, which gives us one way to visualize its construction.

Theorem 6.2.2. *Let G be a group, and suppose $A \subseteq G$ is nonempty. Then (A) exists uniquely. In fact,*

$$(A) = \bigcap_{\substack{J < G \\ J \supseteq A}} J. \tag{6.24}$$

Theorem 6.2.2 states that a way to construct (A) is to collect into a family all the subgroups of G that are supersets of A, and then take the intersection across this family. An important question to ask about the construction in Eq. (6.24) is whether there even exist any subgroups J to use in the intersection. If there are no subgroups of G that contain all elements of A, then Eq. (4.10) is a vacuous construction. However, since $G < G$ and $A \subseteq G$, this is not the case. You'll prove that the construction in Eq. (6.24) satisfies properties U1–U3 and that (A) is unique in Exercise 1.

Equation (6.24) might not actually be the way you would construct (A) for a given $A \subseteq G$. It might be easier to begin with the elements of A and build up (A) by tossing in the needed elements of G until you're sure you're finished.

Example 6.2.3. From Example 6.1.12, and $A = \{i, j\}$, determine (A).

Solution: Clearly, (A) must contain 1, and by property H1, closure of $*$, it must contain $i^2 = -1$ and $i * j = k$. Continuing, $-i, -j, -k \in (A)$ also. Thus, $(A) = Q$. ∎

Constructing (A) in the manner of Example 6.2.3 might be a bit complicated, especially if G is infinite and not abelian, so be careful.

6.2.2 Cyclic subgroups

If $A = \{a\}$ contains only one element, we usually write (a) instead of $(\{a\})$. In this case, we call (a) the subgroup of G generated by a, and a is called a *generator*. A subgroup that has a generator is called a *cyclic subgroup*. The easiest way to see what (a) looks like is to build it from the bottom up, so to speak, and then demonstrate that what you've created satisfies U1–U3. Using the definition of a^n from Section 6.1 (page 212), consider the set

$$S = \{a^n : n \in \mathbb{Z}\}. \quad (6.25)$$

The claim is that $S = (a)$, which we'll verify here by showing S has properties U1–U3, thereby earning the right to be called (a). The details will help you in proving Theorem 6.2.9.

Let $n = 1$ to see that $a \in S$, so that S has property U1. To show that S has property U2, we must show that it has properties H1–H3:

(H1) Pick $x, y \in S$. Then there exist $m, n \in \mathbb{Z}$ such that $x = a^m$ and $y = a^n$. Now $m + n \in \mathbb{Z}$, so that $a^m * a^n = a^{m+n} \in S$. Thus, S is closed under $*$.

(H2) Letting $n = 0$, we have that $e = a^0 \in S$.

(H3) Pick $a^n \in S$. Since Theorem 6.1.13 applies for all $m, n \in \mathbb{Z}$, we have that $(a^n)^{-1} = a^{-n} \in S$. Thus, S is closed under inverses.

To show that S has property U3, suppose $B < G$ and $\{a\} \subseteq B$. We show $S \subseteq B$ by showing $a^n \in B$ for all $n \in \mathbb{Z}$. Since $a \in S$ and B is closed under $*$, it must be that $a^n \in B$ for all $n \in \mathbb{N}$. Certainly $a^0 = e \in B$, and since B is closed under inverses, $a^{-n} \in B$ for all $n \in \mathbb{N}$. Thus, $S \subseteq B$, and we have finished the proof that $(a) = \{a^n : n \in \mathbb{Z}\}$.

Example 6.2.4. In $(\mathbb{R}^+, \cdot, 1, ^{-1})$,

$$(3) = \{3^n : n \in \mathbb{Z}\} = \{\ldots, 1/9, 1/3, 1, 3, 9, \ldots\}.$$

Example 6.2.5. To construct (3) in $(\mathbb{Z}, +, 0, -)$, note that zero is the identity in this context, so that $3^0 = 0$. Constructing 3^n for $n \in \mathbb{N}$ is repeated addition, not repeated multiplication. Thus

$$3^1 = 3, \quad 3^2 = 3+3 = 6, \quad 3^3 = 6+3 = 9, \quad 3^4 = 9+3 = 12, \quad \text{and so on.} \quad (6.26)$$

Constructing 3^{-n} involves negation, not reciprocation, so that

$$\begin{aligned} 3^{-1} &= -3, \\ 3^{-2} &= (3^{-1})^2 = [(-3)+(-3)] = -6, \\ 3^{-3} &= (3^{-1})^3 = [(-6)+(-3)] = -9, \quad \text{and so on.} \end{aligned} \quad (6.27)$$

Thus, $(3) = \{3k : k \in \mathbb{Z}\}$.

Example 6.2.5 suggests that it might be worthwhile to rewrite the exponent definitions in Eqs. (6.13)–(6.15) and the exponent rules in Eqs. (6.16)–(6.18) in what we call their *additive form*. In particular, if $(G, +, 0, -)$ is a group and $a \in G$, the fact that the binary operation is a form of addition suggests that, instead of writing $a^0 = e$, which seems to connote multiplication, we instead write

$$0a = 0. \tag{6.28}$$

We must be careful, though, for the zero on the left-hand side of Eq. (6.28) is an integer, and the zero on the right-hand side is the identity element of the group. By saying $0a = 0$, we mean that the group element a is, in a sense, added to itself zero times, not multiplied by itself, and this is to be defined as the zero of the group. Similarly, the additive form of Eq. (6.16) would be

$$ma + na = (m+n)a. \tag{6.29}$$

You'll write the additive forms of the other exponent definitions and rules in Exercise 2.

The subgroup generated by $a \in G$ might or might not be an infinite set. In Example 6.2.4, the subgroup generated by 3 is an infinite cyclic subgroup. However, in $(\mathbb{C}\setminus\{0+0i\}, \times, 1+0i, ^{-1})$, the subgroup generated by i has a different look. First, $i^0 = 1, i^1 = i, i^2 = -1$, and $i^3 = -i$. But $i^4 = 1$, and 4 is the smallest positive power of i for which this is true. Furthermore, for any $n \in \mathbb{Z}$, $n = 4k+r$ where $0 \leq r \leq 3$, so that $i^n = i^{4k+r} = (i^4)^k \cdot i^r = i^r$. Thus every power of i is an element of $\{i^0, i^1, i^2, i^3\}$, and we've shown that

$$(i) = \{i^n : n \in \mathbb{Z}\} = \{i^0, i^1, i^2, i^3\} = \{1, i, -1, -i\}, \tag{6.30}$$

so that (i) is a finite cyclic subgroup in an infinite group.

This suggests a general result for some cyclic subgroups. Suppose G is a group, and $a \in G$. Suppose there exists some $k \in \mathbb{Z}$ such that $a^k = e$. Since a^{-k} would also be the identity, we may assume $k \in \mathbb{N}$. Of all values of $k \in \mathbb{N}$ for which $a^k = e$, let n be the smallest, and consider the set

$$T = \{a^0, a^1, a^2, \ldots, a^{n-1}\}. \tag{6.31}$$

Since n is the smallest natural number for which $a^n = e$, no $a^k = e$ for $1 \leq k \leq n-1$. Thus e appears exactly once in Eq. (6.31) as a^0. More than that, all elements of T are distinct (Exercise 3). Clearly, $\{a^k : 0 \leq k \leq n-1\} \subseteq \{a^k : k \in \mathbb{Z}\}$, so that $T \subseteq (a)$. In Exercise 3, you'll show $T \supseteq (a)$ to have $T = (a)$.

If n is the smallest natural number for which $a^n = e$, we say that the element a has *order* n, and we denote the order of a by $o(a)$. If $a^n \neq e$ for all $n \in \mathbb{N}$, we say that a has *infinite order*. In Definition 6.1.7, we defined the order of a group as its cardinality. Here, we're defining the order of an element of a group in terms of its powers. By constructing (a) as we've done here, we've demonstrated that these two uses of the term order are tied together.

Theorem 6.2.6. *If G is a group and $a \in G$ has order n, then the subgroup generated by a has order n.*

Naturally, if G is finite and $a \in G$, the subgroup generated by a will be finite.

Example 6.2.7. For $(Q, *, 1, ^{-1})$ from Example 6.1.12,

$$(j) = \{j^n : 0 \leq n \leq 3\} = \{1, j, -1, -j\},$$

and j has order 4 in a group of eight elements.

In Exercises 6 and 7, you'll prove the following.

Theorem 6.2.8. *If G is a group and $a \in G$, then (a) is abelian.*

Theorem 6.2.9. *Suppose G is an abelian group, and $a, b \in G$. Then*

$$(\{a, b\}) = \{a^m b^n : m, n \in \mathbb{Z}\}. \tag{6.32}$$

If G is a group and there exists some $g \in G$ such that $(g) = G$, then G is said to be *cyclic*, and g is called a *generator*. (See Exercise 13 for examples.) There are some nifty little theorems about cyclic groups. By Theorem 6.2.8, a cyclic group is abelian. We'll point out one more characteristic of cyclic groups in Exercise 12, but this and other results will become pretty transparent after we discuss morphisms in Section 6.5.

EXERCISES

1. Prove Theorem 6.2.2 by showing that the construction in Eq. (6.24) satisfies properties U1–U3, and that any two subgroups H_1 and H_2 that both satisfy U1–U3 must be equal.

2. Let $(G, +, 0, -)$ be an additive group. Write Eqs. (6.13)–(6.18) in their additive form.

3. Let G be a group, and $a \in G$ have the property that $a^n = e$, where n is the smallest such natural number. Define $T \subseteq G$ as in Eq. (6.31).

   (a) Show that the elements of T are distinct. That is, if $0 \leq k < l \leq n - 1$, then $a^k \neq a^l$.

   (b) Show $T \supseteq (a)$ to have $T = (a)$.[10]

4. Suppose $x^k = e$ for some $k \in \mathbb{N}$. Show that $o(x) \mid k$.

5. For each of the following groups, determine the subgroup generated by the given element.

   (a) $(\mathbb{Z}, +, 0, -)$; 5
   (b) $(\mathbb{R}\setminus\{0\}, \times, 1, ^{-1})$; 5
   (c) $(S, \oplus, 0, -)$ from Example 6.1.5; 5
   (d) $(S, \oplus, 0, -)$ from Example 6.1.5; 2
   (e) $(S, \oplus, 0, -)$ from Example 6.1.5; 3
   (f) $(S, \oplus, 0, -)$ from Example 6.1.5; 0

---
[10] Pick $a^k \in (a)$ and apply the division algorithm to k and n.

(g) $(Q, \times, 1, ^{-1})$ from Example 6.1.12; j

(h) $(\mathbb{C}\backslash\{0\}, \times, 1, ^{-1})$; $z = 1/\sqrt{2} + (1/\sqrt{2})i$

6. Prove Theorem 6.2.8: If G is a group and $a \in G$, then (a) is abelian.

7. Prove Theorem 6.2.9: Suppose G is an abelian group, and $a, b \in G$. Then

$$(\{a, b\}) = \{a^m b^n : m, n \in \mathbb{Z}\}. \tag{6.33}$$

8. Write Eq. (6.33) in its additive form.

9. For the group $(\mathbb{Z}, +, 0, -)$, describe the following:

   (a) $(4) \cap (6)$

   (b) $(10) \cap (3)$

   (c) $(8) \cap (16)$

   (d) $(\{4, 6\})$

   (e) $(\{4, 7\})$

10. Consider the group $(\mathbb{Z}, +, 0, -)$, and let $a, b \in \mathbb{Z}\backslash\{0\}$. Write $g = \gcd(a, b)$. Show that $(\{a, b\}) = (g)$.

11. Suppose G is a group, $a \in G$, let $m, n \in \mathbb{N}$, and suppose $\gcd(m, n) = g$. Show that $(\{a^m, a^n\}) = (a^g)$.

12. Show that a cyclic group is countable.

13. Are the following groups cyclic? Explain.

    (a) $(\mathbb{Z}, +, 0, -)$

    (b) $(S, \oplus, 0, -)$ from Example 6.1.5

    (c) $(\mathbb{R}\backslash\{0\}, \times, 1, ^{-1})$

    (d) $(Q, \times, 1, ^{-1})$ from Example 6.1.12

    (e) $(\mathbb{Q}, +, 0, -)$

6.3 Integers Modulo *n* and Quotient Groups

In this section we return to Example 6.1.5 to derive it in a formal way from the integers. The result of this construction is called the *integers modulo n* ($n = 6$ for Example 6.1.5). This construction serves as a good illustration of what is called a *quotient group*, which we will discuss as a generalization of the process of deriving the integers modulo *n*.

6.3.1 Integers modulo *n*

In Example 6.1.5, we noted that $(S, \oplus, 0, -)$ is a group, where $S = \{0, 1, 2, 3, 4, 5\}$ and $\oplus$ was defined by Cayley Table 6.1. You might have noticed the similarity of $\oplus$ to

6.3 Integers Modulo n and Quotient Groups

regular addition of integers, except that it was a sort of circular addition. Any sum such as $4 + 5$ that exceeded 5 in $\mathbb{Z}$ was reduced by 6 to guarantee that the sum is in S. This circular summing is sometimes called clock arithmetic, where in this case, the numbers $\{0, 1, 2, 3, 4, 5\}$ can be written around the perimeter of a clock and addition performed in a natural circular way. This example of a group might sound a bit simplistic, but actually, in spite of its simplicity, it is a most important example in group theory. Perhaps we should say *because of* its simplicity its importance in the study of group theory is particularly striking and beautiful.

Here is a standard way of constructing this group by beginning with the integers. It takes a few steps and might seem a bit esoteric, but it's representative of a standard procedure. Study it carefully. Instead of considering the special case of $S = \{0, 1, 2, 3, 4, 5\}$, we consider the general case $\{0, 1, 2, \ldots, n-1\}$.

Step 1. Construct elements of the set. Start with the group $(\mathbb{Z}, +, 0, -)$, and let $n \in \mathbb{N}$ be given. Let

$$(n) = \{kn : k \in \mathbb{Z}\} = \{\ldots, -3n, -2n, -n, 0, n, 2n, 3n, \ldots\} \quad (6.34)$$

be the subgroup of $\mathbb{Z}$ generated by n. Define an equivalence relation on $\mathbb{Z}$ as follows. Define

$$x \equiv_n y \quad \text{if and only if} \quad x - y \in (n). \quad (6.35)$$

That is, $x \equiv_n y$ if and only if $x - y = kn$ for some $k \in \mathbb{Z}$. This is the same definition of equivalence that we used in Example 2.5.5, so proving $\equiv_n$ is an equivalence relation on $\mathbb{Z}$ has already been done. Recall from Exercise 6 in Section 2.7 that $x \equiv_n y$ if and only if x and y have the same remainder when divided by n. Also, recall the equivalence classes generated by $\equiv_n$:

$$[0] = \{\ldots, -2n, -n, 0, n, 2n, \ldots\}$$
$$[1] = \{\ldots, -2n+1, -n+1, 1, n+1, 2n+1, \ldots\}$$
$$[2] = \{\ldots, -2n+2, -n+2, 2, n+2, 2n+2, \ldots\}$$
$$\vdots \quad (6.36)$$
$$[k] = \{\ldots, -2n+k, -n+k, k, n+k, 2n+k, \ldots\}$$
$$\vdots$$
$$[n-1] = \{\ldots, -n-1, -1, n-1, 2n-1, 3n-1, \ldots\}.$$

Notice the very important fact that $[0] = (n)$; that is, the equivalence class of the identity in $(\mathbb{Z}, +, 0, -)$ is the subgroup from which the equivalence is defined. Lump these n equivalence classes into a family and call it $\mathbb{Z}/(n)$, the *integers modulo n*.

$$\mathbb{Z}/(n) = \{[0], [1], [2], \ldots, [n-1]\}. \quad (6.37)$$

Recall that every equivalence class has infinitely many names depending on the representative element by which we choose to address it. For example, in $\mathbb{Z}/(6)$, $[5] = [-1] = [41]$.

Step 2. Define a binary operation on the created set. On the elements of $\mathbb{Z}/(n)$, which are themselves sets, define a form of addition $\oplus_n$ in the following way:

$$[a] \oplus_n [b] = [a +_\mathbb{Z} b]. \tag{6.38}$$

For example, $[4] \oplus_6 [5] = [4+5] = [9] = [3]$, where we choose to refer to the sum as $[3]$ since 3 is a representative element of $[9]$ between 0 and 5.

Notice this new binary operation $\oplus_n$ is a way of combining two sets to produce another set. It's not union or intersection, which up to now are the only binary operations on sets we've seen. Instead, $\oplus_n$ combines $[a]$ and $[b]$ by using representative elements from each to produce a representative element of the set we're calling $[a] \oplus_n [b]$.

Step 3. Show that the created set with its binary operation gives rise to a group. We must show that $(\mathbb{Z}/(n), \oplus_n)$ is a group by showing $\oplus_n$ is well defined, closed, and associative. We must also find an identity element and find inverses for all elements. Clearly, $\oplus_n$ is closed. For if $[a], [b] \in \mathbb{Z}/(n)$, then $a + b \in \mathbb{Z}$. Since the equivalence classes in $\mathbb{Z}/(n)$ partition $\mathbb{Z}$, every integer is in some equivalence class. In particular, $[a+b] \in \mathbb{Z}/(n)$. In Exercise 1, you'll show the rest of the required properties. Showing $\oplus_n$ is well defined is a great example of how one element of a set can have more than one name, and how the resulting ambiguity must not affect how the sums work.

Now we're done. The following Cayley Table 6.39 displays the final product for the particular case $\mathbb{Z}/(6)$:

$\oplus_n$	[0]	[1]	[2]	[3]	[4]	[5]
[0]	[0]	[1]	[2]	[3]	[4]	[5]
[1]	[1]	[2]	[3]	[4]	[5]	[0]
[2]	[2]	[3]	[4]	[5]	[0]	[1]
[3]	[3]	[4]	[5]	[0]	[1]	[2]
[4]	[4]	[5]	[0]	[1]	[2]	[3]
[5]	[5]	[0]	[1]	[2]	[3]	[4]

(6.39)

Take a closer look at the equivalence classes $[0], [1], \ldots, [n-1]$, for there is another standard notation for these sets that we'll use when we derive quotient groups in general. Although we know from Eqs. (6.36) what is in each equivalence class, let's consider a way to visualize an arbitrary $[k]$ in terms of the subgroup of the integers $(n) = [0]$ that motivated this whole mess in the first place. Picture all the integers on the number line in standard fashion, and pretend that each integer is like a key on a piano keyboard. Take countably infinitely many of your friends, and each of you place a finger on the elements of $(n) = [0]$, so that you are pointing to all the multiples of n: $\{\ldots, -2n, -n, 0, n, 2n, \ldots\}$. Now suppose you want to point to all the elements of $[k]$. How can you do it? Everyone in unison should lift his/her finger off the keyboard, and everyone should shuffle over to the right k units, then put his/her finger back down. In other words, to generate all the elements of $[k]$, take all the elements of (n) and add k to each one, so that you're sort of translating each element of (n) through the set of integers by k units. Here's yet another way to say it:

$$[k] = \{x + k : x \in (n)\}. \tag{6.40}$$

Let's create new notation for the construction in Eq. (6.40), writing

$$(n) + k = \{x + k : x \in (n)\}, \tag{6.41}$$

where we're slightly abusing orthodox notation by apparently combining the subgroup (n) with an integer k by using the binary operation defined for $\mathbb{Z}$. This notation is standard, however, and it makes the definition of addition in Eq. (6.38) look like this:

$$[(n) + a] \oplus_n [(n) + b] = (n) + [a + b]. \tag{6.42}$$

With this notation for the equivalence classes in $\mathbb{Z}/(n)$, the Cayley Table, written in its painfully rigorous form, would look like Table 6.43 for the case $n = 6$.

Now let's relax the notation for the specific context of $\mathbb{Z}/(n)$. First, $\mathbb{Z}/(n)$ is usually written $\mathbb{Z}_n$. Also, instead of always writing $(n) + k$, we allow ourselves simply to write k, understanding that we are not talking about a single integer k, but the entire equivalence class of integers of which k is a representative element.

$\oplus_n$	$(n)+0$	$(n)+1$	$(n)+2$	$(n)+3$	$(n)+4$	$(n)+5$
$(n)+0$	$(n)+0$	$(n)+1$	$(n)+2$	$(n)+3$	$(n)+4$	$(n)+5$
$(n)+1$	$(n)+1$	$(n)+2$	$(n)+3$	$(n)+4$	$(n)+5$	$(n)+0$
$(n)+2$	$(n)+2$	$(n)+3$	$(n)+4$	$(n)+5$	$(n)+0$	$(n)+1$
$(n)+3$	$(n)+3$	$(n)+4$	$(n)+5$	$(n)+0$	$(n)+1$	$(n)+2$
$(n)+4$	$(n)+4$	$(n)+5$	$(n)+0$	$(n)+1$	$(n)+2$	$(n)+3$
$(n)+5$	$(n)+5$	$(n)+0$	$(n)+1$	$(n)+2$	$(n)+3$	$(n)+4$

(6.43)

We also generally revert back to the regular addition sign $+$ rather than always writing $\oplus_n$. What this boils down to is that mathematicians get a little lazy and let the familiar notation from $(\mathbb{Z}, +, 0, -)$ also serve for the group $\mathbb{Z}/(n)$ with addition, so that $(\mathbb{Z}_n, +, 0, -)$ is a relaxed notation for $(\mathbb{Z}/(n), \oplus_n, (n) + 0, -)$. The reason this is probably all right is that we are less interested in fancy notation than we are in the internal structure of $\mathbb{Z}_n$ as a group. And it's just as helpful to visualize $\mathbb{Z}_n$ as clock arithmetic with regular old numbers as it is to think in terms of combining equivalence classes of integers. Understood this way, we can then write something like $17 + 9 =_6 2$ and know just what we mean.

6.3.2 Quotient groups

The derivation of $\mathbb{Z}_n$ is affected by the fact that $+$ is a commutative operation on $\mathbb{Z}$. If you've already done Exercise 1, did you notice at what point you exploited this fact?[11] Now let's generalize the program for deriving $\mathbb{Z}_n$ to create a quotient group in a more abstract setting. At first, we're going to restrict ourselves to an abelian group (which we shouldn't have to do). By doing this, we avoid an obstacle in showing that the binary operation on the quotient group is well defined. Then in Section 6.4 we'll address the non-abelian case. For now, simply try not to use the fact that G is abelian anywhere it's not absolutely necessary, and take note of the step(s) in the proof where you do use it. The proofs of all the steps along the way in this derivation are left to you in Exercise 2. As you read them, refer back to their parallel steps in $\mathbb{Z}_n$, and note how the new notation here is a generalization of the $\mathbb{Z}_n$ notation to an arbitrary group.

---

[11] Somewhere in showing $\oplus_n$ is well defined.

Step 1. Given an *abelian* group $(G, *, e, {}^{-1})$ and a subgroup $H < G$, we define an equivalence relation $\equiv_H$ on G in the following way. Define

$$a \equiv_H b \Leftrightarrow a * b^{-1} \in H. \tag{6.44}$$

First, we would need to show that this definition of equivalence is in fact an equivalence relation, so that G is partitioned into equivalence classes. Call the set of equivalence classes G/H, read "G mod H."

Before we complete the construction of the group G/H by defining its binary operation, let's see how the equivalence classes generated by 6.44 look. In $\mathbb{Z}_n$, we have a pretty clear picture of what is in each equivalence class, as illustrated by Eqs. (6.36). In a general group, though, we want a way to visualize the elements of $[g]$ for a given $g \in G$. Therefore we ask what the statement $x \in [g]$ means. In Exercise 2, you'll show that

$$[g] = \{h * g : h \in H\}; \tag{6.45}$$

that is, you'll show $x \in [g]$ if and only if there exists some $h \in H$ such that $x = h * g$. You can think of this as saying that x is an element of G that is scooted over by $*$ exactly g amount from some element of H. The set in Eq. (6.45) is called the *right coset of H generated by g*, and notationally we write

$$H * g = \{h * g : h \in H\}. \tag{6.46}$$

A coset $H * g$ is strikingly similar to that from Eq. (6.40), and can be visualized in a similar way as a translation of all elements of H through G by g units, so to speak. Writing $G/H = \{H * g : g \in G\}$, the set of all cosets of H, we're ready to continue the construction of the quotient group by defining a binary operation.

Step 2. Define a binary operation $*_H$ on G/H in the following way:

$$(H * a) *_H (H * b) = H * (a * b). \tag{6.47}$$

Step 3. Show that $*_H$ is a well-defined, closed, associative binary operation on G/H, find the identity element, and find inverses for elements.

Unless you're unnecessarily sloppy with your algebraic manipulation in completing steps 1–3, there is only one place where you must exploit the fact that G is abelian, and it is the same place where you exploited commutativity of $+$ on $\mathbb{Z}$ in your work with $\oplus_n$. In Section 6.4, we'll return to this construction in the context of an arbitrary group that might not be abelian. However, even then, if the program is going to work, we cannot completely do without some way to switch the order of certain elements. In Section 6.4, we'll guarantee the property we need by requiring H to be a special kind of subgroup of G called a *normal subgroup*.

The set of all cosets with $*_H$ forms the new group, which we call the *quotient group* generated by H. Here's the definition in its entirety.

Definition 6.3.1. Suppose $(G, *, e, {}^{-1})$ is a group (abelian), and let $H < G$. For $a, b \in G$, define an equivalence relation $\equiv_H$ by declaring $a \equiv_H b$ if and only if $a * b^{-1} \in H$. Then

$$G/H = \{H * g : g \in G\} \tag{6.48}$$

with binary operation $*_H$ defined by

$$(H * a) *_H (H * b) = H * (a * b), \qquad (6.49)$$

identity element $H * e = H$, and inverse $(H * a)^{-1} = H * a^{-1}$ is called the *quotient group* of G created by *modding out* the subgroup H. Elements of G/H are called *right cosets* of H.

For equivalence classes in general, we know that $[a] = [b]$ if and only if $a \equiv b$; otherwise, $[a] \cap [b] = \emptyset$. In the context of Definition 6.3.1, these facts become $H * a = H * b$ if and only if $a * b^{-1} \in H$; otherwise, $H * a \cap H * b = \emptyset$. In other words, a and b generate the same coset of H if and only if $a * b^{-1} \in H$; otherwise, the cosets they generate are disjoint.

Example 6.3.2. Let G be the set of all functions $f : \mathbb{R} \to \mathbb{R}$. Define $f + g$ in the following way. For $x \in \mathbb{R}$, let $[f + g](x) = f(x) + g(x)$. Clearly, the function $\mathbf{0}$ defined by $\mathbf{0}(x) = 0$ for all $x \in \mathbb{R}$ is the identity element, and $-f$, defined by $[-f](x) = -f(x)$, is the additive inverse of $f \in G$. Thus $(G, +, \mathbf{0}, -)$ is a group. Let $H \subset G$ be the set of all constant functions, which is clearly a subgroup of G.

What do the elements of the quotient group G/H look like? If $f \in G$ is given, then $H + f = \{h + f : h \in H\}$ is the set of all translations of f up and down in the plane by the constant functions in H. That is, $g \in H + f$ if there exists some constant function $h \in H$ such that $g = h + f$. This might remind you of a fact from calculus. If f is an antiderivative of a function f_1, then $H + f$ is the set of all the antiderivatives of f_1.

6.3.3 Cosets and Lagrange's theorem

Even if G is not an abelian group and the family of cosets of a subgroup of G do not give rise to a quotient group, we can still derive some important results about elements and subgroups of G by looking at the cosets of the subgroup. One thing to keep in mind if G is not abelian is that we have to distinguish between left and right cosets. For $H < G$ and some fixed $g \in G$, we use the notation

$$g * H = gH = \{g * h : h \in H\} \qquad (6.50)$$

$$H * g = Hg = \{h * g : h \in H\} \qquad (6.51)$$

to denote the left and right cosets generated by g, respectively (see Exercise 3). Whether or not these cosets are the same for a particular $g \in G$ is a question for Section 6.4. However, because the results we want to derive in this section are the same for either left or right cosets, we'll look only at right cosets. First, all cosets of a given $H < G$ have the same cardinality (Exercise 4).

Theorem 6.3.3. *If G is a group and $H < G$, then $|H| = |Hg|$ for all $g \in G$.*

Regardless of whether G is of finite or infinite order, the number of cosets of H in G might be finite. If so, we call the number of cosets of H the *index* of H in G, and we denote this number by $(G : H)$. Naturally, if G is finite, then so is $(G : H)$, and the following theorem is immediate as an implication of Theorem 6.3.3.

Theorem 6.3.4 (Lagrange). *If G is a finite group and $H < G$, then the order of H divides the order of G.*

Proof: Since cosets are disjoint and all have the same cardinality as H, it follows that $|G| = (G : H) \times |H|$. ∎

If G is a finite group and $x \in G$, then Lagrange's theorem states that the order of the subgroup generated by x divides $o(G)$. Since the order of (x) is the same as the order of the element x (Theorem 6.2.6), we have $o(x) \mid o(G)$. Beginning here, it is possible to prove several results about elements of a group (see Exercises 5–7).

EXERCISES

1. For the construction $(\mathbb{Z}_n, \oplus_n)$ as discussed on page 221:
   (a) Show that $\oplus_n$ is well defined on $\mathbb{Z}_n$
   (b) Show that $\oplus_n$ is associative
   (c) Determine the identity element
   (d) Determine the inverse of $[k]$.

2. Suppose $(G, *, e, ^{-1})$ is an abelian group and $H < G$. For $a, b \in G$, define $a \equiv_H b$ if $a * b^{-1} \in H$.
   (a) Show that $\equiv_H$ is an equivalence relation on G.
   (b) Show that the equivalence classes generated by $\equiv_H$ are cosets of H. That is, for any given $g \in G$, $[g] = H * g$.
   (c) Define $*_H$ on G/H by $(H * a) *_H (H * b) = H * (a * b)$. Verify that G/H with $*_H$ is a group by showing the following:
      G1: Show that $*_H$ is well defined on G/H.
      G2: Show that $*_H$ is closed on G/H.
      G3: Show that $*_H$ is associative on G/H.
      G4: Show that $H * e$ is the identity element of G/H.
      G5: Show that every $H * a$ has an inverse in G/H.

3. Find all cosets of $H = \{\pm 1, \pm i\}$ in the quaternion group. For each possible $g \in Q$, is $gH = Hg$?

4. Prove Theorem 6.3.3: If G is a group and $H < G$, then $|H| = |Hg|$ for all $g \in G$.

5. Suppose $o(G) = n$, and $g \in G$. Show that $g^n = e$.[12]

6. Suppose G is cyclic of order n. Show that if $k \mid n$, then there exists a subgroup of G of order k.

7. Show that a group of prime order is cyclic.[13]

---

[12] Apply Theorem 6.2.6 and Lagrange's theorem to (g).
[13] Any element except e is a generator.

6.4 Permutation Groups and Normal Subgroups

In this section, we take an in-depth look at a particular class of groups, and two particularly important subgroups that derive from it. Our main purpose here is to introduce the idea of a normal subgroup and to see how it's just the right thing to patch up the hole we left in our derivation of quotient groups by requiring that G be abelian.

6.4.1 Permutation groups

Let A be any nonempty set, and let S be the set of all functions $f : A \xrightarrow[\text{onto}]{1-1} A$. We want to take a close look at the group formed on S with the binary operation of composition. First, note that composition as a binary operation is well defined on S. For if $f_1 = f_2$ and $g_1 = g_2$, then $f_1(a) = f_2(a)$ and $g_1(a) = g_2(a)$ for all $a \in A$. Thus

$$(f_1 \circ g_1)(a) = f_1[g_1(a)] = f_2[g_2(a)] = (f_2 \circ g_2)(a) \tag{6.52}$$

for all $a \in A$, and we have that $f_1 \circ g_1 = f_2 \circ g_2$. Since the composition of one-to-one onto functions is a one-to-one onto function, composition is closed on S. The fact that composition is associative is a mere exercise in the manipulation of parentheses. For if we pick any $a \in A$, then

$$[(f \circ g) \circ h](a) = (f \circ g)[h(a)] = f(g(h(a))) = f[(g \circ h)(a)] = [f \circ (g \circ h)](a). \tag{6.53}$$

Thus $[(f \circ g) \circ h](a) = [f \circ (g \circ h)](a)$ for all $a \in A$, so that $(f \circ g) \circ h = f \circ (g \circ h)$, and $\circ$ is associative on S. Writing $i : A \to A$, the identity function, then for a given $f \in S$ and any $a \in A$,

$$(f \circ i)(a) = f[i(a)] = f(a) \quad \text{and} \quad (i \circ f)(a) = i[f(a)] = f(a). \tag{6.54}$$

Thus $i \circ f = f \circ i = f$ for all $f \in S$ and i is the identity element. Furthermore, by Theorem 3.2.5, every $f \in S$ has its inverse $f^{-1} \in S$. Since

$$f[f^{-1}(a)] = a = i(a) \quad \text{and} \quad f^{-1}[f(a)] = a = i(a) \tag{6.55}$$

for all $a \in A$, we have $f \circ f^{-1} = f^{-1} \circ f = i$. Thus $(S, \circ, i, ^{-1})$ is a group, called the *permutation group* on A, and elements of S are called *permutations* of A.

Example 6.4.1. Let $A = \{1, 2, 3, 4, 5, 6\}$. For this A, the permutation group is denoted S_6, and is called the *symmetric group* on six elements. One way to visualize what $f \in S_6$ does is to imagine six fixed slots, numbered 1–6, with some sort of object in each position. If $f(2) = 5$, this means that the object in slot 2 is moved to slot 5. Thus every element of S_6 does a sort of fruitbasket turnover of the ordered numbers $(1, 2, 3, 4, 5, 6)$. There are several ways to describe notationally a particular $f \in S_6$. One way is to display the image of every element of A as follows:

$$f = \begin{pmatrix} 1 & 2 & 3 & 4 & 5 & 6 \\ 3 & 1 & 2 & 4 & 6 & 5 \end{pmatrix}, \tag{6.56}$$

which means $f(1) = 3$, $f(2) = 1$, $f(3) = 2$, $f(4) = 4$, $f(5) = 6$, and $f(6) = 5$. When we compose two permutations in S_6, we work right to left as usual. For example, if we write

$$g = \begin{pmatrix} 1 & 2 & 3 & 4 & 5 & 6 \\ 3 & 1 & 4 & 6 & 2 & 5 \end{pmatrix}, \tag{6.57}$$

we may calculate $f \circ g$ element by element. To find $(f \circ g)(1)$, we see 1 is first mapped to 3 by g, then 3 is mapped to 2 by f. Thus $(f \circ g)(1) = 2$. Doing the same for the remaining elements yields

$$\begin{aligned} f \circ g &= \begin{pmatrix} 1 & 2 & 3 & 4 & 5 & 6 \\ 3 & 1 & 2 & 4 & 6 & 5 \end{pmatrix} \circ \begin{pmatrix} 1 & 2 & 3 & 4 & 5 & 6 \\ 3 & 1 & 4 & 6 & 2 & 5 \end{pmatrix} \\ &= \begin{pmatrix} 1 & 2 & 3 & 4 & 5 & 6 \\ 2 & 3 & 4 & 5 & 1 & 6 \end{pmatrix}. \end{aligned} \tag{6.58}$$

Calculating $g \circ f$ reveals that $(S_6, \circ, i, ^{-1})$ is not abelian:

$$g \circ f = \begin{pmatrix} 1 & 2 & 3 & 4 & 5 & 6 \\ 4 & 3 & 1 & 6 & 5 & 2 \end{pmatrix}. \tag{6.59}$$

Also,

$$\begin{pmatrix} 1 & 2 & 3 & 4 & 5 & 6 \\ 3 & 1 & 2 & 4 & 6 & 5 \end{pmatrix}^{-1} = \begin{pmatrix} 1 & 2 & 3 & 4 & 5 & 6 \\ 2 & 3 & 1 & 4 & 6 & 5 \end{pmatrix}. \tag{6.60}$$

Another way to describe f from Eq. (6.56) more succinctly is with *cycle notation*, where $f = (132)(56)$ means 1 is mapped to 3, 3 is mapped to 2, and 2 is mapped to 1 for one cycle. The other cycle indicates 5 is mapped to 6, and 6 is mapped to 5. The absence of 4 means it's mapped to itself. Since (132) and (56) are disjoint, it doesn't matter which you write first, nor does it matter whether you think of f as a single permutation or as the composition of (132) and (56). If you want, you can scroll the numbers in a cycle around to make it start with any element in the cycle. Thus $(132) = (213) = (321)$. The identity mapping in S_n is generally written (1). If you compose several permutations in cycle notation that are not disjoint, you can simplify the expression into disjoint cycles.

Example 6.4.2. Simplify $f = (132)(14)(24)(23)(563)$.

Solution: Find $f(1)$ first by tracing 1 through the cycles. Since composition of functions is performed from right to left, we see that 1 is first mapped to 4 by (14), and 4 is not mapped elsewhere by (132). Thus $f(1) = 4$. Now find $f(4)$ by noting that 4 is mapped to 2 by (24), then 2 is mapped to 1 by (132). Thus $f(4) = 1$, and we have completed one cycle (14). Continuing with 2, we see that 2 is mapped to 3 by (23), then 3 is mapped to 2 by (132). Thus $f(2) = 2$, and we can omit it in the cycle notation. Continuing with 3, we see $f(3) = 5$, $f(5) = 6$, and $f(6) = 3$. Thus the simplified expression is $f = (14)(356)$. ∎

How many elements are there in S_6? What about in S_n for any $n \in \mathbb{N}$? If you covered Section 3.4, you know that $|S_n| = n!$. If you omitted that section, however, we'll just point out that the question is equivalent to asking how many ways are there to arrange the elements of $\mathbb{N}_n$. We ask how many possible numbers can go in the first position of the bottom row of 6.56, then the second position, and so on. Multiplying these, we see

$|S_n| = n! = n \cdot (n-1) \cdot (n-2) \cdots 2 \cdot 1$. Thus, S_n is a group on $n!$ elements, and if $n \geq 3$, S_n is not abelian.

Example 6.4.3. For $f = (132)(56)$ in S_6, determine (f).

Solution: By Exercise 3 from Section 6.2 the subgroup generated by f must contain all powers of f.

$$\begin{aligned} f^2 &= f \circ f = (132)(56) \circ (132)(56) = (123), \\ f^3 &= f^2 \circ f = (123) \circ (132)(56) = (56), \\ f^4 &= f^3 \circ f = (56) \circ (132)(56) = (132), \\ f^5 &= f^4 \circ f = (132) \circ (132)(56) = (123)(56), \\ f^6 &= f^5 \circ f = (123)(56) \circ (132)(56) = (1). \end{aligned} \quad (6.61)$$

Thus

$$\begin{aligned} (f) &= \{(1), f, f^2, f^3, f^4, f^5\} \\ &= \{(1), (132)(56), (123), (56), (132), (123)(56)\}. \end{aligned} \quad (6.62)$$

■

6.4.2 The alternating group A_4

Now let's look at an important subgroup of S_n. A permutation of the form (ij) is called a *transposition* because all it does is switch the positions two elements. Notice $(ij)^{-1} = (ij)$ and $(ij) = (ji)$. If $i = 1$ (or $j = 1$, it doesn't matter), then (ij) becomes $(1j)$. If neither i nor j is 1, then

$$(ij) = (1i)(1j)(1i). \quad (6.63)$$

Therefore, if you're playing some game where objects are lined up in positions $1, \ldots, n$ and you want to swap the positions of the objects in positions i and j, it's possible to do it even if you restrict yourself to swapping an object's position only with the object in position 1, only it might take a little longer. However, notice that (ij) and its equivalent in Eq. (6.63) both use an odd number of transpositions.

Even though (ijk) is not a transposition, it can be written as a composition of several transpositions. See if you can figure out how to send i to j, j to k, and k to i by several transpositions involving only $\{i, j, k\}$. A possible answer:

$$(ijk) = (ik)(ij). \quad (6.64)$$

If none of $\{i, j, k\}$ is 1, what does Eq. (6.64) become if we require that all transpositions involve 1, as in Eq. (6.63)? Taking a hint from Eq. (6.63) and applying it to both (ik) and (ij), we can write

$$(ijk) = (1i)(1k)(1i)(1i)(1j)(1i). \quad (6.65)$$

Since $(1i)^{-1} = (1i)$, the two transpositions in the middle of Eq. (6.65) cancel to yield

$$(ijk) = (1i)(1k)(1j)(1i). \quad (6.66)$$

Now $(ijk) = (jki) = (kij)$. Thus, if one of $\{i, j, k\}$ is 1, we may assume $i = 1$, and observe that $(1jk) = (1k)(1j)$. However, notice this one fact. Every form of (ijk) that we have written here involves an even number of transpositions.

Suppose we now consider the cycle $\sigma = (x_1, x_2, \ldots, x_m)$, where no $x_k = 1$. Can you take a hint from Eq. (6.66) and jump right to a similar form for σ that involves only transpositions of the form $(1x_k)$? How about

$$\sigma = (1x_1)(1x_m)(1x_{m-1})(1x_{m-2}) \cdots (1x_2)(1x_1)? \tag{6.67}$$

If some $x_k = 1$, rotate the terms of σ to have $x_1 = 1$. Then Eq. (6.67) works by deleting the transpositions on each end. Be assured there are many other answers. One thing we would like to know is whether the ways of writing σ with transpositions all have something in common.

From all this work, we can see that any $f \in S_n$, once written in cycle notation with disjoint cycles, can be broken down into a composition of transpositions in at least some way. The identity can be thought of as requiring zero transpositions, or if you prefer, we can write $(1) = (12)(12)$ for $n \geq 2$. One characteristic of elements of S_n that we want to point out, but not prove in this text, is that of all possible decompositions of a particular $f \in S_n$ into transpositions, either they will all involve an even number of transpositions, or they will all involve an odd number. This is not a trivial fact to demonstrate. Therefore, we'll just accept it here and leave the proof to your later coursework in algebra. Thus the $n!$ elements of S_n are partitioned into two classes. If $f \in S_n$ always decomposes into an even number of transpositions, then f is called an *even permutation*. Similarly, if $f \in S_n$ always decomposes into an odd number of transpositions, then f is called an *odd permutation*.

Now let's create an important subset of S_n. Let

$$A_n = \{f \in S_n : f \text{ is an even permutation}\}. \tag{6.68}$$

What do you think the relationship between A_n and S_n is? You'll prove the following in Exercise 2.

Theorem 6.4.4. *Let* $n \in \mathbb{N}$. *Then* $A_n < S_n$.

We call A_n the *alternating group* on n elements.

Given that $|S_n| = n!$, how many elements do you think A_n has? Your first thought might be that since every element of S_n is either even or odd, it would only be natural that half would be even and half would be odd, so that $|A_n| = n!/2$. If so, you're right. In Exercise 4 from Section 6.3, you showed that all cosets of a subgroup have the same cardinality. Thus, if you can show that A_n has precisely two cosets, that is, itself and $S_n \setminus A_n$, it will follow that $|A_n| = n!/2$ (Exercise 3).

6.4.3 The dihedral group D_8

Now let us look at an important subgroup of S_4. Consider a rigid square with the numbers 1, 2, 3, 4 etched on its corners, sitting in the xy-plane, where the positions of each corner are also written in the plane (see Fig. 6.1). Consider the following two moves for the square:

1. a rotation $90°$ counterclockwise, a move we'll call ρ (the Greek letter rho); and

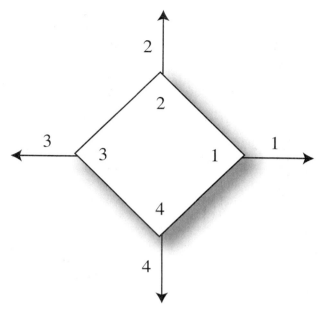

Figure 6.1 A square in the xy plane for the dihedral group.

2. a flip upside down, sending the top corner to the bottom, and vice versa, and leaving the side corners fixed. Call this move ϕ (the Greek letter phi).

The move ρ can be expressed as an element of S_4; specifically, $\rho = (1\,2\,3\,4)$. Visualize ρ and the cycle notation expression of it as sending the corner that initially occupies position 1 in the xy plane to position 2, and so on. Similarly, $\phi = (2\,4)$, which sends the corner initially in position 2 to position 4, and vice versa. If we consider all possible ultimate positions of the square that can result from combining moves ρ and ϕ in any way by composition, we have created a way of visualizing the subgroup of S_4 generated by $\{\rho, \phi\}$. This subgroup of S_4 is called the *dihedral group*, and is denoted D_8.

First, let's note how many elements there are in D_8. We ask how many possible ultimate positions are there for the square that can result from a combination of rotations and flips. Perhaps it's clear that the square can end up either top side up or top side down, and in any one of four states of rotation. Thus eight distinguishable ultimate positions are possible, and $o(D_8) = 8$.

Let's create a Cayley table for D_8 in the following way. Each element of D_8, that is, each ultimate position for the square, can be obtained by doing any necessary rotation first, and then flipping afterward, if necessary. Thus every element of D_8 can be written in the form $\phi^m \rho^n$. As an example, $\phi \rho^3$ would rotate 270° counterclockwise and then flip. This maneuver could be written in cycle notation as $(24)(1234)^3$, which upon simplification becomes $(12)(34)$. Also, since $o(\rho) = 4$ and $o(\phi) = 2$, $\phi^m \rho^n$ can always be written with $0 \leq m \leq 1$ and $0 \leq n \leq 3$. Writing elements of D_8 in this way, we have

$$D_8 = \{i, \rho, \rho^2, \rho^3, \phi, \phi\rho, \phi\rho^2, \phi\rho^3\}. \tag{6.69}$$

To fill in the Cayley table values, we must combine all elements of D_8 and write the compositions in a form from Eq. (6.69). This takes a little bit of work, but yields some useful principles as you work through it. For example, what is $(\phi\rho)\circ(\phi) = \phi\rho\phi$? The first maneuver is a flip, followed by a 90° rotation and flip. Which element of D_8 from Eq. (6.69) is the equivalent of this set of moves? Perhaps you can see that $\phi\rho\phi = (1\,4\,3\,2) = \rho^3$. Table 6.70 contains a few of the entries for the Cayley table of D_8. In Exercise 4, you'll finish out this table, and be pointed to a systematic approach that can save you a lot of time. Remember! In a Cayley table, the entry down the left column is written on the left, and the entry across the top row is written on the right. Since composition is read from right to left, it means that the entry from the top row is actually performed first!

$$
\begin{array}{c|cccccccc}
\circ & i & \rho & \rho^2 & \rho^3 & \phi & \phi\rho & \phi\rho^2 & \phi\rho^3 \\
\hline
i & i & \rho & \rho^2 & \rho^3 & \phi & \phi\rho & \phi\rho^2 & \phi\rho^3 \\
\rho & \rho & \rho^2 & \rho^3 & i & \phi\rho^3 & \phi & & \\
\rho^2 & \rho^2 & \rho^3 & i & \rho & \phi\rho^2 & & & \\
\rho^3 & \rho^3 & i & \rho & \rho^2 & & & & \\
\phi & \phi & & & & & & & \\
\phi\rho & \phi\rho & & & & & & & \\
\phi\rho^2 & \phi\rho^2 & & & & & & & \\
\phi\rho^3 & \phi\rho^3 & & & & & & & \\
\end{array}
\qquad (6.70)
$$

6.4.4 Normal subgroups

Let's return to quotient groups and recall from Section 6.3 how our assumption that G is abelian got us over the hump of showing that the binary operation on G/H from Definition 6.3.1 is well defined. For notational simplicity, write

$$Ha * Hb = H(ab), \qquad (6.71)$$

using juxtaposition for the binary operation on G and in the coset notation. To show that $*$ is well defined on G/H, we would suppose

$$Ha_1 = Ha_2 \quad \text{and} \quad Hb_1 = Hb_2, \qquad (6.72)$$

and try to show from this that

$$H(a_1 b_1) = H(a_2 b_2). \qquad (6.73)$$

The equations in (6.72) are set equalities, and certainly do not mean that $a_1 = a_2$ or $b_1 = b_2$. Instead, they state that if a_1 and a_2 generate the same coset of H, and b_1 and b_2 generate the same coset of H, then $a_1 b_1$ and $a_2 b_2$ will generate the same coset of H. To show Eq. (6.73), we could try to chase an element back and forth between the cosets. Let's try to do that and see two places where we would get stuck in the absence of G being abelian.

Pick $x \in H(a_1 b_1)$. Then x can be written as $x = h_1 a_1 b_1$ for some $h_1 \in H$. We must write $x = k a_2 b_2$ for some $k \in H$ to have $x \in H(a_2 b_2)$. We can make partial progress. Since $h_1 a_1 \in H a_1$, and since $H a_1 = H a_2$, then $h_1 a_1 \in H a_2$, so that $h_1 a_1 = h_2 a_2$ for some $h_2 \in H$. Thus $x = h_2 a_2 b_1$. However, now we're stuck because the only way to replace b_1 with b_2 is by exploiting the fact that $Hb_1 = Hb_2$, which means we must have some element of H to the immediate left of b_1. Although replacing h_1a_1 with h_2a_2 was no problem, we need at least to be able to replace h_2a_2 with a_2h_3 for some $h_3 \in H$. This is a bit like commutativity, but not quite as strong. Then with $x = a_2h_3b_1$, we could exploit $Hb_1 = Hb_2$ to write $x = a_2h_4b_2$ for some $h_4 \in H$. We would almost be home at this point, except that we need to replace a_2h_4 with h_5a_2 for some $h_5 \in H$. If we could get past these two humps, which are really the same hump met in opposite directions, we would have written $x = h_5a_2b_2 \in H(a_2b_2)$, and be done with $\subseteq$. Showing $\supseteq$ would be similar. Therefore, let's make a demand on H that will allow us to get past these humps. The question "we addressed" was whether there exists some $h_3 \in H$ such that $h_2a_2 = a_2h_3$. Thus, instead of making the sweeping requirement that G be abelian, we require that H have the following feature.

Definition 6.4.5. Suppose G is a group and $H < G$. If for all $g \in G$ and $h \in H$ there exists $h_1 \in H$ such that $hg = gh_1$, then H is called a *normal subgroup* of G, and we write $H \triangleleft G$.

Definition 6.4.5 only allows you to swap hg for gh_1. It does not explicitly allow you to swap gh with some h_1g. However, you can show that it does, and you will in Exercise 5.

All our work here shows that $*$ from Eq. (6.71) is well defined if $H \triangleleft G$. Furthermore, your other work from Section 6.3, Exercise 2 where you did not exploit the abelian nature of G completes the proof that G/H is a group, and we arrive at the following:

Theorem 6.4.6. Suppose G is a group and H is a normal subgroup of G. Then G/H with binary operation $*$ defined by $(Ha) * (Hb) = H(ab)$ is a group with identity He and inverses $(Ha)^{-1} = Ha^{-1}$.

6.4.5 Equivalences and implications of normality

There are ways other than Definition 6.4.5 to define normal subgroup, and different authors approach the idea in different ways. As we'll see in Section 6.5, normality can be naturally defined in terms of mappings between groups. However, we've chosen our definition, so any equivalent forms will have to be demonstrated as theorems. If you read the preceding element chasing proof one more time, another way to pinpoint just what kind of subgroup H needs to be might jump out at you. Instead of stating our requirement as the existence of $h_3 \in H$ so that $h_2a_2 = a_2h_3$, we could have said there exists $h_3 \in H$ such that $h_3 = a_2^{-1}h_2a_2$, or simply that $a_2^{-1}h_2a_2 \in H$. That makes the proof of the following immediate:

Theorem 6.4.7. Suppose $H < G$. Then H is normal if and only if for all $h \in H$ and $g \in G$, $g^{-1}hg \in H$.

It's one thing to say that a subgroup is closed under the operations of the group. However, the fact that $g^{-1}hg \in H$ for all $h \in H$ and $g \in G$ adds a new dimension to the closure of H by saying, in a sense, that H cannot be kicked around by things of the form

$g^{-1} \square g$. An expression of the form $g^{-1}hg$ is called a *conjugate* of h. If H is normal, the conjugates of all its elements lie in H. Thus choosing any $h \in H$ and $g \in G$, it might be that $g^{-1}hg$ is different from h, but at least it's still in H. This is the way you'll want to argue the following in Exercise 6.

Theorem 6.4.8. *If $n \geq 1$, then $A_n \triangleleft S_n$.*

If G is not abelian and $H < G$ is not normal, then a conjugate $g^{-1}hg$ might or might not be in H. For example, using $\rho = (1234) \in D_8$ and $(12) \in S_4$,

$$(12)^{-1}(1234)(12) = (12)(1234)(12) = (1342) \notin D_8, \quad (6.74)$$

which by Theorem 6.4.7 proves that $D_8 \not\triangleleft S_4$. On the other hand, using $(123) = (13)(12) \in A_4$ and $(14) \in S_4$,

$$(14)^{-1}(123)(14) = (14)(123)(14) = (234) = (24)(23) \in A_4, \quad (6.75)$$

which is a consequence of the fact that $A_4 \triangleleft S_4$. If G is abelian, conjugation is not particularly interesting.

Theorem 6.4.9. *A group G is abelian if and only if every $h \in G$ has no conjugates other than itself.*

The proof of Theorem 6.4.9 should be immediately clear, for the conditions that $gh = hg$ for all $g, h \in G$ and $g^{-1}hg = h$ for all $g, h \in G$ are identical.

If we think of g as fixed and allow h to take on all values in H, we create what we call a conjugate of the subgroup H. Notationally, we write this as

$$g^{-1}Hg = \{g^{-1}hg : h \in H\}. \quad (6.76)$$

One interesting characteristic of the conjugate of a subgroup is that it is also a subgroup of G. You'll prove the following theorem in Exercise 7:

Theorem 6.4.10. *Let G be a group and $H < G$. Fix $g \in G$. Then $g^{-1}Hg < G$.*

To see an example of a conjugate subgroup, consider $D_8 < S_4$, and let $g = (12)$. For some $\delta \in D_8$, the expression $g^{-1}\delta g$ can be thought of as first switching the numbers in positions 1 and 2 on the square (an illegal move in D_8), doing the δ rotation and/or flip, and then switching whatever numbers are in positions 1 and 2 again. For example,

$$g^{-1}\rho g = (12)(1234)(12) = (1342) \notin D_8. \quad (6.77)$$

Transforming all the elements of D_8 in this way creates another subgroup of S_4 that you might need to play around with to understand. It turns out that $g^{-1}D_8g$ as an algebraic structure is just like D_8, except that the rigid square we described before won't work as a way to visualize it. Instead, picture the numbers $\{1, 2, 3, 4\}$ being pushed from corner to corner by $g^{-1}\rho g$ and $g^{-1}\phi g$ according to Fig. 6.2. The point to be made is that $g^{-1}D_8g \neq D_8$. Granted, they both contain the identity, but except for that overlap, they cut through S_4 in different directions.

If $H \triangleleft G$, the situation is slightly different concerning conjugates of H. The proof of the following theorem should be quick (Exercise 8). It states $H \triangleleft G$ if and only if H has no conjugates other than itself.

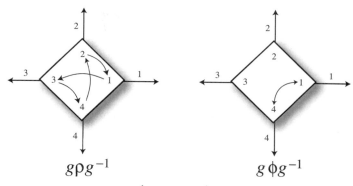

Figure 6.2 Effects of $g^{-1}\rho g$ and $g^{-1}\phi g$ on the square.

Theorem 6.4.11. *Suppose $H < G$. Then H is normal if and only if $g^{-1}Hg = H$ for all $g \in G$.*

In showing that $*$ is well defined on G/H, we supposed $Ha_1 = Ha_2$ to get us from h_1a_1 to h_2a_2. Another feature we might have required of H to get us over the next hump from h_2a_2 to a_2h_3 involves linking the right coset Ha_2 to the left coset a_2H. That is, if we had had that $Ha_2 = a_2H$, our problem would have been solved in precisely the same way. This leads to the following equivalence of normality, which you'll prove in Exercise 9.

Theorem 6.4.12. *Suppose $H < G$. Then H is normal if and only if $Hg = gH$ for all $g \in G$.*

To sum up, notice how our retreat from the requirement that G be abelian motivated the definition of a characteristic of $H < G$ that allows us still to construct the quotient group. And notice the following, which you'll prove in Exercise 11.

Theorem 6.4.13. *If G is abelian, then every $H < G$ is normal.*

Since our definition of normal subgroup was inspired by a retreat from the global condition of G being abelian, let's compare and contrast normality in all its forms to similarly worded statements about abelian groups.

If G is abelian and $H < G$:

1. For all $g, h \in G$, $gh = hg$.
2. For all $g, h \in G$, $g^{-1}hg = h$.
3. For all $g \in G$, $g^{-1}Hg = H$.
4. For all $g \in G$, $Hg = gH$.

If $H \triangleleft G$:

For all $g \in G$ and $h \in H$, there exists $h_1 \in H$ such that $hg = gh_1$.

For all $g \in G$ and $h \in H$, $g^{-1}hg \in H$.

For all $g \in G$, $g^{-1}Hg = H$.

For all $g \in G$, $Hg = gH$.

As a final observation, consider four sets $H \subseteq H_1 \subseteq G \subseteq G_1$, where $H \triangleleft G$ instead of simply $H < G$ as we considered in Section 6.1. Normality of H as a subgroup of G says something about the way elements of H behave in the presence of elements of G, and not simply how they behave among themselves. Thus, if H_1 is also a group, then

$H \triangleleft G$ will imply $H \triangleleft H_1$ as well. For if $g^{-1}hg \in H$ for all $g \in G$, then certainly the same is true for all $g \in H_1$. However, if G_1 is a group, it might be that $H \not\triangleleft G_1$. Even though $g^{-1}hg \in H$ for all $g \in G$, there might be some $g \in G_1 \setminus G$ for which $g^{-1}hg \notin H$. A trivial example of the truth of this is that $D_8 \triangleleft D_8$, but $D_8 \not\triangleleft S_4$. And, although it takes a little playing around to see it clearly, $D_8 \triangleleft A_4$.

EXERCISES

1. For $g = (1\,2)$ and $h = (3\,4)$ in S_4, determine $(\{g, h\})$, the subgroup of S_4 generated by g and h.

2. Prove Theorem 6.4.4: Let $n \in \mathbb{N}$. Then $A_n < S_n$.

3. Assuming $|S_n| = n!$, show that $|A_n| = n!/2$ by showing that A_n has precisely two cosets.[14]

4. Complete Table (6.70), the Cayley table for D_8.[15]

5. Suppose $H \triangleleft G$. Show that for all $g \in G$ and $h \in H$, there exists $h_1 \in H$ such that $gh = h_1 g$.[16]

6. Prove Theorem 6.4.8: If $n \geq 1$, then $A_n \triangleleft S_n$.

7. Prove Theorem 6.4.10: Let G be a group and $H < G$. Fix $g \in G$. Then $g^{-1}Hg < G$.

8. Prove Theorem 6.4.11: Suppose $H < G$. Then H is normal if and only if $g^{-1}Hg = H$ for all $g \in G$.

9. Prove Theorem 6.4.12: Suppose $H < G$. Then H is normal if and only if $Hg = gH$ for all $g \in G$.

10. In S_4, let $f = (123)$ and $g = (14)$. Is $g(f) = (f)g$?

11. Prove Theorem 6.4.13: If G is abelian, then every $H < G$ is normal.

12. In Exercise 8 of Section 6.1, you showed that the center of a group G is a subgroup of G. Prove that the center of G is normal in G.

13. In Exercise 9 of Section 6.1, you showed that the intersection across a family of subgroups of G is itself a subgroup of G. Show that the intersection across a family of normal subgroups of G is normal in G.

6.5 Group Morphisms

Now that we have a basic understanding of groups and some of their internal structure, let's turn our attention outward to a special type of function from a group G to a group

---

[14] When do $f_1, f_2 \in S_n$ generate the same coset of A_n? See the comments that follow Definition 6.3.1.
[15] First convert all expressions of the form $\rho^i \phi$ to their equivalents in the form $\phi \rho^j$. Use these results to take expressions of the form $\phi^s \rho^t \phi^u \rho^v$ and reorder $\rho^t \phi^u$ in the middle. Thus all ρ's will gravitate to the right and all ϕ's to the left.
[16] Apply Definition 6.4.5 to $h^{-1}g^{-1}$.

H. The special feature we want these functions to have is that they preserve the binary operation.

Definition 6.5.1. Suppose $(G, *, e_G, ^{-1})$ and $(H, \cdot, e_H, ^{-1})$ are groups, and suppose $\phi : G \to H$ is a function with the property that $\phi(x * y) = \phi(x) \cdot \phi(y)$ for all $x, y \in G$. Then ϕ is called a *morphism* from G to H. If ϕ is one-to-one, it is called a *monomorphism*. If ϕ is onto, it is called an *epimorphism*. If ϕ is both one-to-one and onto, it is called an *isomorphism*, and we write $G \cong H$, which is read "G is isomorphic to H." If $\phi : G \to G$ is an isomorphism, ϕ is called an *automorphism*.

Example 6.5.2. Let $\mathbb{Z}$ be the group of integers under addition, and E the group of all even integers under addition. Then $\phi_1 : \mathbb{Z} \to E$ defined by $\phi_1(n) = 2n$ is an isomorphism, and $\phi_2(n) : \mathbb{Z} \to E$ defined by $\phi_2(n) = 4n$ is a monomorphism that is not an isomorphism. The function $\phi_3 : \mathbb{Z} \to \mathbb{Z}$ defined by $\phi_3(n) = -n$ is an automorphism (Exercise 1).

Example 6.5.3. Let $\mathbb{Z}$ be the group of integers under addition, let $n \in \mathbb{N}$, and consider the group $(\mathbb{Z}_n, \oplus_n, (n) + 0, -)$. Then $\phi : \mathbb{Z} \to \mathbb{Z}_n$ defined by $\phi(k) = (n) + k$ is an epimorphism that is not an isomorphism (Exercise 2).

Example 6.5.4. Let $G = (\mathbb{R}^+, \times, 1, ^{-1})$ and $H = (\mathbb{R}, +, 0, -)$. Even though we have not discussed logarithms in this text, your work in precalculus reveals that $\phi(x) = \ln x$ is an isomorphism because $\ln x$ is a one-to-one function from $\mathbb{R}^+$ onto $\mathbb{R}$, and satisfies $\phi(xy) = \ln(xy) = \ln x + \ln y = \phi(x) + \phi(y)$.

Example 6.5.5. Let G and H be any groups, and define $\phi : G \to H$ by $\phi(x) = e_H$ for all $x \in G$. Then ϕ is called the *trivial* morphism.

The morphic behavior of $\phi : G \to H$ does not explicitly require that particular elements of G must map to particular elements of H. However, there are some restrictions of this sort inherent in the definition.

Theorem 6.5.6. *If $\phi : G \to H$ is a group morphism, then:*

1. $\phi(e_G) = e_H$.
2. *For all $x \in G$, $\phi(x^{-1}) = [\phi(x)]^{-1}$.*

We'll prove part 1 and leave part 2 to you in Exercise 3.

Proof:

$$e_H \cdot \phi(e_G) = \phi(e_G) = \phi(e_G * e_G) = \phi(e_G) \cdot \phi(e_G). \tag{6.78}$$

By cancellation, $\phi(e_G) = e_H$. ∎

Notice the strange similarity of Eq. (6.78) to Eq. (2.44), where we showed $a \cdot 0 = 0$ for all $a \in \mathbb{R}$. With part 1 of Theorem 6.5.6 as the root of an induction argument for $n \geq 0$, then with part 2 to take care of $n < 0$, you can show the following (Exercise 4).

Theorem 6.5.7. *If $\phi : G \to H$ is a group morphism and $g \in G$, then for all $n \in \mathbb{Z}$,*

$$\phi(g^n) = [\phi(g)]^n. \tag{6.79}$$

The statement $G \cong H$ in Definition 6.5.1 looks like a form of equivalence. In fact, you can show the following in Exercise 6.

Theorem 6.5.8. *The relation $\cong$ in Definition 6.5.1 is an equivalence relation on the set of all groups.*

The statement $G \cong H$ is a statement about the existence of a one-to-one function from G onto H with the additional property that it preserves the binary operation. Thus most of the work in proving that $\cong$ has properties E1–E3 has been done in Chapter 3 and will require only references to applicable theorems. However, in each property E1–E3, something will have to be shown about the morphic behavior of the functions involved.

The word *isomorphism* has a connotation to it that deserves pointing out. For G to be isomorphic to H means that G and H are essentially the same group, in the sense that all the elements from one can be swapped one for one with those in the other and the internal relationships between them as expressed by $*$ are retained by $\cdot$. By giving a new name to every $x \in G$, namely $\phi(x)$, and swapping the binary operation symbol in G for the symbol in H, we have effectively dressed $(G, *, e_G, {}^{-1})$ in the clothing of $(H, \cdot, e_H, {}^{-1})$. Thus, to be able to map every element of G to a unique and distinct element of H, exhausting all elements of H and preserving the binary operation, is to show that, as far as their structure as groups is concerned, they are identical. Here are some illustrations of this principle that you'll prove in Exercise 7.

Theorem 6.5.9. *Suppose $\phi : G \to H$ is an isomorphism. Then,*

1. *If G is abelian, then H is abelian.*

2. *If G is cyclic, then H is cyclic.*

Instead of writing $\text{Rng}(\phi)$, we usually write $\phi(G)$. Subgroups of the domain and codomain are related in the following theorem, which you'll prove in Exercise 9.

Theorem 6.5.10. *Suppose $\phi : G \to H$ is a group morphism. Then*

1. $\phi(G) < H$.

2. *If $N \triangleleft G$, then $\phi(N) \triangleleft \phi(G)$.*

3. *If $N \triangleleft H$, then $\phi^{-1}(N) \triangleleft G$.*

Notice that part 2 of Theorem 6.5.10 does not say that $\phi(N) \triangleleft H$. If ϕ is not onto, it's possible that $\phi(N)$ is normal in $\phi(G)$ but not normal in H.

Theorem 6.5.6 ensures that e_G always maps to e_H under a group morphism. It might be that other elements of G also map to the identity in H. In Example 6.5.3, every integer of the form kn where $k \in \mathbb{Z}$ maps to $(n) + 0$. In Example 6.5.5, every element of G maps to e_H. We give a name to the set of all elements in G that map to e_H under a morphism.

Definition 6.5.11. Suppose $\phi : G \to H$ is a group morphism. Then $\phi^{-1}(e_H) = \{x \in G : \phi(x) = e_H\}$ is called the *kernel* of ϕ and is denoted $\text{Ker}(\phi)$. If $\text{Ker}(\phi) = \{e_G\}$, we say that the kernel of ϕ is *trivial*.

One important feature of $\text{Ker}(\phi)$ is the following, which you'll prove in Exercise 10.

Theorem 6.5.12. *If $\phi : G \to H$ is a group morphism, then $\text{Ker}(\phi) \triangleleft G$.*

An interesting property of group morphisms is that you can sometimes learn a lot about the behavior of ϕ across all of G by looking at its behavior at certain places in G. Clearly, if ϕ is a monomorphism, then $\text{Ker}(\phi)$ is trivial. Interestingly, the converse of this is also true (Exercise 11).

Theorem 6.5.13. *Suppose $\phi : G \to H$ is a group morphism. Then ϕ is one-to-one if and only if $\text{Ker}(\phi) = \{e_G\}$.*

Thus if $\phi^{-1}(e_H)$ has only one element, then $\phi^{-1}(y)$ has only one element for every $y \in \phi(G)$. We can go even further and show that

$$|\phi^{-1}(y)| = |\text{Ker}(\phi)| \qquad (6.80)$$

for all $y \in \phi(G)$, so that all pre-image sets of individual elements in $\phi(G)$ have the same cardinality. Probably the easiest way to do this is to exploit the fact that $\text{Ker}(\phi) \triangleleft G$ in order to prove something even stronger than Eq. 6.80. Theorem 6.5.14 says that the pre-image of each $y \in \phi(G)$ is simply one of the cosets of $\text{Ker}(\phi)$. Since all cosets of a subgroup have the same cardinality (Theorem 6.3.3), Eq. (6.80) follows. In the next theorem, we use left cosets to keep the notation uncluttered. You'll prove it in Exercise 12.

Theorem 6.5.14. *If $\phi : G \to H$ is a group morphism, then for all $y \in \phi(G)$, there exists $a \in G$ such that $\phi^{-1}(y) = a \, \text{Ker}(\phi)$.*

If you've caught on to what's happening here, you might have observed that any group morphism $\phi : G \to H$ makes some interesting statements about the internal structure of G. The morphism ϕ gives rise to a normal subgroup of G, namely $\text{Ker}(\phi)$. From there all of our theory from Section 6.4 applies to present us with a quotient group $G/\text{Ker}(\phi)$, with its binary operation $a \, \text{Ker}(\phi) * b \, \text{Ker}(\phi) = (ab) \, \text{Ker}(\phi)$, identity $\text{Ker}(\phi)$ and inverses $[a \, \text{Ker}(\phi)]^{-1} = a^{-1} \, \text{Ker}(\phi)$. Thus any time you're given a group G and can manage to find a morphism into some other group H, you've managed to find a normal subgroup of G and motivate a quotient group from it.

Also, if you're like a lot of people at your stage of the mathematical game, you wish there were a better way to see what's going on inside this new quotient group $G/\text{Ker}(\phi)$ than by visualizing cosets whacking each other around. Well, there is another way to visualize $G/\text{Ker}(\phi)$, for it is essentially the same as (isomorphic to) $\phi(G)$. We'll get to a full-blown statement and proof of that soon.

A group morphism $\phi : G \to H$ gives rise to a normal subgroup of G and a quotient group from it. Now let's go the other way: Given a group G and any $N \triangleleft G$, we can *create* a group H and an epimorphism $\phi : G \to H$ such that $\text{Ker}(\phi) = N$. It doesn't involve much we haven't seen before, mostly just some observations. You'll provide the only missing detail in Exercise 13.

Theorem 6.5.15. *If $(G, *, e, ^{-1})$ is a group and $N \triangleleft G$, then $\phi : G \to G/N$ defined by $\phi(x) = Nx$ is an epimorphism whose kernel is N.*

Proof: From our previous work, ϕ is defined on all of G and is well defined. Also, ϕ is onto because every $Nx \in G/N$ is generated by $x \in G$, and $\phi(x) = Nx$. We must show that ϕ behaves morphically and satisfies $\text{Ker}(\phi) = N$. Pick $x, y \in G$. Since the binary operation on G/N is well defined,

$$\phi(xy) = N * (xy) = Nx * Ny = \phi(x)\phi(y). \tag{6.81}$$

Finally, from Exercise 13, $\text{Ker}(\phi) = N$. ∎

So we can have it both ways. Two groups and a morphism give rise to a normal subgroup of the domain. And a group with a normal subgroup gives rise to another group and a morphism onto it. Here's the final tie. Whichever you start with and use to create the other, the range of the morphism and the quotient group of G are isomorphic. That is, $\phi(G)$ and $G/\text{Ker}(\phi)$ are essentially the same group, as the following theorem states. For notational simplicity, we'll assume that a given $\phi : G \to H$ is onto so that we don't have to distinguish between H and $\phi(G)$.

Theorem 6.5.16. *Suppose $\phi : G \to H$ is a group epimorphism. Then*

$$G/\text{Ker}(\phi) \cong H. \tag{6.82}$$

We'll supply only a skeleton of the proof here, leaving most of the details to you in Exercise 14. Before we begin the proof, look at Fig. 6.3, which illustrates all the groups and mappings involved. First, there are the given groups G and H with the epimorphism $\phi : G \to H$. Now create a carbon copy of G, but collect all the elements of $\text{Ker}(\phi)$ and hogtie them together into a single entity in the sketch of $G/\text{Ker}(\phi)$. Similarly, go to each coset of $\text{Ker}(\phi)$, take all of its elements, and lump them together into a single entity in $G/\text{Ker}(\phi)$ to create a visualization of $G/\text{Ker}(\phi)$. We have a link between G and $G/\text{Ker}(\phi)$, and that is the mapping that sends $x \in G$ to $x \, \text{Ker}(\phi)$. Now it might be that ϕ is one-to-one, or maybe not. But the extent to which ϕ collapses elements of G down to single elements of H is precisely the extent to which elements of G clump together into cosets in $G/\text{Ker}(\phi)$ (Theorem 6.5.14), which itself depends upon the size of the kernel. The task is to show that $G/\text{Ker}(\phi)$ and H are isomorphic by finding the required mapping between them. We'll imagine mapping an element of $G/\text{Ker}(\phi)$ to an element of H, and the way we decide where to map a chosen coset $x \, \text{Ker}(\phi) \in G/\text{Ker}(\phi)$ is by grabbing some element of $x \, \text{Ker}(\phi)$, say x, and sending the whole coset to $\phi(x)$ in H. In the proof here, we'll use $\cdot$ to represent the binary operation in H.

Proof: We must find a one-to-one, onto function $\psi : G/\text{Ker}(\phi) \to H$ such that

$$\psi[x \, \text{Ker}(\phi) * y \, \text{Ker}(\phi)] = \psi[x \, \text{Ker}(\phi)] \cdot \psi[y \, \text{Ker}(\phi)] \tag{6.83}$$

for all $x \, \text{Ker}(\phi), y \, \text{Ker}(\phi) \in G/\text{Ker}(\phi)$. Define $\psi : G/\text{Ker}(\phi) \to H$ by

$$\psi[x \, \text{Ker}(\phi)] = \phi(x). \tag{6.84}$$

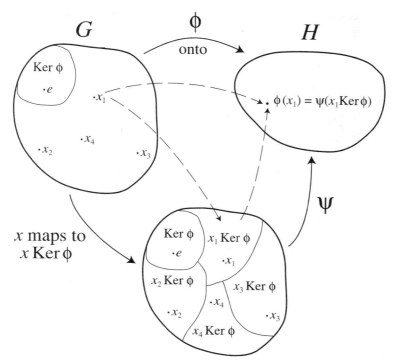

Figure 6.3 Isomorphism between $G/\mathrm{Ker}(\phi)$ and H.

By Exercise 14, ψ is well defined on all of $G/\mathrm{Ker}(\phi)$, is one-to-one and onto, and satisfies Eq. (6.83). ∎

As a final note, the theorems from this section provide us with another logical equivalence to normality of a subgroup. If $N \triangleleft G$, then there exists a group H and an epimorphism $\phi : G \to H$ such that $\mathrm{Ker}(\phi) = N$. Also, if $\phi : G \to H$ is an epimorphism, $\mathrm{Ker}(\phi) \triangleleft G$. Thus, given a group G and $N < G$, N is normal in G if and only if there exists some group H and some morphism $\phi : G \to H$ such that $N = \mathrm{Ker}(\phi)$.

EXERCISES

1. Verify the claims from Example 6.5.2 concerning ϕ_1, ϕ_2, and ϕ_3.

2. Show that ϕ from Example 6.5.3 is an epimorphism that is not an isomorphism.

3. Prove part 2 of Theorem 6.5.6: If $\phi : G \to H$ is a group morphism, then for all $x \in G$, $\phi(x^{-1}) = [\phi(x)]^{-1}$.

4. Prove Theorem 6.5.7: If $\phi : G \to H$ is a group morphism and $g \in G$, then for all $n \in \mathbb{Z}$, $\phi(g^n) = [\phi(g)]^n$.

5. Restate Theorems 6.5.6 and 6.5.7 in their additive form.

6. Prove Theorem 6.5.8: The relation $\cong$ in Definition 6.5.1 is an equivalence relation on the set of all groups.

7. Prove Theorem 6.5.9: Suppose $\phi : G \to H$ is an isomorphism. Then,

    (a) If G is abelian, then H is abelian.

    (b) If G is cyclic, then H is cyclic.

8. Let G be a group, and define $\phi : G \to G$ by $\phi(g) = g^{-1}$. Assuming ϕ is a one-to-one function from G onto itself, show that ϕ is an automorphism of G if and only if G is abelian.

9. Prove Theorem 6.5.10: Suppose $\phi : G \to H$ is a group morphism. Then

    (a) $\phi(G) < H$.

    (b) If $N \triangleleft G$, then $\phi(N) \triangleleft \phi(G)$.

    (c) If $N \triangleleft H$, then $\phi^{-1}(N) \triangleleft G$.

10. Prove Theorem 6.5.12: If $\phi : G \to H$ is a group morphism, then $\text{Ker}(\phi) \triangleleft G$.

11. Prove Theorem 6.5.13: Suppose $\phi : G \to H$ is a group morphism. Then ϕ is one-to-one if and only if $\text{Ker}(\phi) = \{e_G\}$.

12. Prove Theorem 6.5.14: If $\phi : G \to H$ is a group morphism, then for all $y \in \phi(G)$, there exists $a \in G$ such that $\phi^{-1}(y) = a \, \text{Ker}(\phi)$.

13. Finish the proof of Theorem 6.5.15 by showing that $\text{Ker}(\phi) = N$.

14. Finish the proof of Theorem 6.5.16 by showing that $\psi : G/\text{Ker}(\phi) \to H$ defined by Eq. (6.84) satisfies the following:

    (a) ψ is defined on all of $G/\text{Ker}(\phi)$.

    (b) ψ is well defined.

    (c) ψ is one-to-one.

    (d) ψ is onto.

    (e) For all $x \, \text{Ker}(\phi), y \, \text{Ker}(\phi) \in G/\text{Ker}(\phi)$,

    $$\psi[x \, \text{Ker}(\phi) * y \, \text{Ker}(\phi)] = \psi[x \, \text{Ker}(\phi)] \cdot \psi[y \, \text{Ker}(\phi)].$$

15. Construct notation and Cayley tables to determine (up to isomorphism) all groups on five or fewer elements.

16. Find isomorphisms between the four-element groups you found in Exercise 15 and the following.

    (a) The multiplicative group $S = \{\pm 1 \pm i\}$.

    (b) The subgroup of S_4 in Exercise 1 from Section 6.4.

17. Describe all cyclic groups.

18. Show that $D_8 \not\cong Q$, where Q is the quaternion group from Section 6.1.

CHAPTER 7

Rings

We can create algebraic structures of greater complexity than a group by endowing a set with two binary operations and laying down some assumptions about how these operations behave, both on their own and in relation to each other. In this chapter, we'll look at several such structures. Before we do, some explanation is in order about how we're going to proceed, for the theory of rings is full of all kinds of details that make a road map very helpful.

First, in Section 7.1, we'll define the most general algebraic structure with two binary operations, a *ring*, construct several important examples, and define *subring*. In Section 7.2, we'll look at several properties that the most general rings share. At the same time, we'll make a passing reference to *fields*, the most specialized kind of ring we'll study. One particularly important class of rings can be created by what we call *adjoining* an element to a given ring. We devote Section 7.3 to this class of examples. In Section 7.4, we dive down inside a ring to look at specialized substructures of a general ring. Ideals, principal ideals, prime ideals, and maximal ideals are special types of substructures we'll see there. In Sections 7.5–7.7, we'll study four increasingly specialized kinds of rings: integral domains; unique factorization domains; principal ideal domains; and Euclidean domains. Each class of these structures is a proper subset of the class that comes before it, so as we progress, we'll demonstrate (or at least refer to) examples that illustrate this. For example, we'll see a ring that's not an integral domain, an integral domain that's not a unique factorization domain, etc. In Section 7.8 we'll look at ring morphisms, and finally, in Section 7.9, we'll build quotient rings.

7.1 Rings and Subrings

7.1.1 Rings defined

The simplest structure with two binary operations, and therefore where we begin, is called a *ring*. Because the assumptions we make about ring operations so closely resemble those for addition and multiplication on $\mathbb{Z}$, it's common to use the notations $+$ and either $\cdot$ or juxtaposition for the two operations, and shamelessly call them addition and multiplication, respectively, even though by doing so we run some slight risks. One is that we might inadvertently think that some of the rings we create are more like the integers than the assumptions justify, because the WOP of $\mathbb{N}$ makes $\mathbb{Z}$ a very special kind of ring. Thus we have to be careful not to bring any excess baggage from our understanding of the integers that the general ring assumptions do not imply. On the other hand, being able to envision $\mathbb{Z}$ as a sort of quintessential example of a ring means that some of the results we proved for the integers will translate directly over to any ring and be clear to us right away. Thus, some of the theorems we'll state in this section will require very little in the way of a new proof, but mimic those for $\mathbb{Z}$ with very little or no variation. The second risk we run in using notation already associated with the integers is that it might stifle our imagination when we try to create new and interesting rings. There are quite a number of very interesting rings that creative minds have concocted from interesting sets and definitions of equality, addition, and multiplication. We'll see several. Here is the definition of ring, along with an enumeration of its defining characteristics:

Definition 7.1.1. Suppose R is a nonempty set endowed with binary operations $+$ (addition) and $\cdot$ (multiplication), such that R is an abelian group under $+$, with additive identity 0 and additive inverse operation $-$, and $\cdot$ is an associative binary operation that distributes over $+$ from the left and right; that is, $a(b+c) = ab + ac$ and $(b+c)a = ba + ca$ for all $a, b, c \in R$. Then the algebraic structure $(R, +, 0, -, \cdot)$ is called a *ring*. If multiplication is a commutative operation, then R is called a *commutative ring*. If there exists a *nonzero* identity element for multiplication, we denote such an element e, and it is called a *unity element* or simply a *unity*.

Before we make some comments about Definition 7.1.1, let's spell out the essential features of a ring in glorious detail:

(R1) Addition is well defined on R (G1).

(R2) Addition is closed on R (G2).

(R3) Addition is associative (G3).

(R4) There exists $0 \in R$ such that $a + 0 = a$ for all $a \in R$ (G4).

(R5) For all $a \in R$, there exists $-a \in R$ such that $a + (-a) = 0$ (G5).

(R6) Addition is commutative (abelian).

(R7) Multiplication is well defined on R (G1).

(R8) Multiplication is closed on R (G2).

(R9) Multiplication is associative (G3).

(R10) For all $a, b, c \in R$, $a(b+c) = ab + ac$ and $(b+c)a = ba + ca$.

Definition 7.1.1 requires that, if a ring R has a unity element e, it must be nonzero. The only reason for this is that $e = 0$ causes R to collapse to $\{0\}$, just as it does in $\mathbb{R}$

if $1 = 0$. Since the trivial ring $\{0\}$ is not particularly exciting in its internal structure, and since an occasional general result about rings with unity would not apply to $\{0\}$, we simply insist for convenience that $e \neq 0$.

Definition 7.1.1 does not mention the existence of multiplicative inverses. Certainly if a ring has a unity, some elements other than e itself might have a multiplicative inverse. We'll wait until Section 7.2 before we address their existence.

7.1.2 Examples of rings

We can only hope that many our examples of groups and their binary operations can serve as the raw material from which to construct rings, or at least that new settings we'll create will involve binary operations whose basic features are transparent enough to make our path through the verification of properties R1–R10 quick and relatively painless. Here are a few examples of rings. We'll return to all of them several times in later sections.

Example 7.1.2. The integers with addition and multiplication form a commutative ring with unity element 1. Similarly, $\mathbb{Q}, \mathbb{R}$, and $\mathbb{C}$ under the same operations are commutative rings with unity. The even integers are a commutative ring without unity.

Example 7.1.3. For $n \geq 2$, consider $\mathbb{Z}_n$ with addition $\oplus_n$ defined as in Eq. (6.42). Define multiplication $\otimes_n$ in an analogous way:

$$[(n) + a] \otimes_n [(n) + b] = (n) + ab. \tag{7.1}$$

Table 6.2 illustrates the behavior of $\otimes_n$ in $\mathbb{Z}_6$. From all our work in Section 6.3, properties R1–R6 are satisfied. Properties R7–R10 must be shown (Exercise 1). Additionally, multiplication is commutative, and there exists a multiplicative identity. Thus, $(\mathbb{Z}_n, \oplus_n, (n) + 0, -, \otimes_n)$ is a commutative ring with unity $(n) + 1$.

For several reasons it would be a shame not to introduce an example of a matrix ring at this point. First, as a mathematical structure, matrices are very important in theoretical and applied mathematics. They are the bread and butter of linear algebra and of many highly computational processes that have only become practical since computers have been with us. For our purposes, they are a storehouse of examples that exhibit all kinds of interesting behaviors.

Example 7.1.4. A *matrix* is a two-dimensional array, typically of real numbers:

$$A = \begin{bmatrix} a_{1,1} & a_{1,2} & \cdots & a_{1,n} \\ a_{2,1} & a_{2,2} & \cdots & a_{2,n} \\ \vdots & \vdots & \ddots & \vdots \\ a_{m,1} & a_{m,2} & \cdots & a_{m,n} \end{bmatrix} \tag{7.2}$$

The matrix A in Eq. 7.2 is said to have dimensions m by n, and the set of all $m \times n$ matrices with real number entries is denoted $\mathbb{R}_{m \times n}$. If A has dimensions $m \times n$, we sometimes write it as $A_{m \times n}$ if we need to display its dimensions. Notice how entries of A are tagged with a row and a column number, and in that order. The commas in the subscripts are annoying to write and probably not necessary, so we'll omit them if we can get away with it and still be clear. Thus a_{42} means the entry down in row 4 and across in column 2. Most of the matrices we'll use will be 2×2 because we don't need to be more general than that to construct some interesting examples.

Before we even discuss binary operations on sets of matrices, we need first to define what it means for two matrices to be equal. We'll say two matrices $A_{m_1 \times n_1}$ and $B_{m_2 \times n_2}$ are equal if two conditions are satisfied. First, the dimensions of A must be the same as those of B; that is, $m_1 = m_2$ and $n_1 = n_2$. Second, all their corresponding entries must be equal: $a_{jk} = b_{jk}$ for all $1 \le j \le m_1 = m_2$ and $1 \le k \le n_1 = n_2$. This definition is clearly an equivalence relation because properties E1–E3 are satisfied for the dimensions and all the real number entries in the matrices involved reveal that our definition of matrix equality also satisfies properties E1–E3.

Addition of matrices requires that they have the same dimensions, and $A + B$ is merely the matrix whose entries are the sums of the corresponding entries from A and B. For the 2×2 case, writing $A = \begin{bmatrix} a_{11} & a_{12} \\ a_{21} & a_{22} \end{bmatrix}$ and $B = \begin{bmatrix} b_{11} & b_{12} \\ b_{21} & b_{22} \end{bmatrix}$, we have

$$A + B = \begin{bmatrix} a_{11} + b_{11} & a_{12} + b_{12} \\ a_{21} + b_{21} & a_{22} + b_{22} \end{bmatrix}. \tag{7.3}$$

Matrix multiplication is more complicated than addition. In general, in order for the product AB to be defined, A and B do not have to be the same dimensions, but there is some relationship between their dimensions that must be satisfied. Since two square matrices ($n \times n$) can always be multiplied to produce another $n \times n$ matrix, we'll define multiplication only for this case. The 2×2 case should get the point across, but we'll explain it in more general language. Writing

$$A = \begin{bmatrix} a & b \\ c & d \end{bmatrix} \quad \text{and} \quad B = \begin{bmatrix} e & f \\ g & h \end{bmatrix}, \tag{7.4}$$

we define AB in the following way. To calculate entry (j, k) (row j, column k) in the product, mentally highlight row j of A and column k of B. Mentally run your fingers across row j of A and down column k of B, multiplying the pairs $a_{j1}b_{1k}$, $a_{j2}b_{2k}$, etc., and add up these products. This sum of product pairs is entry (j, k) in AB. Thus, for A and B in Eq. (7.4),

$$AB = \begin{bmatrix} ae + bg & af + bh \\ ce + dg & cf + dh \end{bmatrix}. \tag{7.5}$$

If you create two 2×2 matrices pretty much at random, you'll probably see that matrix multiplication defined this way is not commutative. With these definitions of matrix addition and multiplication, let's consider

$$\mathbb{R}_{2 \times 2} = \left\{ \begin{bmatrix} a & b \\ c & d \end{bmatrix} : a, b, c, d \in \mathbb{R} \right\} \tag{7.6}$$

and show that it's a noncommutative ring with a unity element. That matrix addition is well defined (R1) is transparent and notationally tedious. If $A = B$ and $C = D$, then showing $A + C = B + D$ amounts to nothing more than applying the fact that addition is well defined on $\mathbb{R}$ to each entry in $A + C$ and $B + D$. Similar drudgery reveals that matrix multiplication is well defined, so property R7 holds. Closure of addition and multiplication in $\mathbb{R}$ means that all entries in $A + B$ and AB are real numbers, so that properties R2 and R8 hold. Associativity and commutativity of addition in $\mathbb{R}$ means properties R3 and R6 hold. The matrix denoted $\mathbf{0}$, with all zero entries, is the additive identity, and

$$-\begin{bmatrix} a & b \\ c & d \end{bmatrix} = \begin{bmatrix} -a & -b \\ -c & -d \end{bmatrix}. \tag{7.7}$$

The only remaining properties are R9 and R10, associativity of multiplication, and left and right distributivity. These are not immediately obvious, but verifying them with all the necessary notation is about as exciting as counting stripes on the highway, and you're about as likely to make a mistake.

Thus $\mathbb{R}_{2\times 2}$ is a noncommutative ring. Does it have a unity element? Why yes it does. Writing

$$I_{n\times n} = \begin{bmatrix} 1 & 0 & 0 & \cdots & 0 \\ 0 & 1 & 0 & \cdots & 0 \\ 0 & 0 & 1 & \cdots & 0 \\ \vdots & \vdots & \vdots & \ddots & \vdots \\ 0 & 0 & 0 & \cdots & 1 \end{bmatrix}, \tag{7.8}$$

the matrix with ones down what we call the *main diagonal* and zeroes elsewhere, you can see that $I_{2\times 2}A = AI_{2\times 2} = A$ for all $A \in \mathbb{R}_{2\times 2}$. Finally, by exactly the same reasoning already used here, $\mathbb{R}_{n\times n}$ is a noncommutative ring with unity element $I_{n\times n}$, as are $\mathbb{Q}_{n\times n}$ and $\mathbb{Z}_{n\times n}$.

Here's one more example of a ring that we can create from two given rings R and S. Verifying that this creation is a ring is both tedious and transparent, but you should at least mentally walk through the steps (or at least the first few of them) to see that it satisfies properties R1–R10. Rather than going crazy with notation to distinguish operations and elements of R from those of S, writing expressions like 0_R and $+_S$, we'll assume your acquired level of mathematical sophistication makes them unnecessary. Just make sure you notice which set all the operations are being performed in.

Example 7.1.5. Suppose R and S are rings, and consider

$$R \times S = \{(r, s) : r \in R, s \in S\}, \tag{7.9}$$

the Cartesian product of R and S. Define $(r_1, s_1) = (r_2, s_2)$ in $R \times S$ if $r_1 = r_2$ and $s_1 = s_2$. Define addition and multiplication in $R \times S$ by

$$(r_1, s_1) + (r_2, s_2) = (r_1 + r_2, s_1 + s_2) \quad \text{and} \quad (r_1, s_1) \cdot (r_2, s_2) = (r_1 r_2, s_1 s_2). \tag{7.10}$$

Then $R \times S$ is a ring under the operations defined in Eqs. (7.10). The zero element of $R \times S$ is $(0, 0)$, and $-(r, s) = (-r, -s)$. If R and S each have a unity element, then so does $R \times S$, and $e_{R\times S} = (e_R, e_S)$.

7.1.3 Subrings

If R is a ring and $S \subseteq R$ is also a ring under the same operations, we say that S is a *subring* of R. Demonstrating $S \subseteq R$ is a subring, some of properties R1–R10 are inherited from R, while some must be shown for S. Take a look at properties R1–R10 again and note for yourself which ones must be demonstrated for S. Here they are:

(S1) S is closed under addition (R2, H1).

(S2) S contains the additive identity (R4, H2).

(S3) S is closed under additive inverses (R5, H3).

(S4) S is closed under multiplication (R8, H1).

Thus a subring S of a ring R is merely a subgroup of the additive group that is also closed under multiplication. We call $\{0\}$ the *trivial* subring, and all subrings other than $\{0\}$ and R itself are called *proper* subrings. If R has a unity element e, it is not necessary that $e \in S$ in order for S to be a subring of R.

Example 7.1.6. The set of even integers is a subring of $\mathbb{Z}$. It is also a subring of $\mathbb{Q}$ and $\mathbb{R}$. The integers are a subring of $\mathbb{Q}$ and of $\mathbb{R}$. The rationals are a subring of $\mathbb{R}$.

Example 7.1.7. $\mathbb{Z}_{2\times 2}$ is a subring of $\mathbb{R}_{2\times 2}$.

Example 7.1.8. Call a square matrix *diagonal* if its only nonzero entries lie on the main diagonal. Let $D_{2\times 2}$ be the subset of $\mathbb{Z}_{2\times 2}$ consisting of the diagonal matrices. Then $D_{2\times 2}$ is a subring of $\mathbb{Z}_{2\times 2}$ (Exercise 2).

Example 7.1.9. In this example, we create a subring of $\mathbb{Q}$ that we'll return to in Section 7.6. Let $\mathbb{Q}_{OD}$ be the subset of $\mathbb{Q}$ whose denominators are odd. There is more than one way to denote elements of $\mathbb{Q}_{OD}$. The form

$$\left\{ \frac{m}{2n+1} : m, n \in \mathbb{Z} \right\} \tag{7.11}$$

is an obvious way, but another useful way to denote the set is to exploit the prime factorization of the numerator and denominator, isolating 2 to keep it separate from all the other primes involved. This works for all elements except zero, which we'll throw in separately.

$$\mathbb{Q}_{OD} = \{0\} \cup \left\{ \pm \frac{2^n p_1 p_2 \cdots p_r}{q_1 q_2 \cdots q_s} : n \in \mathbb{W} \text{ and } p_i, q_i \text{ are } odd \text{ primes} \right. \tag{7.12}$$
$$\left. \text{for all } 1 \leq i \leq r \text{ and } 1 \leq i \leq s \right\}.$$

How much repetition there is among the p_i and q_i doesn't matter. And notice that the form of an element in Eq. (7.12) includes 1 by letting $n = 0$, $r = s = 1$, and $p_1 = q_1$. Verifying that $\mathbb{Q}_{OD}$ has properties S1–S4 is quick. Using either form (7.11) or (7.12), we see that both addition and multiplication are closed because the product of denominators of two elements is odd. That $\mathbb{Q}_{OD}$ contains 0 and additive inverses is obvious. Thus $\mathbb{Q}_{OD}$ is a subring of $\mathbb{Q}$.

If you're required to show that a particular $S \subseteq R$ is a subring of R, you might save yourself some time if you exploit the fact that S1–S3 are merely the subgroup properties H1–H3 applied to R as an abelian additive group. If you already know that $(S, +, 0, -)$ is a subgroup of $(R, +, 0, -)$, then a lot of your work in showing S is a subring of R is already done. Keep that in mind when you prove the following in Exercise 4.

Theorem 7.1.10. *Suppose $\mathcal{F}$ is a family of subrings of a ring R. Then $\cap_{S \in \mathcal{F}} S$ is a subring of R.*

Although subrings are interesting in their own right, they are bland when compared to the special kind of subring called an *ideal*. Ideals come in all kinds of interesting flavors, and we'll taste several in Section 7.4.

EXERCISES

1. For $\mathbb{Z}_n$ in Example 7.1.3, show the following:
   (a) Addition and multiplication satisfy properties R7–R10.
   (b) Multiplication is commutative.
   (c) There exists a multiplicative identity.

2. Show that the set $D_{2 \times 2}$ in Example 7.1.8 is a subring of $\mathbb{Z}_{2 \times 2}$.

3. Show that the set of all rational numbers with even numerator and odd denominator is a subring of $\mathbb{Q}_{OD}$.

4. Prove Theorem 7.1.10: Suppose $\mathcal{F}$ is a family of subrings of a ring R. Then $\cap_{I \in \mathcal{F}} I$ is a subring of R.

5. Define e_R to be a right unity for a ring R if $r \cdot e_R = r$ for all $r \in R$. Similarly, define e_L to be a left unity if $e_L \cdot r = r$ for all $r \in R$. Let

$$L = \left\{ \begin{bmatrix} a & 0 \\ b & 0 \end{bmatrix} : a, b \in \mathbb{R} \right\}. \tag{7.13}$$

   (a) Show L is a ring by showing it is a subring of $\mathbb{R}_{2 \times 2}$.
   (b) Find an element of L that is a right unity but not a left unity.
   (c) Show that this right unity is not unique.

7.2 Ring Properties and Fields

7.2.1 Ring properties

Because a ring is an abelian group under its addition operation, all the properties of abelian groups that you proved in Chapter 6 apply to addition. With regard to multiplication and its interaction with addition, it would probably be a good idea to swing back by Section 2.3 and point out those theorems we proved for $\mathbb{R}$ that used only its ring properties. Many theorems will then translate very similarly over to a general ring. The only difference is that multiplication might not be commutative, so we have to state and prove certain theorems in two-sided language to get the full strength. We'll state them here, with appropriate comments along the way. You'll prove some of them in Exercises 1 and 2. The corollaries should be mere observations.

Theorem 7.2.1. *If R is a ring, then $a \cdot 0 = 0 \cdot a = 0$ for all $a \in R$.*

Theorem 7.2.1 implies that a ring with unity will not contain 0^{-1}.

Theorem 7.2.2. *If R is a ring and $a, b \in R$, then*

1. $(-a)b = -(ab)$,
2. $a(-b) = -(ab)$,
3. $(-a)(-b) = ab$.

Corollary 7.2.3. *If R is a ring with unity e, then $(-e)a = a(-e) = -a$ for all $a \in R$.*

If R has a unity element, the next theorem follows immediately from Corollary 7.2.3 in the same way that Corollary 2.3.8 follows from Corollary 2.3.7 for $\mathbb{R}$. However, if R has no unity element, such a proof won't work. Not to worry. Since a ring is an abelian group with respect to addition, Eq. (6.18) applies, which says for the $n = -1$ case that in an abelian group $(G, *, e, ^{-1})$, $(a * b)^{-1} = a^{-1} * b^{-1}$. Since the operation we're working with in R is addition, that translates to the following additive form.

Theorem 7.2.4. *If R is a ring, then $-(a + b) = (-a) + (-b)$ for all $a, b \in R$.*

The distributive property extends nicely in a general ring to yield the following result analogous to Exercises 3 and 4 from Section 2.4.

Theorem 7.2.5. *If R is a ring and $a, b_1, b_2, \ldots, b_n \in R$, then*

$$a \sum_{k=1}^{n} b_k = \sum_{k=1}^{n} (ab_k). \tag{7.14}$$

Theorem 7.2.6. *If R is a ring and $a_1, a_2, \ldots, a_m, b_1, b_2, \ldots, b_n \in R$, then*

$$\left(\sum_{j=1}^{m} a_j\right)\left(\sum_{k=1}^{n} b_k\right) = \sum_{j=1}^{m}\left(\sum_{k=1}^{n} a_j b_k\right). \tag{7.15}$$

Just as in a group, if a ring has a unity element, it can have only one. Your proof from Exercise 4 in Section 6.1 will probably translate directly over to a general ring pretty much word for word.

Theorem 7.2.7. *If R is a ring with unity, then the unity element is unique.*

Even though elements of a ring with unity are not assumed to have multiplicative inverses, it might be that some of them do. If for $x \in R$ there exists $y \in R$ such that $xy = yx = e$, then x is called a *unit* of R. Notice that Theorem 7.2.1 implies that zero is not a unit. In Exercise 3 you'll find the units of certain rings.

Divisibility in a ring is defined in pretty much the same way it is for $\mathbb{Z}$, except that we must distinguish between left and right divisors.

Definition 7.2.8. Let R be a ring, $a, b \in R$, and $a \neq 0$. Then a is called a *left divisor* of b if there exists nonzero $k \in R$ such that $ak = b$. Similarly, a is called a *right divisor* of b if there exists nonzero $k \in R$ such that $ka = b$. If R is a commutative ring and there exists nonzero $k \in R$ such that $ak = b$, we say simply that a is a *divisor* of b, or that *a divides b*, and we write $a \mid b$.

If a divides b, it does not necessarily mean that the $k \in R$ such that $ak = b$ or $ka = b$ is unique. In Exercise 4, you'll show that this is true. In a more specialized ring we'll study in Section 7.5, however, we will have uniqueness.

Having defined divisors, we can now define what it means for an element of a ring to be prime, although we won't really look into any of its properties until Section 7.5. To motivate the term by returning to $\mathbb{N}$, a prime p has exactly two distinct natural number divisors, 1 and p. By definition, 1 is not called prime. Thus if $p \in \mathbb{N}$ is prime and $p = ab$ is any factorization of p where $a, b \in \mathbb{N}$, then either $a = 1$ or $b = 1$. In the ring of integers, primes are extended to include the negatives of the primes in $\mathbb{N}$. Thus in $\mathbb{Z}$, for p to be prime means that if $p = ab$ is any factorization of p where $a, b \in \mathbb{Z}$, then either a or b is ± 1. If you've already done Exercise 3a, you showed that ± 1 are the units in $\mathbb{Z}$. In a general ring, a prime element is defined by using the language of divisors and units.

Definition 7.2.9. Suppose R is a ring, and $p \in R$ is not a unit. Then p is said to be prime if every factorization $p = ab$ with $a, b \in R$ implies either a or b is a unit in R.

Definition 7.2.9 automatically excludes zero from being prime, for $0 \cdot 0 = 0$, and 0 is not a unit.

In Exercises 2 and 6 from Section 2.3, you proved multiplicative cancellation for nonzero real numbers and the principle of zero products. In a ring, these properties do not necessarily apply, either as part of the definition or as logical consequences of it. These properties will reappear in Section 7.5, though, when we discuss integral domains. From Definition 7.2.8, if $ab = 0$ while neither a nor b is zero, then a and b are called *divisors of zero* or *zero divisors*. In Exercise 5, you'll find some zero divisors in certain rings, and in Exercise 7, you'll prove the following:

Theorem 7.2.10. *If a is a divisor of zero in a ring with unity, then a is not a unit.*

With Theorem 7.2.10, you can prove the following (Exercise 8):

Theorem 7.2.11. *If R is a ring with unity and $x \in R$ is a unit, then the $y \in R$ satisfying $xy = yx = e$ is unique.*

Many of the results in Section 2.3 involved ordering of real numbers as measured by $<$. Since a ring doesn't necessarily have any such way of comparing its elements, none of these results have meaning in a ring without such a basis for comparison being defined. Thus, rings do not necessarily have positive and negative elements, there is not necessarily a way to measure the size of elements as absolute value does in $\mathbb{R}$, and there's not necessarily a way to make the WOP applicable to subsets. Just because the equation $x^2 = -1$ has no solution in $\mathbb{R}$ does not mean that the equation $x^2 = -e$ cannot have a solution.

Since a ring with its addition operation is an abelian group, we can apply the additive forms of the recursive definitions in Eqs. (6.13–6.15) as you wrote them in Exercise 2 from Section 6.2:

$$0a = 0, \tag{7.16}$$

$$(n+1)a = na + a, \tag{7.17}$$

$$(-n)a = -(na). \tag{7.18}$$

As we stated in Section 6.2, it's important to keep in mind what is an integer and what is a ring element in these equations.

Example 7.2.12. In $\mathbb{Z}_{2\times 2}$, let $A = \begin{bmatrix} 1 & 2 \\ -1 & 0 \end{bmatrix}$. Then

$$0A = 0 \begin{bmatrix} 1 & 2 \\ -1 & 0 \end{bmatrix} = \begin{bmatrix} 0 & 0 \\ 0 & 0 \end{bmatrix}, \tag{7.19}$$

$$6A = \overbrace{\begin{bmatrix} 1 & 2 \\ -1 & 0 \end{bmatrix} + \cdots + \begin{bmatrix} 1 & 2 \\ -1 & 0 \end{bmatrix}}^{6 \text{ times}} = \begin{bmatrix} 6 & 12 \\ -6 & 0 \end{bmatrix}, \tag{7.20}$$

$$-6A = -(6A) = -\begin{bmatrix} 6 & 12 \\ -6 & 0 \end{bmatrix} = \begin{bmatrix} -6 & -12 \\ 6 & 0 \end{bmatrix}. \tag{7.21}$$

Writing Eqs. (6.16–6.18) in their additive forms as you also did in Section 6.2, we have for all $a, b \in R$ and $m, n \in \mathbb{Z}$

$$ma + na = (m+n)a, \tag{7.22}$$

$$m(na) = (mn)a, \tag{7.23}$$

$$n(a+b) = na + nb. \tag{7.24}$$

Again, be careful to distinguish Eqs. (7.22–7.24) from the associative and distributive properties that characterize R and $\mathbb{Z}$ individually.

Example 7.2.13. For $A = \begin{bmatrix} 1 & 2 \\ -1 & 0 \end{bmatrix}$ and $B = \begin{bmatrix} 5 & 0 \\ 2 & 2 \end{bmatrix}$:

$$4A + 2A = 4 \begin{bmatrix} 1 & 2 \\ -1 & 0 \end{bmatrix} + 2 \begin{bmatrix} 1 & 2 \\ -1 & 0 \end{bmatrix} = \begin{bmatrix} 4 & 8 \\ -4 & 0 \end{bmatrix} + \begin{bmatrix} 2 & 4 \\ -2 & 0 \end{bmatrix} = \begin{bmatrix} 6 & 12 \\ -6 & 0 \end{bmatrix}$$

$$= 6 \begin{bmatrix} 1 & 2 \\ -1 & 0 \end{bmatrix} = 6A = (4+2)A, \tag{7.25}$$

$$4(2A) = 4 \begin{bmatrix} 2 & 4 \\ -2 & 0 \end{bmatrix} = \begin{bmatrix} 8 & 16 \\ -8 & 0 \end{bmatrix} = 8 \begin{bmatrix} 1 & 2 \\ -1 & 0 \end{bmatrix} = 8A, \tag{7.26}$$

$$4(A+B) = 4\left(\begin{bmatrix} 1 & 2 \\ -1 & 0 \end{bmatrix} + \begin{bmatrix} 5 & 0 \\ 2 & 2 \end{bmatrix}\right) = 4\begin{bmatrix} 6 & 2 \\ 1 & 2 \end{bmatrix} = \begin{bmatrix} 24 & 8 \\ 4 & 8 \end{bmatrix}$$

$$= \begin{bmatrix} 4 & 8 \\ -4 & 0 \end{bmatrix} + \begin{bmatrix} 20 & 0 \\ 8 & 8 \end{bmatrix} = 4\begin{bmatrix} 1 & 2 \\ -1 & 0 \end{bmatrix} + 4\begin{bmatrix} 5 & 0 \\ 2 & 2 \end{bmatrix} = 4A + 4B. \quad (7.27)$$

Even though the multiplication operation in a ring is not necessarily accompanied by an identity and inverses for elements, we can use the multiplicative forms of the definitions of a^n in a limited way. For any $a \in R$, we begin by defining $a^1 = a$ and $a^{n+1} = a^n \cdot a$ for $n \geq 1$. If R has a unity element e, we also define $a^0 = e$, but only for $a \neq 0$. If a is a unit of R, we can define $a^{-n} = (a^{-1})^n$ for $n \in \mathbb{N}$.

Example 7.2.14. In $\mathbb{Z}_{10}$ from Example 7.1.3,

$$3^0 = 1, \quad 3^1 = 3, \quad 3^2 = 3 \otimes_n 3 = 9, \quad 3^3 = 9 \otimes_n 3 = 7, \quad \text{etc.} \quad (7.28)$$

$$5^0 = 1, \quad 5^1 = 5, \quad 5^2 = 5 \otimes_n 5 = 5, \quad 5^3 = 5 \otimes_n 5 = 5, \quad \text{etc.} \quad (7.29)$$

Since 3 is a unit in $\mathbb{Z}_{10}$, and $3^{-1} = 7$, we have

$$3^{-1} = 7, \quad 3^{-2} = (3^{-1})^2 = 7^2 = 9, \quad 3^{-3} = 7^3 = 3, \quad \text{etc.} \quad (7.30)$$

Example 7.2.15. In $\mathbb{Z}_{2\times 2}$, let $A = \begin{bmatrix} 1 & 2 \\ -1 & 0 \end{bmatrix}$. Then

$$A^0 = I = \begin{bmatrix} 1 & 0 \\ 0 & 1 \end{bmatrix}$$

$$A^1 = \begin{bmatrix} 1 & 2 \\ -1 & 0 \end{bmatrix}$$

$$A^2 = A \times A = \begin{bmatrix} -1 & 2 \\ -1 & -2 \end{bmatrix}$$

$$A^3 = A^2 \times A = \begin{bmatrix} -3 & -2 \\ 1 & -2 \end{bmatrix}, \quad \text{etc.}$$

With these definitions of a^n for appropriate $n \in \mathbb{Z}$, and by arguments exactly like those in Exercise 6 from Section 2.4, we have the following:

Theorem 7.2.16. *Suppose R is a ring, $a, b \in R$, and $m, n \in \mathbb{N}$. Then*

$$a^m \cdot a^n = a^{m+n} \quad (7.31)$$

$$(a^m)^n = a^{mn}. \quad (7.32)$$

If R has a unity element and $a \neq 0$, Eqs. 7.31 and 7.32 hold for $m, n \in \mathbb{W}$. If a is a unit, Eqs. (7.31) and (7.32) hold for all $m, n \in \mathbb{Z}$. Furthermore, if R is commutative,

$$(ab)^n = a^n b^n \quad (7.33)$$

for all $n \in \mathbb{Z}$ for which a^n and b^n are defined.

One big difference in the way we mentally visualize $\mathbb{Z}$ and $\mathbb{Z}_n$ is that $\mathbb{Z}$ extends out indefinitely along the numberline in both directions, whereas $\mathbb{Z}_n$ is circular. If we generate both $\mathbb{Z}$ and $\mathbb{Z}_n$ by considering $1, 1+1, 1+1+1$, etc., no expression of the form $n1 = \sum_{k=1}^{n} 1$ ever produces a sum of zero in $\mathbb{Z}$, but $n1 = \sum_{k=1}^{n} 1 = 0$ in $\mathbb{Z}_n$. In a general ring with unity element e, whether or not some expression $ne = \sum_{k=1}^{n} e = 0$ ever occurs motivates a term.

Definition 7.2.17. Suppose R is a ring with unity, and suppose there exists $n \in \mathbb{N}$ such that $ne = 0$. Then the smallest such n for which this holds is called the *characteristic* of R, and is denoted char R. If no such $n \in \mathbb{N}$ exists, then R is said to have characteristic zero.

In $\mathbb{Z}_n$ the fact that $n \equiv_n 0$, or $n1 = 0$ is zero in $\mathbb{Z}_n$ means that char $\mathbb{Z}_n \leq n$. On the other hand, if $m \in \mathbb{N}$ and $m \equiv_n 0$, that is, if $m1$ is zero in $\mathbb{Z}_n$, then $m = kn$ for some $k \in \mathbb{N}$, so that $m \geq n$. Thus char $\mathbb{Z}_n \geq n$, so that we have proved the following theorem:

Theorem 7.2.18. *If $n \in \mathbb{N}$, then* char $\mathbb{Z}_n = n$.

If adding the unity element to itself n times produces a sum of zero, then the same is true for all elements of the ring (Exercise 11).

Theorem 7.2.19. *Let R be a ring with unity where* char $R = n \neq 0$. *Then $nx = 0$ for all $x \in R$.*

7.2.2 Fields defined

It might seem strange to introduce our next term at this point, but it turns out to be more convenient as we progress through the theory of rings. A *field* is a special kind of ring, and its defining characteristics make it the most specialized kind of ring we'll study in this text. Although we won't delve deeply into a general theory of fields, we will notice a few of their characteristics that are easy to pick up along our way.

Definition 7.2.20. Suppose K is a commutative ring with unity, with the property that every nonzero element has a multiplicative inverse. Then K is called a *field*.

In addition to properties R1–R10, a field K must have the following features:

(K11) There exists $e \in K \setminus \{0\}$ such that $e \cdot k = k$ for all $k \in K$ (G4).

(K12) For all $k \in K \setminus \{0\}$, there exists $k^{-1} \in K \setminus \{0\}$ such that $k \cdot k^{-1} = e$ (G5).

(K13) Multiplication is commutative (abelian).

Notice that properties K11–K13 complete the requirements for $K \setminus \{0\}$ to be an abelian group under multiplication. Thus a shorthand way of defining a field K is to say that K is an abelian group under addition, and $K \setminus \{0\}$ is an abelian group under multiplication.

Example 7.2.21. $\mathbb{Q}$ and $\mathbb{R}$ are fields. Also, $\mathbb{C}$ is a field. In Section 6.1, we showed that $\mathbb{C}$ with addition is an abelian group, and in Exercise 1 from Section 6.1, you showed that $\mathbb{C}\setminus\{0+0i\}$ with multiplication is an abelian group.

EXERCISES

1. Prove Theorem 7.2.1: If R is a ring, then $a \cdot 0 = 0 \cdot a = 0$ for all $a \in R$.

2. Prove Theorem 7.2.2: If R is a ring and $a, b \in R$, then:
   (a) $(-a)b = -(ab)$,
   (b) $a(-b) = -(ab)$,
   (c) $(-a)(-b) = ab$.

3. Find, with verification, all units in the following rings:
   (a) $\mathbb{Z}$
   (b) $\mathbb{Z}_{12}$
   (c) $\mathbb{Z}_7$
   (d) $D_{2\times 2}$ from Example 7.1.8

4. Find *nonzero* elements a, b, k in each of the following rings where $ak = b$ but k is not unique.
   (a) $\mathbb{Z}_{12}$
   (b) $\mathbb{Z}_{2\times 2}$

5. Find a zero divisor in each of the following rings.
   (a) $\mathbb{Z}_n$ for some strategically chosen n
   (b) $\mathbb{Z}_{2\times 2}$

6. Let R and S be rings. Find all units and zero divisors in $R \times S$, in terms of the units and zero divisors of R and S individually.

7. Prove Theorem 7.2.10: If a is a divisor of zero in a ring with unity, then a is not a unit.

8. Prove Theorem 7.2.11: If R is a ring with unity and $x \in R$ is a unit, then the $y \in R$ satisfying $xy = yx = e$ is unique.

9. Suppose R is a ring with unity element e. Show $(me)(ne) = (mn)e$ for all $m, n \in \mathbb{N}$.

10. What is char $(\mathbb{Z}_4 \times \mathbb{Z}_{18})$? Explain.

11. Prove Theorem 7.2.19: Let R be a ring with unity where char $R = n \neq 0$. Then $nx = 0$ for all $x \in R$.[1]

---

[1] Use Theorem 7.2.5.

7.3 Ring Extensions

There is a very important type of algebraic structure that we create from a given algebraic structure by tossing in a new element, stirring well, and letting the mixture expand into another algebraic structure of the same type. It's called the process of *adjoining* an element, to create what is called an *extension* of the original structure. In this section we want to get acquainted with the creation of extensions by adjoining elements to *commutative* rings. In principle, there are really only two types of ring extensions that can result from adjoining an element. We'll begin with a very specific example of the first type, but instead of building it up as an extension of a certain ring in the most rigorous way, we'll just lay the whole structure out there, define equality and the operations, and show that what we've presented is a ring. But don't worry. We'll make up for our lax introduction of this ring in Section 7.8, where we'll see a more rigorous way to construct it. After we've presented our example of the first type of ring extension, we'll point out how other extensions of the same type can be created by precisely the same reasoning. Finally, we'll construct the canonical example of the second type. We'll use these constructions over and over throughout the rest of Chapter 7.

7.3.1 Adjoining roots of ring elements

Example 7.3.1. Let $S = \{a + b\sqrt[3]{2} + c\sqrt[3]{4} : a, b, c \in \mathbb{Z}\}$, the set of all integer linear combinations of $\{1, \sqrt[3]{2}, \sqrt[3]{4}\}$. First, we define $x = a_1 + b_1\sqrt[3]{2} + c_1\sqrt[3]{4}$ to be equal to $y = a_2 + b_2\sqrt[3]{2} + c_2\sqrt[3]{4}$ if $a_1 = a_2$, $b_1 = b_2$, and $c_1 = c_2$. Notice that this definition is an equivalence relation because it's just an application of integer equality in triplicate.[2] Define addition $\oplus$ and multiplication $\odot$ in a natural way, based on the extended distributive property and the behavior of $\sqrt[3]{2}$ in the real numbers:

$$(a + b\sqrt[3]{2} + c\sqrt[3]{4}) \oplus (d + e\sqrt[3]{2} + f\sqrt[3]{4}) = (a + d) + (b + e)\sqrt[3]{2} + (c + f)\sqrt[3]{4}$$

$$(a + b\sqrt[3]{2} + c\sqrt[3]{4}) \odot (d + e\sqrt[3]{2} + f\sqrt[3]{4}) = (ad + 2bf + 2ce) + (ae + bd + 2cf)\sqrt[3]{2}$$
$$+ (af + be + cd)\sqrt[3]{4}. \qquad (7.34)$$

Because $\oplus$ and $\odot$ on S have the familiar behavior we expect when viewed within the context of the real numbers, we could simply use $+$ and $\cdot$. Just remember that a single

---

[2] This definition of equality will raise all kinds of concerns in the mind of your professor because of something you'll probably just assume without any basis. You're probably thinking that our definition of equality here coincides exactly with equality in $\mathbb{R}$, so that two expressions in S are equal if and only if they are equal in $\mathbb{R}$. Clearly, if $x = y$ as we've defined equality for S, then $x = y$ in $\mathbb{R}$. But just because $a_1 + b_1\sqrt[3]{2} + c_1\sqrt[3]{4} = a_2 + b_2\sqrt[3]{2} + c_2\sqrt[3]{4}$ in $\mathbb{R}$, we cannot conclude immediately that $a_1 = a_2$, $b_1 = b_2$, and $c_1 = c_2$. If you crank out $5,096,516,652 - 5184\sqrt[3]{2} + 91,047,715,794\sqrt[3]{4}$ and $2,669,624,714 + 130,936,500,093\sqrt[3]{2} - 11,347,811,196\sqrt[3]{4}$ on a TI-85 calculator, it appears they might be equal in $\mathbb{R}$. They're not, and it is indeed true that equality in $\mathbb{R}$ implies equality in S. That is, if there's a way to write a real number in the form $a + b\sqrt[3]{2} + c\sqrt[3]{4}$, then there's only one way to do it. To prove this, you would need to know more about $\sqrt[3]{2}$ and $\sqrt[3]{4}$ as real numbers and their relationship to each other in the context of $\mathbb{Z}$. The term is *linear independence*, and you'll see it in linear algebra.

entity in S is of the form $a + b\sqrt[3]{2} + c\sqrt[3]{4}$, and includes partial forms like $1, 6\sqrt[3]{2}$, or $4 - \sqrt[3]{4}$ by letting certain coefficients be zero.

From Eqs. (7.34) and closure of integer addition and multiplication, the closure of $\oplus$ and $\odot$ are immediately obvious. Furthermore, S contains the additive identity $0 + 0\sqrt[3]{2} + 0\sqrt[3]{4}$, additive inverses $(-a) + (-b)\sqrt[3]{2} + (-c)\sqrt[3]{4}$, and unity element $1 + 0\sqrt[3]{2} + 0\sqrt[3]{4}$. Furthermore, all remaining ring properties are assumed for all of $\mathbb{R}$, and are therefore inherited by S. Since $\odot$ is commutative, S is a commutative ring with unity element.

The notation we use for the ring in Example 7.3.1 is $\mathbb{Z}[\sqrt[3]{2}]$. This notation is meant to denote that the ring of integers has had an additional element $\sqrt[3]{2}$ thrown in, or *adjoined*, as we say. Adjoining an element to an algebraic structure is obviously different from unioning it onto the set. Instead of merely tossing it in as one more additional element, we toss it in, then combine it by $+$ and $\cdot$ with itself and all other elements to expand into a ring. Thus you can see that the presence of $\sqrt[3]{4}$ in $\mathbb{Z}[\sqrt[3]{2}]$ is necessary so that we have closure of multiplication: $(\sqrt[3]{2})^2 = \sqrt[3]{4}$. However, the fact that $(\sqrt[3]{2})^3 = 2 \in \mathbb{Z}$ means that expressions of the form $a + b\sqrt[3]{2} + c\sqrt[3]{4}$ are all that are necessary. By reasoning similar to that in Example 7.3.1, we could begin with $\mathbb{Z}$ (or $\mathbb{Q}$), choose $n \in \mathbb{N}$ and $x \in \mathbb{Z}$, define equality, addition, and multiplication, and show all ring properties (plus commutativity) for

$$\mathbb{Z}[\sqrt[n]{x}] = \{a_0 + a_1\sqrt[n]{x} + a_2\sqrt[n]{x^2} + \cdots + a_{n-1}\sqrt[n]{x^{n-1}} : a_i \in \mathbb{Z}\}. \quad (7.35)$$

The form of elements of $\mathbb{Z}[\sqrt[n]{x}]$ in Eq. (7.35) assumes that n is the smallest natural number such that $x^n \in \mathbb{Z}$, so that none of the terms $\sqrt[n]{x}, \ldots, \sqrt[n]{x^{n-1}}$ is an integer.

For example, $\mathbb{Q}[\sqrt{5}] = \{a + b\sqrt{5} : a, b \in \mathbb{Q}\}$ fits the form of Eq. (7.35). There's no reason x in Eq. 7.35 cannot be negative, and $\mathbb{Z}[\sqrt{-1}] = \mathbb{Z}[i] = \{a + bi : a, b \in \mathbb{Z}\}$ is called the *Gaussian integers*. It turns out that $\mathbb{Z}[\sqrt{-5}]$ is an important ring, and we'll take a look at it in Section 7.5. Finally, $\mathbb{R}[i] = \mathbb{C}$. Notice how $\mathbb{Z}$ is the subring of $\mathbb{Z}[\sqrt[3]{2}]$ consisting of all elements $a + b\sqrt[3]{2} + c\sqrt[3]{4}$ where $b = c = 0$.

For the class of rings in the next theorem, we'll need to have a handle on the units when we get to Section 7.5. The proof of the next theorem is important to work through because you'll need to mimic the algebraic manipulation in some of our later work.

Theorem 7.3.2. *Suppose $p \in \mathbb{N}$ is prime. Then the only units in $\mathbb{Z}[\sqrt{-p}]$ are ± 1.*

Proof: Suppose $a + b\sqrt{-p}$ is a unit. We show that $b = 0$ and $a = \pm 1$ by supposing

$$(a + b\sqrt{-p})(c + d\sqrt{-p}) = 1 \quad (7.36)$$

and drawing conclusions about a, b, c, d. Multiplying out the terms in Eq. (7.36) and using the definition of equality in $\mathbb{Z}[\sqrt{-p}]$ yields

$$ac - pbd = 1 \quad \text{and} \quad ad + bc = 0. \quad (7.37)$$

Squaring each side of the equations in (7.37) yields

$$a^2c^2 - 2pabcd + p^2b^2d^2 = 1 \quad \text{and} \quad a^2d^2 + 2abcd + b^2c^2 = 0. \quad (7.38)$$

Multiplying the second equation in (7.38) by p and adding the two equations yields

$$a^2c^2 + pa^2d^2 + p^2b^2d^2 + pb^2c^2 = 1$$
$$a^2(c^2 + pd^2) + pb^2(pd^2 + c^2) = 1$$
$$(a^2 + pb^2)(c^2 + pd^2) = 1. \tag{7.39}$$

Each factor in Eq. (7.39) is a positive integer because the components are squared and $p > 0$. Thus by Exercise 10f from Section 2.3,

$$a^2 + pb^2 = 1 \quad \text{and} \quad c^2 + pd^2 = 1. \tag{7.40}$$

Furthermore, since $p \geq 2$, it must be that $b = d = 0$, so that $a = \pm 1$. ∎

7.3.2 Polynomial rings

The other type of extension we want to create might seem fundamentally different from the preceding ones, but the principle is really the same. It just has one notable difference in that the new element we adjoin is, in a sense, more foreign to the original ring than numbers like $\sqrt{-5}$ are to $\mathbb{Z}$. The relationship of $\sqrt{-5}$ to $\mathbb{Z}$ is characterized by the fact that $(\sqrt{-5})^2 = -5 \in \mathbb{Z}$, or if you prefer, $(\sqrt{-5})^2 + 5 = 0$. Similarly, if $x = \sqrt{1 + \sqrt[3]{2}}$, then $(x^2 - 1)^3 - 2 = 0$. Thus, as with $\sqrt{-5}$, there is some way to manipulate x using only the ring elements and operations to produce zero. The term that describes this relationship of $\sqrt{-5}$ and $\sqrt{1 + \sqrt[3]{2}}$ to the integers is *algebraic*, and numbers that are not algebraic are called *transcendental*. For example, π is transcendental over $\mathbb{Z}$ because there is no way to combine π and any finite set of integers using the ring operations a finite number of times to produce zero. Strict definitions of these terms will come in your later work in algebra. For now, we simply construct an example where the symbol we adjoin is transcendental over $\mathbb{Z}$ because we define the ring and the behavior of the symbol to make it so.

Let R be a commutative ring, and write $R[t]$ to mean the set of all polynomials in the variable t, where the coefficients are elements of R. That is,

$$R[t] = \{a_n t^n + a_{n-1} t^{n-1} + \cdots + a_1 t + a_0 : n \in \mathbb{W}, a_k \in R \text{ for all } k,$$
$$\text{and } a_n \neq 0 \text{ if } n \neq 0\}. \tag{7.41}$$

The reason we insist on $a_n \neq 0$ for $n \neq 0$ is that it would be kind of silly to suggest that the highest power term of an element of $R[t]$ is $a_n t^n$ when its coefficient a_n wipes it out. However, if $n = 0$, we do want to allow for $a_0 = 0$, the zero polynomial. We address an arbitrary element of $R[t]$ as f, as if it were a function. However, we won't write it as $f(t)$, because we're not interested at this point in an element of $R[t]$ primarily as an expression into which we substitute numeric values for t, but more as a string of symbols whose behavior in the ring we're constructing involves the symbol t merely as a way to describe how elements of $R[t]$ add and multiply. First, we define $f = a_m t^m + \cdots + a_0$ and $g = b_n t^n + \cdots + b_0$ to be equal if $m = n$ and $a_k = b_k$ for all $1 \leq k \leq n$, which is clearly an equivalence relation. Concerning the binary operations on $R[t]$, define addition and multiplication of two elements in the familiar way of adding and multiplying two polynomials. Notationally it's very ugly to state the definitions of addition and multiplication formally, but you're certainly familiar with the way they're done. Assuming this, let's check that all ring properties R1–R10 are satisfied.

First of all, the fact that R has properties R1–R3 and R6 means that $R[t]$ does, too. Letting $n = 0$ and $a_0 = 0$ reveals that $f = 0$ (viewed as a polynomial in $R[t]$ and not as a mere element of R) is the additive identity (R4). The existence of additive inverses in R makes property R5 clear. Considering the way polynomials multiply, properties R7–R10 call on all the similar properties of both addition and commutative multiplication in R. Showing these is a messy task, but not difficult. Furthermore, multiplication in $R[t]$ is commutative because R is commutative, and if R has a unity element e, the polynomial e is the unity element in $R[t]$. Thus $R[t]$ is a commutative ring and has a unity element if R does. The new ring $R[t]$ is called the *polynomial ring* over R.

If we were to write an element of $\mathbb{Z}[\sqrt[3]{2}]$ as $c(\sqrt[3]{2})^2 + b\sqrt[3]{2} + a$, we could say that elements of $\mathbb{Z}[\sqrt[3]{2}]$ are three-term polynomials in the symbol $\sqrt[3]{2}$. The fundamental difference between $\mathbb{Z}[\sqrt[3]{2}]$ and $R[t]$ is that the three-term polynomials $c(\sqrt[3]{2})^2 + b\sqrt[3]{2} + a$ are all that is necessary to have closure of the ring operations when $\sqrt[3]{2}$ is adjoined to $\mathbb{Z}$. The behavior of $\sqrt[3]{2}$, that is, the fact that $(\sqrt[3]{2})^3 = 2 \in \mathbb{Z}$, means that it's not necessary to have terms of the form $(\sqrt[3]{2})^n$ for $n \geq 3$. However, in $R[t]$, polynomials can be of any *degree*, whatever that means.

7.3.3 Degree of a polynomial

If $f \in R[t]$ is written as $f = a_n t^n + \cdots + a_0$, where $a_n \neq 0$, we define the *degree* of f to be n, and we write $\deg f = n$. This definition does not assign a degree to the zero polynomial, so we won't assign a degree to it. Some authors define $\deg 0 = -\infty$. Even though $-\infty$ is not a real number, this degree assignment can serve as a way to make theorems involving $\deg f$ hold for the zero polynomial. Instead of assigning a degree to the zero polynomial and making it a special case in theorems, we will agree that polynomials whose degree we're working with are always nonzero polynomials. In $R[t]$, the degree of a polynomial is a measure of its size, like $|x|$ in $\mathbb{R}$, even though its properties do not really jibe with the norm properties N1–N3 (beginning on page 57) and Theorem 2.3.22. But as we'll see later, it gives us at least some way to apply the WOP to $R[t]\setminus\{0\}$. Here are some properties of polynomial degree that you'll prove in Exercise 5.

Theorem 7.3.3. *Suppose R is a commutative ring and $f, g \in R[t]$ are nonzero polynomials. Then the following hold:*

1. *If $\deg f = \deg g = n$, then $\deg(f + g) \leq n$.*

2. *If $\deg f > \deg g$, then $\deg(f + g) = \deg f$.*

3. $\deg(fg) \leq \deg f + \deg g$.

It's easy to see why part 1 in Theorem 7.3.3 is an inequality, for if $f = 2t^2 + 1$ and $g = -2t^2 + 4t$ are elements of $\mathbb{Z}[t]$, then $f + g = 4t + 1$. However, the fact that part 3 is an inequality instead of an equation might seem strange. The existence of zero divisors in R allows for this odd behavior in $R[t]$. For example, in $\mathbb{Z}_6[t]$,

$$(2t^2 + 3)(3t + 3) = 6t^3 + 6t^2 + 9t + 9 = 3t + 3. \qquad (7.42)$$

Thus, the degree of a product can be strictly less than the sum of the degrees of the factors. Equation (7.42) also illustrates that $f \mid g$ is possible for some f and g where $\deg f > \deg g$. These unfamiliar idiosyncracies can happen because of the presence of zero divisors in R. In Section 7.5, these behaviors will go away when we look at the polynomial ring over an integral domain.

EXERCISES

1. Using i, j, k as in the quaternion group (Example 6.1.12), construct the ring extension $\mathbb{Z}[i, j, k]$, defining equality, addition, and multiplication, then showing that all ring properties R1–R10 are satisfied. Is it commutative?

2. Finding the units in a ring amounts to solving the equation $xy = 1$. In $\mathbb{Z}_6$, the only units are 1 and 5, so the equation $xy \equiv_6 1$ implies $x, y \in \{1, 5\}$. Use this fact and the technique in the proof of theorem 7.3.2 find all 16 units in $\mathbb{Z}_6[\sqrt{2}]$.

3. Find, with verification, all units in $\mathbb{Z}[i]$.

4. Prove that the following commutative rings with unity are fields by showing that every nonzero element is a unit.
   (a) $\mathbb{Q}[\sqrt{2}]$
   (b) $\mathbb{Q}[i]$

5. Prove Theorem 7.3.3: Suppose R is a commutative ring and $f, g \in R[t]$ are nonzero polynomials. Then the following hold:
   (a) If $\deg f = \deg g = n$, then $\deg(f + g) \leq n$.
   (b) If $\deg f > \deg g$, then $\deg(f + g) = \deg f$.
   (c) $\deg(fg) \leq \deg f + \deg g$.

6. Calculate $(2t^2 + 4t + 1)(3t^3 + 3t + 4)$ in $\mathbb{Z}_5[t]$, then in $\mathbb{Z}_6[t]$.

7. Give an example of a ring R and $f \in R[t]$ such that f is a divisor of zero.

7.4 Ideals

7.4.1 Definition and examples

In the same way a normal subgroup is a special kind of subgroup that exhibits a characteristic stronger than closure, we define a special class of subring where we have something stronger than closure of multiplication:

Definition 7.4.1. Suppose R is a ring and $I \subseteq R$ is closed under addition, contains 0, and is closed under additive inverses. Suppose I also has the property that $rx \in I$ for all $x \in I$ and $r \in R$. Then I is called a *left ideal* of R. Similarly, if $xr \in I$ for all $x \in I$ and $r \in R$, then I is called a *right ideal*. If I is both a left and right ideal, it is called a *two-sided ideal*. If R is commutative, there is no distinction between left and right ideals, and we will simply use the term *ideal*. Also, for simplicity, if R is not commutative, then we will refer simply to an *ideal* to denote an arbitrary left or right ideal.

Since R is an abelian group with respect to its addition operation, I is an additive subgroup, even a normal one because R is abelian as an additive group. What makes an ideal, say a left ideal, more than a subring is that it is more than simply closed with regard to multiplication. A left ideal has properties S1–S4, but property S4 is replaced with the stronger property that $rx \in I$ for all $x \in I$ and $r \in R$. Here are the defining properties of a left (or right) ideal:

(Y1) I is closed under addition (S1).

(Y2) I contains the additive identity (S2).

(Y3) I is closed under additive inverses (S3).

(Y4) For all $x \in I$ and $r \in R$, $rx \in I$ (or $xr \in I$).

If I is a left ideal of R, we might say that I is impenetrable from the left against multiplication, even if the thing multiplied by is outside I. An ideal *absorbs*, if you will, multiplication from the left. Similarly for a right ideal.

In a ring R, the ideal $\{0\}$ is called the *trivial* ideal. Also, R is an ideal of itself. All other ideals besides $\{0\}$ and R are called *proper* ideals.

Example 7.4.2. Let S be the set of all polynomials in $\mathbb{Z}[t]$ whose constant term is even. Then S is an ideal in $\mathbb{Z}[t]$ (Exercise 1).

Example 7.4.3. In the ring $\mathbb{Z}$, use the notation $(n) = \{kn : k \in \mathbb{Z}\}$ in precisely the same way we did in Section 6.3. We've already shown that (n) is a subgroup of the additive group, so to show that (n) is an ideal in $\mathbb{Z}$, the only thing that must be shown is that it absorbs multiplication. But this is clearly true, for if $kn \in (n)$ and $m \in \mathbb{Z}$, then $m(kn) = (mk)n \in (n)$.

Example 7.4.3 is a special case of a general type of ideal in a ring R with unity. Starting with any element $a \in R$, we simply create a left (or right) ideal by multiplying a on the left (or right) by every $r \in R$. That leads you right to the following theorem, which you'll prove in Exercise 2.

Theorem 7.4.4. *Suppose R is a ring and $a \in R$. Then the set*

$$Ra = \{ra : r \in R\} \tag{7.43}$$

is a left ideal in R.

By similar reasoning, $aR = \{ar : r \in R\}$ is a right ideal in R. If R does not have a unity element, then the ideals Ra and aR might not contain a.

Example 7.4.5. Let E be the ring of even integers. Then
$$6E = \{\ldots, -24, -12, 0, 12, 24, \ldots\}. \tag{7.44}$$

Example 7.4.6. For the polynomial ring $\mathbb{Z}[t]$, $(3t+2)\mathbb{Z}[t]$ is the ideal of all multiples of $f = 3t + 2$ in $\mathbb{Z}[t]$. Since $\mathbb{Z}[t]$ has unity, $3t + 2 \in (3t+2)\mathbb{Z}[t]$.

A result analogous to Theorem 7.1.10 holds for ideals, but we must distinguish between left and right (Exercise 5).

Theorem 7.4.7. *Suppose $\mathcal{F}$ is a family of left (or right) ideals of a ring R. Then $\cap_{I \in \mathcal{F}} I$ is a left (or right) ideal of R.*

In Exercise 13, you'll show that the intersection of a left ideal and a right ideal need not be either a left or right ideal. The example you'll use to illustrate this requires Theorem 7.4.11 and some familiarity with Example 7.4.12. In a commutative ring, where there is no difference between left and right ideals, Theorem 7.4.7 states that the intersection of a family of ideals is an ideal.

The union of two ideals is not necessarily an ideal (Exercise 7). However, there is a theorem that will come in handy in Section 7.6 that says something about the union across a special family of ideals (Exercise 8).

Theorem 7.4.8. *Suppose $\{I_n\}_{n \in \mathbb{N}}$ is a family of left (or right) ideals of a ring with the property that $I_n \subseteq I_{n+1}$ for all $n \in \mathbb{N}$. Then $\cup_{n=1}^{\infty} I_n$ is a left (or right) ideal.*

7.4.2 Generated ideals

Analogous to the subgroup generated by a subset of a group, we can define a similar term for ideals.

Definition 7.4.9. Suppose R is a ring and $A \subseteq R$ is nonempty. Suppose $I \subseteq R$ has the following properties:

(U1) $A \subseteq I$.

(U2) I is a left ideal of R.

(U3) If J is a left ideal of R and $A \subseteq J$, then $I \subseteq J$.

Then I is called a *left ideal generated by A*, and is denoted $(A)_l$.

We can similarly define a *right ideal generated by A* and denote it $(A)_r$. If R is commutative, then there is no distinction between $(A)_l$ and $(A)_r$, so we denote such an ideal (A) and call it the *ideal generated by A*. Does $(A)_l$ exist? If so, is it unique? And, if so, what does it look like? Here's your answer (Exercise 9).

Theorem 7.4.10. *Suppose R is a ring and $A \subseteq R$ is nonempty. Let $\mathcal{F}$ be the family of all left ideals of R that contain all elements of A. Then $(A)_l$ exists uniquely and can be written as*

$$(A)_l = \bigcap_{I \in \mathcal{F}} I. \tag{7.45}$$

If the path we traveled when we discussed subgroups generated by $A \subseteq G$ suggests a direction for us to go from here, we would consider the left ideal generated by a single element $a \in R$ and show that the top-down form of $(a)_l$ in Theorem 7.4.10 is equivalent to a form that can be built from the bottom up, by starting with a and building up to a subset of R that has properties U1–U3. Let's do this now. In order for this program to work, R must have a unity element.

Theorem 7.4.11. *Suppose R is a ring with unity, and $a \in R$. Then $(a)_l = Ra$ (the construction in Eq. (7.43)) and $(a)_r = aR$.*

Proof: We prove for $(a)_l$ only by showing Ra satisfies properties U1–U3. The proof for $(a)_r$ is similar. First, since R has a unity e, $a = ea \in Ra$, so that U1 is satisfied. Second, from Theorem 7.4.4, Ra is an ideal of R, so that U2 is satisfied. Finally, suppose J is any left ideal of R that contains a, and pick any $x \in Ra$. Then $x = ra$ for some $r \in R$. But since J is an ideal of R and $a \in J$, it must be that $ra \in J$, so that $Ra \subseteq J$. Thus U3 is satisfied and $(a)_l = Ra$. ∎

Example 7.4.12. Describe the left and right ideals of $\mathbb{Z}_{2 \times 2}$ generated by $\begin{bmatrix} 2 & 0 \\ 0 & 3 \end{bmatrix}$.

Solution: We'll describe the left ideal here and then you'll describe the right ideal in Exercise 11. For $\begin{bmatrix} a & b \\ c & d \end{bmatrix} \in \mathbb{Z}_{2 \times 2}$,

$$\begin{bmatrix} a & b \\ c & d \end{bmatrix} \begin{bmatrix} 2 & 0 \\ 0 & 3 \end{bmatrix} = \begin{bmatrix} 2a & 3b \\ 2c & 3d \end{bmatrix}. \tag{7.46}$$

Thus the left ideal is the set of all 2×2 matrices whose first column entries are even, and whose second column entries are multiples of three. ∎

Regardless of whether the presence of a unity allows $(a)_l$ to be written in the form Ra, the left ideal generated by a single element $a \in R$ is called the *principal left ideal generated by a*. If R does not have a unity element, then Ra won't contain a. Thus the construction of (a) would be a bit more complicated than that in Theorem 7.4.11. When we work with (a), we'll always be in the context of a ring with unity. In Exercise 14, you'll construct a form analogous to Eq. (7.43) for $(A)_l$, where $A = \{a_1, a_2, \ldots, a_n\}$. To simplify the notation, we'll write $(A) = (a_1, \ldots, a_n)$.

If R is a commutative ring with unity, there is an important link between the existence of proper ideals and what kind of ring R is. We'll prove one direction of the next theorem, and you'll prove the other in Exercise 16.

Theorem 7.4.13. *Suppose R is a commutative ring with unity. Then R is a field if and only if it has no proper ideals.*

Proof: Suppose R is a commutative ring with unity e.

($\Leftarrow$) Suppose R has no proper ideals. Choose any nonzero $x \in R$. We show that x has a multiplicative inverse in R. Since $x \neq 0$ and R has no proper ideals, $(x) = R$, so that $e \in (x)$. With Theorem 7.4.11, $(x) = Rx$, so that there exists $r \in R$ such that $e = rx$. Thus x has a multiplicative inverse in R and R is therefore a field.

($\Rightarrow$) (Exercise 16.) ∎

Suppose I is a proper ideal of a ring R, and $a \in R \backslash I$. Similar to the way we adjoin an element to a ring to create an extension, we can construct a left or right ideal of R that contains a and all elements of I. Our next theorem addresses the construction for the left-sided case, and you'll prove it in Exercise 19. We'll need this construction in Section 7.9 in the context of a commutative ring with unity.

Theorem 7.4.14. *Let R be a ring, I an ideal of R, and fix $a \in R \backslash I$. Let*

$$J = \{ra + i : r \in R, i \in I\}. \tag{7.47}$$

Then J is a left ideal of R.

7.4.3 Prime ideals

If $p \in \mathbb{N}$ is prime and $p \mid ab$, then either $p \mid a$ or $p \mid b$ (Theorem 2.7.15). Another way to say this is: If $pk_1 = ab$, then either $pk_2 = a$ or $pk_3 = b$. Using the language of the ideal generated by p and Theorem 7.4.11, yet another way to say this is: If $ab \in (p)$, then either $a \in (p)$ or $b \in (p)$. In a general ring, we assign a term to any *proper* ideal with this special property, whether or not the ideal is principal.

Definition 7.4.15. Suppose R is a ring, and P is a proper ideal of R with the property that $ab \in P$ implies either $a \in P$ or $b \in P$. Then P is called a *prime ideal* of R.

In Exercise 20 you'll determine whether certain ideals are prime. Although we haven't looked into properties of prime elements in a ring yet, it deserves to be said right away that we must be careful about jumping to conclusions about prime ideals and principal ideals generated by prime elements. In Section 7.5, we'll see that if a principal ideal (a) is a prime ideal, then a must be a prime element. However, just because a is a prime element, it does *not* mean that (a) is a prime ideal. Example 7.5.17 will illustrate an example of how this can happen. In Section 7.6, where we look at principal ideal domains, we'll see that a prime ideal will always be generated by a prime element. (See Exercise 20 for examples.)

7.4.4 Maximal ideals

Any proper ideal of a ring R is contained in a larger ideal, namely R itself. But if an ideal is as large as it can be without actually being all of R, then we call it *maximal*. Here's the definition. Since we're primarily interested in maximal ideals in a commutative ring, we'll set the definition in that context.

Definition 7.4.16. Suppose R is a commutative ring, and I is a proper ideal with the property that if J is any ideal such that $I \subseteq J \subseteq R$, then either $J = I$ or $J = R$. Then I is called a *maximal* ideal.

We can visualize a maximal ideal M in the following way: If M is a maximal ideal, then for any $r \in R \setminus M$, the only ideal of R that contains r and all elements of M is R itself.

Example 7.4.17. If $p \in \mathbb{N}$ is prime, then the ideal it generates in $\mathbb{Z}$ is maximal (Exercise 21).

Example 7.4.18. In $\mathbb{Z}$, (4) is not maximal, for it is a proper subset of the ideal of even integers.

Example 7.4.19. In $\mathbb{Z}_{2 \times 2}$, the right ideal $I = \left\{ \begin{bmatrix} 2a & 2b \\ 2c & 2d \end{bmatrix} : a, b, c, d \in \mathbb{Z} \right\}$ is not maximal, for $J = \left\{ \begin{bmatrix} 2a & 2b \\ c & d \end{bmatrix} : a, b, c, d \in \mathbb{Z} \right\}$ is also a right ideal (Exercise 22).

We see right away that there is a relationship between prime and maximal ideals in commutative rings with unity (Exercise 23).

Theorem 7.4.20. *If M is a maximal ideal in a commutative ring with unity, then M is prime.*

The reason that the ring in Theorem 7.4.20 has to have a unity element can be seen by letting R be the even integers and $M = (4) = \{\ldots, -8, -4, 0, 4, 8, \ldots\}$. Since $2 \times 2 \in (4)$, then (4) is not prime. However, (4) is maximal (Exercise 24).

About now you should be asking for either a theorem claiming that prime ideals in commutative rings with unity are also maximal (so that the terms are logically equivalent) or an example of a prime ideal that is not maximal. Well, they're not logically equivalent, and if you've done the exercises up to this point, you have got verification right before your eyes (Exercise 25). When we restrict ourselves to principal ideal domains in Section 7.6, we'll see that prime ideals are also maximal.

EXERCISES

1. Show that S defined in Example 7.4.2 is an ideal in $\mathbb{Z}[t]$.

2. Prove Theorem 7.4.4: Suppose R is a ring and $a \in R$. Then the set

$$Ra = \{ra : r \in R\} \tag{7.48}$$

is a left ideal in R.

3. Let $M = \left\{ \begin{bmatrix} a & b \\ 0 & 0 \end{bmatrix} : a, b \in \mathbb{Z} \right\}$. Show that M is a right ideal in $\mathbb{Z}_{2\times 2}$ but not a left ideal.

4. Let R be a commutative ring and Z the set of all zero divisors in R. What is wrong with the following proof that $Z \cup \{0\}$ is an ideal in R?

   Proof: Suppose R is a commutative ring and Z the zero divisors in R.

   (Y1) Let $z_1, z_2 \in Z \cup \{0\}$. If $z_1 = z_2 = 0$, then clearly $z_1 + z_2 = 0 \in Z \cup \{0\}$. If precisely one of z_1, z_2 is zero, then without loss of generality, $z_1 = 0$ and $z_2 \neq 0$. Then z_2 is a zero divisor, so there exists nonzero $a \in R$ such that $z_1 a = 0$. Thus $(z_1 + z_2)a = z_2 a = 0$, so that $z_1 + z_2$ is a zero divisor. If neither z_1 nor z_2 is zero, then there exist nonzero $a_1, a_2 \in R$ such that $z_1 a_1 = z_2 a_2 = 0$. Thus
   $$(z_1 + z_2)a_1 a_2 = z_1 a_1 a_2 + z_2 a_2 a_1 = 0 a_2 + 0 a_1 = 0,$$
   so that $z_1 + z_2$ is a zero divisor. In any case $z_1 + z_2 \in Z \cup \{0\}$, so that $Z \cup \{0\}$ is closed under addition.

   (Y2) By definition, $0 \in Z \cup \{0\}$.

   (Y3) Let $z \in Z \cup \{0\}$. If $z = 0$, then $-z = 0 \in Z \cup \{0\}$. If $z \neq 0$, then there exists nonzero $a \in R$ such that $za = 0$. Thus $(-z)(a) = -za = -0 = 0$, so that $-z$ is a zero divisor. In either case $-z \in Z \cup \{0\}$.

   (Y4) Let $z \in Z \cup \{0\}$, and $r \in R$. If $z = 0$, then $rz = 0 \in Z \cup \{0\}$. If $z \neq 0$, then there exists nonzero $a \in R$ such that $za = 0$. Since $a(rz) = r(az) = 0$, it follows that rz is a zero divisor.

   Since $Z \cup \{0\}$ satisfies properties Y1–Y4, $Z \cup \{0\}$ is an ideal in R. ∎

5. Prove the left-sided case of Theorem 7.4.7: Suppose $\mathcal{F}$ is a family of left ideals of a ring R. Then $\bigcap_{I \in \mathcal{F}} I$ is a left ideal of R.

6. In $\mathbb{Z}$, what is $(6) \cap (15)$?

7. Demonstrate a ring R and two ideals I_1 and I_2 such that $I_1 \cup I_2$ is not an ideal in R.

8. Prove the left-sided case of Theorem 7.4.8: Suppose $\{I_n\}_{n \in \mathbb{N}}$ is a family of left ideals of a ring R with the property that $I_n \subseteq I_{n+1}$ for all $n \in \mathbb{N}$. Then $\bigcup_{n=1}^{\infty} I_n$ is a left ideal in R.

9. Prove Theorem 7.4.10: Suppose R is a ring and $A \subseteq R$ is nonempty. Let $\mathcal{F}$ be the family of all left ideals of R that contain all elements of A. Then $(A)_l$ exists uniquely and can be written as
$$(A)_l = \bigcap_{I \in \mathcal{F}} I. \tag{7.49}$$

10. In $\mathbb{Z}[t]$, describe (t).

11. Describe the right ideal in Example 7.4.12.

12. Find the principal left and right ideals generated by $A = \begin{bmatrix} 1 & 0 \\ 0 & 0 \end{bmatrix}$ in $\mathbb{Z}_{2\times 2}$.

13. Let $M = \left\{ \begin{bmatrix} a & 0 \\ b & 0 \end{bmatrix} : a, b \in \mathbb{Z} \right\}$. Let $r = \begin{bmatrix} 2 & 0 \\ 3 & 0 \end{bmatrix}$, and consider the following subsets of M:

$$rM = \{rx : x \in M\} \quad (7.50)$$
$$Mr = \{xr : x \in M\}. \quad (7.51)$$

   (a) Show that M is a ring by showing it is a subring of $\mathbb{Z}_{2\times 2}$.
   (b) By Theorem 7.4.4, the sets in Eqs. (7.50) and (7.51) are ideals in M. Show that neither is a subset of the other.
   (c) Show that $rM \cap Mr$ is neither a left nor right ideal in M.

14. Suppose R is a ring with unity, and $A = \{a_1, a_2, \ldots, a_n\}$ is a subset of R. Construct a form of $(A)_l$ that is analogous to $(a)_l = Ra$ and show that it satisfies U1–U3.

15. Show that the ideal defined in Example 7.4.2 is actually $(2, t)$.

16. Prove the $\Rightarrow$ direction of Theorem 7.4.13: If R is a field, then it has no proper ideals.

17. Suppose R is a ring with unity element, and suppose $a \in R$ is a unit. Show that $(a)_l = (a)_r = R$.

18. In the ring of integers, let $a, b \in \mathbb{Z}\setminus\{0\}$ and $g = \gcd(a, b)$. Show that $(a, b) = (g)$.

19. Prove Theorem 7.4.14: Let R be a ring, I an ideal of R, and fix $a \in R\setminus I$. Let $J = \{ra + i : r \in R, i \in I\}$. Then J is a left ideal of R.

20. Determine with proof whether each of the following ideals is prime:

   (a) (6) and (7) in $\mathbb{Z}$.
   (b) $(2, t)$ in $\mathbb{Z}[t]$.
   (c) (t) in $\mathbb{Z}[t]$.
   (d) The left ideal from Exercise 12.

21. Show that if $p \in \mathbb{N}$ is prime, then the ideal it generates in $\mathbb{Z}$ is maximal.

22. Verify that I and J in Example 7.4.19 are right ideals of $\mathbb{Z}_{2\times 2}$.

23. Prove Theorem 7.4.20: If M is a maximal ideal in a commutative ring with unity, then M is prime.

24. Show that (4) is maximal in the ring of even integers.

25. Demonstrate an example of a prime ideal that is not maximal.[3]

26. In $\mathbb{Z} \times \mathbb{Z}$, is $(5) \times (3)$ prime and/or maximal? Verify your answer.

7.5 Integral Domains

If R is a *commutative* ring *with unity* that has no zero divisors, R is called an *integral domain*, or *domain* for short. The absence of zero divisors means that the principle of

---
[3] You shouldn't have to look far.

zero products applies in a domain by definition. Thus not only is $\mathbb{Z}$ a ring, it's also a domain. Since all nonzero elements of a field are units, a field contains no zero divisors and is therefore a domain. Thus $\mathbb{Q}$, $\mathbb{R}$, and $\mathbb{C}$ are domains. In Exercise 5 from Section 7.1, you found zero divisors in $\mathbb{Z}_n$ for at least some n. Which values of n cause $\mathbb{Z}_n$ to be a domain will follow from Theorems 7.5.1 and 7.5.7. You'll show the following (Exercise 1) with the help of a result from Section 2.7:

Theorem 7.5.1. *If $p \in \mathbb{N}$ is prime, then $\mathbb{Z}_p$ is a domain.*

Example 7.5.2. $\mathbb{Z}[\sqrt[3]{2}]$ and $\mathbb{Q}_{OD}$ are domains, for as subrings of $\mathbb{R}$, they are commutative, contain 1, and have no zero divisors.

In a domain, the fact that there are no zero divisors makes the following true (Exercise 2):

Theorem 7.5.3. *If D is a domain, and if $ac = bc$ and $c \neq 0$, then $a = b$.*

With multiplicative cancellation in hand, you can prove the following (Exercise 3).

Theorem 7.5.4. *Suppose D is a domain and $a \mid b$ in D. Then the $k \in D$ such that $ak = b$ is unique.*

Then with Theorem 7.5.4, the following term becomes meaningful.

Definition 7.5.5. Suppose D is a domain, $a, b \in D$, and a is not a unit. If $a \mid b$, where $ak = b$ and k is not a unit, then a is called a *proper* divisor of b.

If R is a ring in which Theorem 7.5.4 doesn't apply, such as those in Exercise 4 from Section 7.1, then Definition 7.5.5 cannot be unambiguously applied to all pairs $a, b \in D$. For it's possible to have $ak_1 = b$ and $ak_2 = b$ in a ring R where k_1 is a unit in R but k_2 is not. Here's an illustration. In $\mathbb{Z}_{2 \times 2}$,

$$\begin{bmatrix} 1 & 1 \\ 1 & 1 \end{bmatrix} \begin{bmatrix} 1 & 1 \\ 0 & 1 \end{bmatrix} = \begin{bmatrix} 1 & 2 \\ 1 & 2 \end{bmatrix} = \begin{bmatrix} 1 & 1 \\ 1 & 1 \end{bmatrix} \begin{bmatrix} 2 & 1 \\ -1 & 1 \end{bmatrix}. \tag{7.52}$$

Now $\begin{bmatrix} 1 & 1 \\ 0 & 1 \end{bmatrix}$ is a unit in $\mathbb{Z}_{2 \times 2}$, for

$$\begin{bmatrix} 1 & 1 \\ 0 & 1 \end{bmatrix} \begin{bmatrix} 1 & -1 \\ 0 & 1 \end{bmatrix} = \begin{bmatrix} 1 & -1 \\ 0 & 1 \end{bmatrix} \begin{bmatrix} 1 & 1 \\ 0 & 1 \end{bmatrix} = \begin{bmatrix} 1 & 0 \\ 0 & 1 \end{bmatrix}, \tag{7.53}$$

while $\begin{bmatrix} 2 & 1 \\ -1 & 1 \end{bmatrix}$ is not a unit in $\mathbb{Z}_{2 \times 2}$ (Exercise 4). Finding the multiplicative inverse of a matrix is something you'll see in linear algebra.

Multiplicative cancellation in a domain is a logical consequence of the principle of zero products. We could have defined a domain as a commutative ring with unity where multiplicative cancellation holds for nonzero elements, then shown that the principle of zero products follows from that. In a commutative ring, the principle of zero products and multiplicative cancellation are logically equivalent, as the following theorem says (Exercise 5).

7.5 Integral Domains

Theorem 7.5.6. *Suppose R is a commutative ring with unity such that $ac = bc$ implies $a = b$ for all $a, b, c \in R$ with $c \neq 0$. Then R is a domain.*

If a commutative ring with unity has nonzero characteristic, then the only way it can be a domain is if the characteristic is prime (Exercise 6).

Theorem 7.5.7. *If D is a domain and $\operatorname{char} D = n \neq 0$, then n is prime.*

With Theorems 7.2.18, 7.5.1, and 7.5.7, we see that $\mathbb{Z}_n$ is a domain if and only if n is prime. With the following theorem, which you'll prove in Exercise 7, it follows that $\mathbb{Z}_p$ is a field if and only if p is prime.

Theorem 7.5.8. *A finite integral domain is a field.*

Proving Theorem 7.5.8 requires very little that hasn't already been done. Since a domain is a commutative ring with unity, the only thing left in showing it's a field is the existence of multiplicative inverses. With Theorem 7.5.8, we see that $\mathbb{Z}_p$ is a field if p is prime.

In the domain $\mathbb{Z}$, we showed in Theorem 2.7.11 that if $a \mid b$ and $b \mid a$, then $a = \pm b$. And, coincidentally, ± 1 are the units of $\mathbb{Z}$. In a general domain, we say that a and b are *associates* if $a \mid b$ and $b \mid a$. Since this definition does not apply to zero, we declare zero to be an associate of itself. Notice that this declaration and the definition of associate prevent zero from having any other associates.

Example 7.5.9. In $\mathbb{Z}[i]$, $1 + i$ and $1 - i$ are associates (Exercise 9).

You'll show the following in Exercise 10:

Theorem 7.5.10. *Suppose D is a domain, and $a, b \in D$. Then a and b are associates if and only if there exist units $u, v \in D$ such that $a = ub$ and $b = va$.*

You should feel an equivalence relation coming on (Exercise 11).

Theorem 7.5.11. *Let D be a domain, and define $a \sim b$ if a is an associate of b. Then $\sim$ is an equivalence relation on D.*

And what do you think the equivalence class of the unity element consists of? You'll prove the following in Exercise 12:

Theorem 7.5.12. *If D is a domain and $\sim$ is the equivalence relation of association, then $[e]$ is the set of units of D.*

If a and b are associates, then there ought to be some senses in which they are interchangeable. One example of how this is true is in $\mathbb{Z}$, where $(6) = (-6)$. Associates generate the same principal ideal. The best way to show this is first to prove the following (Exercise 13). Its corollary is immediate.

Theorem 7.5.13. *Suppose D is a domain and $a, b \in R$. Then $a \mid b$ if and only if $(b) \subseteq (a)$.*

Corollary 7.5.14. *Suppose D is a domain and $a, b \in D$. Then $(a) = (b)$ if and only if a and b are associates.*

If a is a proper divisor of b, then by Theorems 7.5.4 and 7.5.10, a and b are not associates. So $(b) \subseteq (a)$ by Theorem 7.5.13, but $(a) \neq (b)$ by Corollary 7.5.14. Thus we have the following:

Corollary 7.5.15. *If D is a domain and a is a proper divisor of b in D, then $(b) \subset (a)$.*

Now let's look at the relationship between prime elements and prime ideals. First, the following holds in a domain (Exercise 14).

Theorem 7.5.16. *If D is a domain and (p) is a prime ideal in D, then p is prime in D.*

But what about the converse? If $p \in D$ is a prime element, must it generate a prime ideal? In $\mathbb{Z}$, the answer is yes. Thanks to Theorem 2.7.15, if $p \in \mathbb{N}$ is prime, then (p) is a prime ideal in $\mathbb{Z}$. In $\mathbb{Z}$, however, Theorem 2.7.15 depends on the existence of gcds, which itself depends on the WOP. There are some domains where strangely enough a prime element generates an ideal that isn't prime, and where a prime p can satisfy $p \mid ab$, while $p \nmid a$ and $p \nmid b$. The next example presents some of the building blocks of an integral domain to illustrate these possibilities. You'll provide most of the details in Exercise 16. We'll return to this example in Section 7.6.

Example 7.5.17. The ring $\mathbb{Z}[\sqrt{-5}]$ is a domain because it's a subring of $\mathbb{C}$ and contains 1. However, we can show the following:

1. 3 and $2 \pm \sqrt{-5}$ are prime in $\mathbb{Z}[\sqrt{-5}]$.
2. $2 \pm \sqrt{-5} \notin (3)$.
3. There exists a prime element $p \in \mathbb{Z}[\sqrt{-5}]$ such that (p) is not a prime ideal.
4. There exist $a, b, p \in \mathbb{Z}[\sqrt{-5}]$ where p is prime in D, $p \mid ab$, but $p \nmid a$ and $p \nmid b$.

First note

$$9 = 3 \cdot 3 = (2 + \sqrt{-5})(2 - \sqrt{-5}). \tag{7.54}$$

We show here that 3 is prime in $\mathbb{Z}[\sqrt{-5}]$ and leave all the other claims for you to verify in Exercise 16. Suppose

$$3 = (a + b\sqrt{-5})(c + d\sqrt{-5}), \tag{7.55}$$

where $a, b, c, d \in \mathbb{Z}$. We'll show that one of these two factors must be a unit, which by Theorem 7.3.2 must be ± 1. Multiplying out the factors and using the definition of equality in $\mathbb{Z}[\sqrt{-5}]$ yields

$$ac - 5bd = 3 \quad \text{and} \quad ad + bc = 0. \tag{7.56}$$

Proceeding as in the Proof of Theorem 7.3.2, we square each equation, multiply the latter by 5, add and factor to have

$$(a^2 + 5b^2)(c^2 + 5d^2) = 9. \tag{7.57}$$

Since both factors in Eq. (7.57) are positive integers, both must be in the set $\{1, 3, 9\}$. Considering each possibility reveals either a contradiction or the fact that one of the factors is a unit. Thus 3 is prime in $\mathbb{Z}[\sqrt{-5}]$.

In $D[t]$, the polynomial ring over a domain, the degree of a product of polynomials behaves in a more predictable way than in $R[t]$ for a ring R. You'll prove the following theorems in Exercises 17 and 18.

Theorem 7.5.18. *Suppose D is a domain and $f, g \in D[t]$ are nonzero polynomials. Then*

1. $\deg(fg) = \deg f + \deg g$.
2. *If $f \mid g$, then $\deg f \leq \deg g$.*

Theorem 7.5.19. *If D is a domain, then so is $D[t]$.*

In Exercise 19 you'll determine precisely which polynomials in $D[t]$ are units. Then you'll see more clearly what it means for a polynomial in $D[t]$ to be prime. In a polynomial ring, we generally use the word *irreducible* instead of *prime* to refer to such a polynomial.

Example 7.5.20. In Exercise 20, you'll show that $f = t^2 - 2$ is irreducible in $\mathbb{Z}[t]$. However, if we write $\mathbb{Z}[\sqrt{2}, t]$ to mean the polynomial ring over the domain $\mathbb{Z}[\sqrt{2}]$, then f is reducible.

Example 7.5.21. Let $f = 2t^2 - 4$. In Exercise 19, you'll look into the reducibility of f in $\mathbb{Z}[t]$ and $\mathbb{Q}[t]$.

If $f = a_n t^n + \cdots + a_0$ is a polynomial in $\mathbb{Z}[t]$ such that $\deg f \geq 1$, we call $\gcd(a_n, a_{n-1}, \ldots, a_0)$ the *content* of f. If the content of f is one, we say that f is *primitive*. Insight from Example 7.5.21 should make your proof of the following immediate (Exercise 21).

Theorem 7.5.22. *Suppose $f \in \mathbb{Z}[t]$ is irreducible and $\deg f \geq 1$. Then f is primitive.*

Example 7.5.21 and Exercise 19 reveal that Theorem 7.5.22 does not apply in $\mathbb{Q}[t]$. That is, if a polynomial with integer coefficients is irreducible when viewed as an element $\mathbb{Z}[t]$, then it is primitive. However, $2t + 6$ is irreducible in $\mathbb{Q}[t]$, but has content 2.

EXERCISES

1. Prove Theorem 7.5.1: If $p \in \mathbb{N}$ is prime, then $\mathbb{Z}_p$ is a domain.[4]

2. Prove Theorem 7.5.3: If D is a domain, and if $ac = bc$ and $c \neq 0$, then $a = b$.

3. Prove Theorem 7.5.4: Suppose D is a domain and $a \mid b$ in D. Then the $k \in D$ such that $ak = b$ is unique.

4. Show that $\begin{bmatrix} 2 & 1 \\ -1 & 1 \end{bmatrix}$ is not a unit in $\mathbb{Z}_{2 \times 2}$.

[4] See Theorem 2.7.15.

5. Prove Theorem 7.5.6: Suppose R is a commutative ring with unity such that $ac = bc$ implies $a = b$ for all $a, b, c \in R$ with $c \neq 0$. Then R is a domain.

6. Prove Theorem 7.5.7: If D is a domain and char $D = n \neq 0$, then n is prime.[5]

7. Prove Theorem 7.5.8: A finite integral domain is a field.[6]

8. Find multiplicative inverses for all nonzero elements of $\mathbb{Z}_7$.

9. Show that $1 + i$ and $1 - i$ are associates in $\mathbb{Z}[i]$.

10. Prove Theorem 7.5.10: Suppose D is a domain, and $a, b \in D$. Then a and b are associates if and only if there exist units $u, v \in D$ such that $a = ub$ and $b = va$.[7]

11. Prove Theorem 7.5.11: Let D be a domain, and define $a \sim b$ if a is an associate of b. Then $\sim$ defines an equivalence relation on D.

12. Prove Theorem 7.5.12: If D is a domain and $\sim$ is the equivalence relation of association, then $[e]$ is the set of units of D.

13. Prove Theorem 7.5.13: Suppose D is a domain and $a, b \in D$ are nonzero. Then $a \mid b$ if and only if $(b) \subseteq (a)$.

14. Prove Theorem 7.5.16: If D is a domain and (p) is a prime ideal in D, then p is prime in D.

15. Is 2 prime in $\mathbb{Z}[\sqrt{2}]$? Explain.

16. Verify the remaining claims from Example 7.5.17:

    (a) $2 + \sqrt{-5}$ is prime in $\mathbb{Z}[\sqrt{-5}]$. (The proof for $2 - \sqrt{-5}$ would be similar.)

    (b) $2 + \sqrt{-5} \notin (3)$. (The proof for $2 - \sqrt{-5}$ would be similar.)

    (c) There exists a prime element $p \in \mathbb{Z}[\sqrt{-5}]$ such that (p) is not a prime ideal.

    (d) There exist $a, b, p \in \mathbb{Z}[\sqrt{-5}]$ where p is prime in D, $p \mid ab$, but $p \nmid a$ and $p \nmid b$.

17. Prove Theorem 7.5.18: Suppose D is a domain and $f, g \in D[t]$ are nonzero polynomials. Then

    (a) $\deg(fg) = \deg f + \deg g$.

    (b) If $f \mid g$, then $\deg f \leq \deg g$.

18. Prove Theorem 7.5.19: If D is a domain, then so is $D[t]$.

19. Let D be a domain whose set of units is U, and let K be a field.

    (a) What are the units of $D[t]$?

    (b) What are the units in $K[t]$?

    (c) Is $f = 2t^2 - 4$ reducible in $\mathbb{Z}[t]$? In $\mathbb{Q}[t]$?

20. Explain why $f = t^2 - 2$ is irreducible in $\mathbb{Z}[t]$ but not in $\mathbb{Z}[\sqrt{2}, t]$.

---

[5] See Exercise 9 from Section 7.2.
[6] To find a^{-1}, define $f(x) = ax$ and apply Exercise 4 from Section 3.3.
[7] Don't forget zero.

21. Prove Theorem 7.5.22: Suppose $f \in \mathbb{Z}[t]$ is irreducible and $\deg f \geq 1$. Then f is primitive.

22. Suppose R is a ring and let $r \in R$ be nonzero. Define $f : R \to R$ by $f(x) = rx$. Show that f is not necessarily one-to-one. What additional condition on R guarantees f is one-to-one? Prove.

23. Suppose R and S are both domains. Does it follow that $R \times S$ is a domain?

7.6 UFDs and PIDs

7.6.1 Unique factorization domains

According to Theorems 2.4.8 and 2.7.18, every natural number $n \geq 2$ has a unique factorization into natural number primes. When we consider $\mathbb{Z}$, the implication of these theorems should be clear. For $n \in \mathbb{Z}$, if $n \geq 2$, we lose uniqueness in most cases in that we could introduce some negative signs here and there. But this is the only way we lose uniqueness. Even then, the prime factors in two different factorizations can be paired up as associates of each other ($\pm p$), allowing us to say that $n \geq 2$ has a prime factorization in $\mathbb{Z}$ that is unique up to order and association of the factors. For $n \leq -2$, applying the same principle to $-n$ allows us to say that every nonzero integer that is not a unit has a factorization into prime integers, and this factorization is unique up to order and association of the factors. Some domains have this same property: that every nonzero element that is not a unit can be written as a product of prime elements of the domain, uniquely up to order and association of the factors. A domain with this feature is called a *unique factorization domain*, or UFD for short. Right away we see that a field is a UFD, for every nonzero element is a unit.

In our progression from more general to more specialized rings, UFDs are the point where we can show that any two nonzero elements have a greatest common divisor, unique up to association. We'll adapt the definition of gcd in Definition 2.7.12 to make it applicable to a general domain, then take a passing glance at how you show existence and some sort of uniqueness of gcd in a UFD. The first place we actually need the existence of the gcd is in a PID, where it is a fairly easy thing to show.

An important example of a UFD that we'll merely make reference to right now is $\mathbb{Z}[t]$. To verify that $\mathbb{Z}[t]$ is a UFD takes some mathematical machinery that we won't create until Section 7.7, where we study $\mathbb{Q}[t]$ in some depth. While $\mathbb{Z}[t]$ is a subring of $\mathbb{Q}[t]$, it is not as specialized a ring because there are limitations on coefficients in $\mathbb{Z}[t]$ that do not apply in $\mathbb{Q}[t]$. However, there are some features $\mathbb{Q}[t]$ has that we need to apply to elements of $\mathbb{Z}[t]$ to show it's a UFD.

Since a UFD is by definition a domain, we do not need a theorem claiming that a UFD is a domain. However, not all domains are UFDs. For example, $\mathbb{Z}[\sqrt{-5}]$ is a domain, and Eq. (7.54) shows that 9 has two distinct factorizations into prime elements in $\mathbb{Z}[\sqrt{-5}]$. Thus $\mathbb{Z}[\sqrt{-5}]$ is not a UFD.

With a very minor adjustment in the wording, we can define gcd in a domain.

Definition 7.6.1. Suppose D is a domain, and let $a, b \in D$ be nonzero. Suppose $g \in D$ has the following characteristics:

(D1) $g \mid a$ and $g \mid b$.

(D2) If h is any element of D with the properties that $h \mid a$ and $h \mid b$, then it must be that $h \mid g$ also.

Then g is called a *greatest common divisor* of a and b, and is denoted $\gcd(a, b)$.

In Section 2.7, we said that a practical way you might find $\gcd(a, b)$ in $\mathbb{N}$ is by breaking down a and b into their prime factorizations, taking the appropriate number of 2's, 3's, etc. and building the gcd from that. We didn't prove that such a trick produces a natural number that satisfies D1–D2 because the WOP provided an easier and more useful way. Alas, in a general UFD, proving the existence of $\gcd(a, b)$, unique up to association, must be done by exploiting the unique prime factorizations of a and b in that somewhat sloppy way. We'll state the theorem here, followed by some details of the proof that might make the notation minimally sloppy. You'll finish the proof in Exercise 1.

Theorem 7.6.2. Suppose D is a UFD, and $a, b \in D$ are nonzero. Then there exists $g \in D$ that satisfies D1–D2, and if g_1 and g_2 both satisfy D1–D2, then g_1 and g_2 are associates.

Here are some suggestions on how to prove Theorem 7.6.2 and keep the notation from getting outrageously complicated. If we break a and b down into prime factorizations, then let $\{p_1, p_2, \ldots, p_n\}$ be all the primes that appear in either a or b, we can write

$$a = p_1^{\alpha_1} p_2^{\alpha_2} \cdots p_n^{\alpha_n} \quad \text{and} \quad b = p_1^{\beta_1} p_2^{\beta_2} \cdots p_n^{\beta_n}, \tag{7.58}$$

where some of the α_k and β_k might be zero. If we let $\gamma_k = \min\{\alpha_k, \beta_k\}$ for all $1 \leq k \leq n$, we claim that $g = p_1^{\gamma_1} p_2^{\gamma_2} \cdots p_n^{\gamma_n}$ satisfies D1–D2. Since $\gamma_k = \min\{\alpha_k, \beta_k\}$, we know that $\gamma_k \leq \alpha_k$ and $\gamma_k \leq \beta_k$ for all k. This should make it easy to show that g has property D1. To show that g has property D2, suppose $h \mid a$ and $h \mid b$. Then there exist $k_1, k_2 \in D$ such that $hk_1 = a$ and $hk_2 = b$. If the unique factorization of h is written as $h = q_1^{\delta_1} q_2^{\delta_2} \cdots q_m^{\delta_m}$, then the prime factorizations of hk_1 and hk_2 must agree with Eqs. (7.58), so that some possible reordering of the p_k allows us to say that $q_k = p_k$ for all $1 \leq k \leq m \leq n$. If $m < n$, then letting $\delta_k = 0$ for $m + 1 \leq k \leq n$ gives us $h = p_1^{\delta_1} \cdots p_n^{\delta_n}$. Now, the fact that $h \mid a$ implies that $\delta_k \leq \alpha_k$ for all $1 \leq k \leq m$. Similarly, $h \mid b$ implies that $\delta_k \leq \beta_k$ for all $1 \leq k \leq n$. Thus $\delta_k \leq \gamma_k$ for all $1 \leq k \leq n$, and by constructing the appropriate $l \in D$, we can show $hl = g$. Thus g has property D2. Showing that g is unique up to association is surprisingly easy. In fact, the way you proved uniqueness of $\gcd(a, b)$ in $\mathbb{N}$ should translate directly over to D to imply that any g_1 and g_2 that satisfy D1–D2 must be associates.

7.6.2 Principal ideal domains

Since a domain is commutative and has a unity element, left and right ideals are identical, and the principal ideal generated by $a \in D$ can always be written as Da (or aD).

Furthermore, from Exercise 14 in Section 7.4, if $A = \{a_1, \ldots, a_n\}$, then

$$(A) = \{d_1 a_1 + d_2 a_2 + \cdots + d_n a_n : d_k \in D \text{ for all } 1 \leq k \leq n\}. \quad (7.59)$$

Let's write this form of (A) in Eq. (7.59) as $Da_1 + Da_2 + \cdots + Da_n$.

In some domains, principal ideals are the only ones there are. Every ideal will have a single generator. If D is a domain such that every ideal in D is principal, then D is called a *principal ideal domain*, or a PID for short. Right away we see that a field K is a PID, for Exercise 16 from Section 7.4 says that the only ideals of K are $\{0\}$ and K itself, the former generated by zero, the latter by e. You've already seen an example of a domain that is not a PID, but before we point it out, we'll concentrate on understanding why some of the domains we're acquainted with are PIDs, and results for PIDs in general. Then the example of a domain (actually a UFD) that is not a PID will point itself out. First let's review the integers a bit.

After we defined divisibility and gcd in $\mathbb{Z}$ in Chapter 2, we showed that $\gcd(a, b)$ exists for all $a, b \in \mathbb{Z} \setminus \{0\}$ (Theorem 2.7.13). Furthermore, it is the smallest positive element of

$$S = \{ma + nb : m, n \in \mathbb{Z}\}. \quad (7.60)$$

Thus our proof that $\gcd(a, b)$ exists in $\mathbb{Z}$ depended on the WOP of $\mathbb{N}$ to give us a number that we could show satisfies properties D1–D2. Furthermore, in showing that this smallest element of S divides both a and b, you employed the division algorithm. And you proved the division algorithm with the help of the WOP.

Notice that S in Eq. (7.60) is precisely $\mathbb{Z}a + \mathbb{Z}b$, the ideal generated by $\{a, b\}$. Thus, $\gcd(a, b)$ is the smallest element of (a, b). Moreover, in Exercise 18 of Section 7.4, you showed that $\{a, b\}$ and $\gcd(a, b)$ generate the same ideal in $\mathbb{Z}$, so that the ideal generated by a two-element set in $\mathbb{Z}$ is really principal after all. Once again, it's the WOP at work, leading you to the smallest positive element of an ideal, and allowing you to show that it generates the whole ideal. By looking in this same direction, you can show the following in Exercise 2:

Theorem 7.6.3. *The integers are a PID.*

The way you'll attack this deserves a comment. To show any particular domain D is a PID, you must show that every ideal I can be written as Da, where $a \in I$ is a generator of I that you must find. If $I = \{0\}$, that's a piece of cake. However, if I is an arbitrary ideal with a nonzero element, then it has a positive element. Then the WOP of $\mathbb{N}$ comes in and gives you a generator for I on a silver platter (details notwithstanding). In $\mathbb{Z}$, the WOP is the basis for its being a PID and for its containing gcds. However, the WOP depends on $<$ as a measure of size of elements in $\mathbb{N}$.

Another example of a PID is $\mathbb{Q}[t]$. Rather than show this directly, we'll show in Section 7.7 that $\mathbb{Q}[t]$ is a Euclidean domain, and that all Euclidean domains are PIDs.

Example 7.6.4. The ring $\mathbb{Q}_{OD}$ from Example 7.1.9 is a PID. In Exercise 3, you'll prove this by applying the WOP in a creative way to an arbitrary ideal in $\mathbb{Q}_{OD}$ to find a generator.

Now let's suppose we're working in a domain where we don't necessarily have a measure of element size, so that we have no way to apply the WOP. With gcd defined as in Definition 7.6.1, and with the assumption that all ideals are principal, you can prove the following (Exercise 4):

Theorem 7.6.5. *Let D be a PID, and $a, b \in D$ be nonzero. Then there exists $\gcd(a, b) \in D$.*

Instead of going to (a, b) and taking its smallest element to be $\gcd(a, b)$ as you did in $\mathbb{Z}$, you simply use the fact that (a, b) has *some* generator. Since $(a, b) = Da + Db$, we have the following:

Corollary 7.6.6. *If D is a PID and $a, b \in D$ are nonzero, then there exist $m, n \in D$ such that $\gcd(a, b) = ma + nb$.*

Unfortunately, Theorem 7.6.5 says nothing about the uniqueness of $\gcd(a, b)$ but there is something close (Exercise 5).

Theorem 7.6.7. *If g_1 and g_2 are both greatest common divisors of a and b in a PID, then g_1 and g_2 are associates.*

Thus even if there are several gcds of a and b, they're all unit multiples of each other. If g is one gcd of a and b, then everything in its equivalence class of associates is, too. If $[e]$ is the set of gcds of a and b, then we say a and b are *relatively prime*, and note that if a and b are relatively prime, then there exists a linear combination such that $ma + nb = e$.

Now let's find that example of a domain that is not a PID. In any commutative ring, a maximal ideal is prime (Theorem 7.4.20). However, there exist prime ideals that are not maximal, such as (t) in $\mathbb{Z}[t]$ (Exercise 25 from Section 7.4). And (t) is a proper subset of $(2, t)$, the ideal from Example 7.4.2 of all polynomials with even constant term. The following theorem lets us close in for the kill (Exercise 6).

Theorem 7.6.8. *Suppose D is a PID and I is a prime ideal in D. Then I is maximal.*

Corollary 7.6.9. *$\mathbb{Z}[t]$ is not a PID.*

Proof: $(2, t)$ is prime but not maximal in $\mathbb{Z}[t]$. ∎

By Theorem 7.6.8, if an ideal in a domain is prime but not maximal, the domain is not a PID. Since $(2, t)$ is prime but not maximal in $\mathbb{Z}[t]$, then $\mathbb{Z}[t]$ is not a PID. In Exercise 7 you'll show explicitly that $(2, t)$ is not principal by demonstrating that any supposed generator fails. In Section 7.7, we'll show that $\mathbb{Z}[t]$ is a UFD. Now we want to show that a PID is a UFD. We have to take several steps to get there, but fortunately a lot of the uniqueness work has already been done for $\mathbb{N}$ in Section 2.7 and translates over to a PID almost word for word. The sticky part in showing existence of prime factorizations in a PID stems from the fact that the defining characteristic of a PID concerns its ideals, while the defining characteristic of a UFD concerns its elements. The link between them is the following: All ideals in a PID have a generator, and the way these ideals contain each other as subsets is tied by Theorem 7.5.13 to divisibility of their generators. Since in a PID $a \mid b$ if and only if $(a) \supseteq (b)$, the way an element breaks down into factors is directly linked to the way principal ideals stack. So let's make an important observation about the way ideals in a PID *cannot* stack.

Let D be a PID and $\{I_n\}_{n\in\mathbb{N}}$ a family of ideals such that $I_n \subseteq I_{n+1}$ for all $n \in \mathbb{N}$. By Theorem 7.4.8, $I = \cup_{n=1}^{\infty} I_n$ is an ideal. Since D is a PID, I is principal, and has a generator $a \in I$. Now since $a \in \cup_{n=1}^{\infty} I_n$, there exists $n \in \mathbb{N}$ such that $a \in I_n$. We claim that for all $k \geq n$, $I_k = I_n$. For if this is not true, then there exists some $k > n$ such that $I_k \setminus I_n$ is nonempty. Let $x \in I_k \setminus I_n$. Since $x \in I_k$, then $x \in I$, so that $x = ay$ for some $y \in D$. Also, since I_n is an ideal and $a \in I_n$, then $ay \in I_n$. But $ay = x$, so $x \in I_n$, which is a contradiction. Thus it is impossible that $I_k \supset I_n$.

The upshot of all this is that if you're in a PID and have a set of hypothesis conditions that allows you to create a family of ideals $\{I_n\}_{n=1}^{\infty}$ where $I_n \subset I_{n+1}$ for all $n \in \mathbb{N}$, then you've produced a contradiction, and at least some of the hypothesis conditions are false in a PID.

Now let's show that every element of a PID can be written as a product of prime elements, and that this factorization is unique up to order and association of the factors. To prove the former, suppose D is a PID, and suppose there exists $a \in D$ that cannot be written as a product of primes. Then a is not prime itself. Thus there exist $a_1, b_1 \in D$ such that $a = a_1 b_1$ and neither a_1 nor b_1 is a unit. Furthermore, since a cannot be written as a product of primes, then a_1 or b_1 (or both) cannot either. Without loss of generality, we may assume it's a_1, which means that a_1 is not prime. And notice, since b_1 is not a unit, $(a) \subset (a_1)$ by Corollary 7.5.15.

Now since a_1 is not prime, there exist $a_2, b_2 \in D$ such that $a_1 = a_2 b_2$ and neither a_2 nor b_2 is a unit. Once again, since a_1 cannot be written as a product of primes, either a_2 or b_2 cannot either. Assume it's a_2, and note that $(a_1) \subset (a_2)$ because b_2 is not a unit. Continuing in the same way, we can generate $\{a_n\}_{n\in\mathbb{N}} \subseteq D$ such that $(a_n) \subset (a_{n+1})$, which is a contradiction. Thus the assumption that there exists $a \in D$ that cannot be written as a product of primes is false, and all elements of D have a factorization into primes. With all this said, we have proved the following:

Theorem 7.6.10. *If D is a PID and $a \in D$ is nonzero and not a unit, then there exist primes $p_1, p_2, \ldots, p_n \in D$ such that $a = p_1 p_2 \cdots p_n$.*

To show uniqueness of the prime factorization, almost all the work translates directly over from our work in Section 2.7. The first step is analogous to Theorem 2.7.14, but the proof comes out a little differently because primes in a PID are defined in language that differs from that in $\mathbb{N}$.

Theorem 7.6.11. *If D is a PID, and $a, p \in D$ are such that p is prime and $a \neq 0$, then either $p \mid a$ or p and a are relatively prime.*

> *Proof:* Let $g = \gcd(a, p)$. Since $g \mid p$ and p is prime, then either g is a unit, or it's an associate of p. If g is a unit, then writing $am + np = g$ and multiplying both sides through by g^{-1} reveals that a and p are relatively prime. If g is an associate of p, then $gu = p$ for some unit u. The fact that $g \mid a$ means that $pu^{-1} \mid a$, so that $p \mid a$. ∎

With Theorem 7.6.11 in hand, the proofs of the next two theorems become identical to those for Theorems 2.7.15 and 2.7.17:

Theorem 7.6.12. *If D is a PID and $a, b, p \in D$ are such that p is prime, a and b are nonzero and $p \mid ab$, then either $p \mid a$ or $p \mid b$.*

Theorem 7.6.13. *Suppose D is a PID and $p, a_1, a_2, \ldots, a_n \in D$ are such that p is prime and $a_k \neq 0$ for all $1 \leq k \leq n$. If $p \mid a_1 a_2 \cdots a_n$, then there exists k ($1 \leq k \leq n$) such that $p \mid a_k$.*

The uniqueness result is analogous to Theorem 2.7.18, but the proof doesn't come out exactly the same because of possible association of the factors. You'll provide the new proof in Exercise 10. Exercise 9 will come in handy along the way.

Theorem 7.6.14. *If D is a PID and $a \in D$ is nonzero and not a unit, then the prime factorization of a from Theorem 7.6.10 is unique up to order and association of the factors.*

With Theorems 7.6.10 and 7.6.14, we have the following:

Theorem 7.6.15. *A PID is a UFD.*

Finally, as a direct result of Theorem 7.6.12, we have the following, which you'll prove in Exercise 11.

Theorem 7.6.16. *If D is PID and $p \in D$ is prime, then (p) is a prime ideal.*

EXERCISES

1. Prove Theorem 7.6.2: Suppose D is a UFD, and $a, b \in D$ are nonzero. Then there exists $g \in D$ that satisfies D1–D2, and if g_1 and g_2 both satisfy D1–D2, then g_1 and g_2 are associates.

2. Prove Theorem 7.6.3: The integers are a PID.

3. Show that $\mathbb{Q}_{OD}$ is a PID.[8]

4. Prove Theorem 7.6.5: Let D be a PID, and $a, b \in D$ be nonzero. Then there exists $\gcd(a, b) \in D$.

5. Prove Theorem 7.6.7: If g_1 and g_2 are both greatest common divisors of a and b in a PID, then g_1 and g_2 are associates.

6. Prove Theorem 7.6.8: Suppose D is a PID and I is a prime ideal in D. Then I is maximal.

7. Show that $(2, t)$ is not a principal ideal of $\mathbb{Z}[t]$.

8. Prove Theorem 7.6.12: If D is a PID and $a, b, p \in D$ are such that p is prime and $p \mid ab$, then either $p \mid a$ or $p \mid b$.

9. Show that if D is a domain, $p \in D$ is prime and $u \in D$ is a unit, then pu is also prime.[9]

---

[8] Of all nonzero elements of an ideal I, pick one with the smallest power of 2 in the numerator. Use this element as a generator.

[9] If $pu = ab$ is a factorization of pu, then $a(bu^{-1})$ is a factorization of p.

10. Prove Theorem 7.6.14: If D is a PID and $a \in D$ is nonzero and not a unit, then the prime factorization of a from Theorem 7.6.10 is unique up to order and association of the factors.

11. Prove Theorem 7.6.16: If D is PID and $p \in D$ is prime, then (p) is a prime ideal.

7.7 Euclidean Domains

7.7.1 Definition and properties

Let's return to the division algorithm for $\mathbb{Z}$ (Theorem 2.7.6) to expand it and cast it in somewhat different language. In writing $b = aq + r$, we insisted that a be positive, so that $0 \leq r < a$ is meaningful. We did not have to make this restriction on a. Obviously $a = 0$ won't work, but we could have allowed for $a < 0$ to have a somewhat broader theorem. If we wanted to extend Theorem 2.7.6 to include the case $a < 0$, it would look something like this:

Theorem 7.7.1 (Extended Division Algorithm). *Suppose $a, b \in \mathbb{Z} \setminus \{0\}$. Then there exist unique $q, r \in \mathbb{Z}$ such that $b = aq + r$ and $0 \leq r < |a|$.*

Extending the division algorithm as stated in Theorem 2.7.6 to include the case $a < 0$ is fairly easy to do. For if $a < 0$, then $-a > 0$, and by Theorem 2.7.6, there exist $q_1, r_1 \in \mathbb{Z}$ such that $b = (-a)q_1 + r_1$ and $0 \leq r < -a$. Letting $q_2 = -q_1$ and $r_2 = r_1$, we have $b = aq_2 + r_2$, and $0 \leq r_2 = r_1 < -a = |a|$. Your proof of uniqueness of q and r from Theorem 2.7.6 probably did not depend on the sign of a, and would therefore apply for this case, too.

With Theorem 7.7.1, we're using absolute value as a measure of the size of integers, and saying that any two nonzero integers can be related by breaking one of them (b) down into a certain multiple (q) of the other (a), with some possible stuff left over (r), but where the stuff left over is smaller in size than a. There are other important domains where some notion of size can be imposed on the elements, and something akin to the division algorithm using that measure of size works for all nonzero elements. Since the division algorithm does not involve the zero of the domain, we don't insist that zero be assigned a measure of size. Here is the definition.

Definition 7.7.2. Suppose D is an integral domain, and suppose there exists a function $d : D \setminus \{0\} \to \mathbb{W}$ with the property that $d(a) \leq d(ab)$ for all $a, b \in D$, which is called a *valuation*. Suppose also that, for all $a, b \in D \setminus \{0\}$, there exist $q, r \in D$ such that $b = aq + r$, and either $r = 0$ or $d(r) < d(a)$. Then D is called a *Euclidean domain* (ED).

Definition 7.7.2 deserves a few comments. First, demonstrating that a domain is an ED requires the creation of a valuation d that assigns a nonnegative integer size to all nonzero elements of the domain. There might be more than one such valuation, but we'll point out some features that d has as a result of the requirement that $d(a) \leq d(ab)$

for all $a, b \in D\setminus\{0\}$. Second, given two nonzero elements a and b, either b must be a multiple of a, in which case $r = 0$, or just the right distance from a multiple of a so that writing $b = aq + r$ can be done with r of sufficiently small size.

Example 7.7.3. The integers form an ED. Letting $d(a) = |a|$, we have that $|a| \leq |ab|$ for all $a, b \in \mathbb{Z}\setminus\{0\}$. With Theorem 7.7.1 we have that $\mathbb{Z}$ is an ED.

Example 7.7.4. A field K is an ED. If we let $d : K\setminus\{0\} \to \mathbb{W}$ be defined by $d(x) = 1$ for all $x \in K\setminus\{0\}$, we see that $d(x) = d(xy)$ for all $x, y \in K\setminus\{0\}$. Furthermore, for $x, y \in K\setminus\{0\}$, we may let $q = y/x$ to have $x = qy$, so that $r = 0$ is always possible.

Divisibility in a field is trivial because nonzero elements are all multiples of each other. Theorem 7.7.6 will reveal that any valuation on $K\setminus\{0\}$ would have to be constant.

There is one more classic example of an ED that we want to mention without providing any of the proof. The Gaussian integers $\mathbb{Z}[i]$ are an ED, and the function $d : \mathbb{Z}\setminus\{0 + 0i\} \to \mathbb{W}$ defined by $d(a + bi) = a^2 + b^2$ can be shown to be a valuation that works.

Before we spend some quality time with one more very important example of an ED, let's derive some results about EDs in general. First, let's see how to show that an ED is a PID. If D is an ED and we pick any ideal I, we must find some $a \in I$ where every element of I is a multiple of a. But that shouldn't be too hard, for an element $a \in I$ such that $d(a)$ is minimal would probably serve nicely as a generator for I, and the division algorithm on D ought to be just the right tool to enable us to show it. All that said, you'll prove the following in Exercise 1.

Theorem 7.7.5. *An ED is a PID.*

If D is an ED, it is by definition an integral domain. Thus, it has a unity element e. If we pick any $a \in D$, it follows that $d(e) \leq d(ea) = d(a)$, so that $d(e)$ is minimal among all values of d. This fact and the division algorithm should come in handy when you prove the $\Rightarrow$ direction of the following in Exercise 2.

Theorem 7.7.6. *If D is an ED, then $d(u) = d(e)$ if and only if u is a unit.*

Theorem 7.7.6 sheds a little more light on some things we already know. First, since absolute value can serve as a valuation on $\mathbb{Z}$, we see that the units in $\mathbb{Z}$ are precisely the values of x for which $|x| = |1|$, namely ± 1. Also, since every nonzero element of a field K is a unit, any valuation d must satisfy $d(x) = d(e)$ for all $x \in K\setminus\{0\}$. Thus, d must be constant. Conversely, if a valuation on an ED is constant, then every nonzero element is a unit, so that the ED is a field.

As we've progressed from rings to domains to UFDs to PIDs to EDs, we claimed that there are examples of one structure that do not qualify as an example of the next most restrictive structure. In every case up to now, we've provided an example with complete proof, except for $\mathbb{Z}[t]$, which we'll address later in this section. So about now you should be asking, "Where's my example of a PID that's not an ED?" Well, this is the only place in our progression from more general to more specialized rings where we're going to present an example of a PID that's not an ED, and give only a loose explanation of how this would be shown. The classic example of a PID that is not an ED was first constructed by T. Motzkin in 1949, very recently indeed by mathematical standards. It is $\mathbb{Z}[(1+\sqrt{-19})/2]$, the set of all expressions of the form $m + n(1+\sqrt{-19})/2)$, where $m, n \in \mathbb{Z}$. For convenience, let's write $\alpha = (1+\sqrt{-19})/2$, and discuss how one goes about showing $\mathbb{Z}[\alpha]$ is a PID but not an ED.

First let's address the fact that $\mathbb{Z}[\alpha]$ is not an ED. We do this by showing that every ED has a certain feature, then showing that $\mathbb{Z}[\alpha]$ does not have this feature. If D is any ED that is not a field, then there will exist nonzero, nonunit elements. If d is a valuation on D, then a nonzero, nonunit element x will satisfy $d(x) > d(e)$. Among all elements of D, let a be a nonzero, nonunit element for which $d(a)$ is minimal. Such an element is called a *universal side divisor*. By the division algorithm on D, any $x \in D$ can be written as $x = aq + r$, where either $r = 0$ or $d(r) < d(a)$. Since $d(a)$ is minimal among all nonzero, nonunit elements of D, we may say that either $r = 0$ or $d(r) = d(e)$. That is, if a is a universal side divisor, then every $x \in D$ may be written as $x = aq + r$, where either $r = 0$ or r is a unit. Every ED that is not a field will have universal side divisors because the valuation won't be constant.

The next thing we would need to know is what the units are in $\mathbb{Z}[\alpha]$. It turns out that the only units in $\mathbb{Z}[\alpha]$ are ± 1. So let's suppose that $\mathbb{Z}[\alpha]$ were an ED. Since the only units in $\mathbb{Z}[\alpha]$ are ± 1, $\mathbb{Z}[\alpha]$ is not a field. Thus there exists a universal side divisor $a \in \mathbb{Z}[\alpha]$, and every $x \in \mathbb{Z}[\alpha]$ can be written as $x = aq + r$, where $r \in \{0, \pm 1\}$. In particular, $x = 2$ must be writable in this way. So $aq = 2 - r$, and the only possible values of $2 - r$ are $\{1, 2, 3\}$. Thus, a divides at least one of $\{1, 2, 3\}$ but, since a is not a unit, $a \nmid 1$. With some work, we could show that the only divisors of 2 are $\{\pm 1, \pm 2\}$ and the only divisors of 3 are $\{\pm 1, \pm 3\}$. Thus, $a \in \{\pm 2, \pm 3\}$. If we let $x = \alpha$, however, it turns out that there is no $q \in \mathbb{Z}[\alpha]$ for which $\alpha = aq + r$, given that $a \in \{\pm 2, \pm 3\}$ and $r \in \{0, \pm 1\}$. This is a contradiction, so $\mathbb{Z}[\alpha]$ is not an ED.

Now let's address the fact that $\mathbb{Z}[\alpha]$ is a PID. First, let's take the defining characteristic of an ED and state it first in the original way, then in an altered form. A domain D is an ED if there exists a valuation d on $D\setminus\{0\}$ such that for all nonzero $a, b \in D$:

1. There exist $q, r \in D$ such that $b = aq + r$ and either $r = 0$ or $d(r) < d(a)$.

2. Either b is in the ideal generated by a, or it is possible to subtract from b some q multiple of a to produce an element $r = b - aq$ whose valuation is smaller than that of a, that is, $d(r) < d(a)$.

Let's weaken this second form a bit. Suppose D is a domain with a valuation such that for all nonzero $a, b \in D$, either b is in the ideal generated by a, or there is some *linear combination* of a and b whose valuation is smaller than that of a. That is, either $b \in (a)$ or there exist nonzero $m, n \in D$ such that $d(ma + nb) < d(a)$. This property is sort of like the division algorithm, but not quite as strong. In the event that b is not a multiple of a, we don't insist that some multiple of a can be subtracted from b to produce an element of small valuation, but only that some linear combination of a and b has sufficiently small valuation. A valuation with this feature is called a *Dedekind-Hasse norm*, and if a domain is such that there exists a Dedekind-Hasse norm, then we can show the following (Exercise 3):

Theorem 7.7.7. *Suppose D is a domain with a valuation d, and with the property that, for all nonzero $a, b \in D$, either $b \in (a)$, or there exist $m, n \in D$ such that $d(ma + nb) < d(a)$. Then D is a PID.*

The trick then is to show that $\mathbb{Z}[\alpha]$ entertains a Dedekind-Hasse norm, so that $\mathbb{Z}[\alpha]$ is a PID. A function $d : D \setminus \{0\} \to \mathbb{R}$ that works is $d(a + b\alpha) = a^2 + ab + 5b^2$, for which we won't provide any details.

7.7.2 Polynomials over a field

An important example of an ED is $\mathbb{Q}[t]$, and we want to address that now. Every nonzero polynomial $f \in \mathbb{Q}[t]$ has a nonnegative degree, and $\deg f$ is precisely the valuation we want to use. From Theorem 7.5.18, since $\deg(fg) = \deg f + \deg g$, and since $\deg g \geq 0$, we have that $\deg f \leq \deg(fg)$ for all $f, g \in \mathbb{Q}[t]$. Showing $\mathbb{Q}[t]$ is an ED then boils down to proving a sort of division algorithm on $\mathbb{Q}[t]$. Here is precisely the theorem we need, with uniqueness of q and r to boot. The technique we'll use to get the proof off the ground should look surprisingly familiar. You'll provide a few of the details in Exercise 4:

Theorem 7.7.8. *Suppose $f, g \in \mathbb{Q}[t]$ are nonzero polynomials. Then there exist unique $q, r \in \mathbb{Q}[t]$ such that $g = fq + r$ and either $r = 0$ or $\deg r < \deg f$.*

Proof: Choose $f, g \in \mathbb{Q}[t]$, both nonzero polynomials, and define

$$S = \{g - fq : q \in \mathbb{Q}[t]\}. \tag{7.61}$$

If the zero polynomial is in S, then we have $g = fq$ for some $q \in \mathbb{Q}[t]$. Otherwise we may let r be any polynomial in S whose degree is minimal. Then $r = g - fq$ for some $q \in S$, and since $\mathbb{Q}[t]$ is closed under addition and multiplication, we have that $r \in \mathbb{Q}[t]$. Thus, we have $g = fq + r$. To show $\deg r < \deg f$, suppose $\deg r \geq \deg f$. Then we may write

$$f = a_m t^m + \cdots + a_0 \quad \text{and} \quad r = b_n t^n + \cdots + b_0, \tag{7.62}$$

where $m \leq n$, and neither a_m nor b_n is zero. By Exercise 4, we may create a nonzero element of S whose degree is strictly less than $\deg r$, which is a contradiction. Also by Exercise 4, q and r are unique. ∎

Having shown that $\mathbb{Q}[t]$ is an ED opens a floodgate of interesting facts about this very important ring. First, unlike $\mathbb{Z}[t]$, $\mathbb{Q}[t]$ is a PID, so every nontrivial ideal has a nonzero generator f, and $\deg f$ is of minimum degree among all elements of (f). Furthermore, from our comments after the statement of Theorem 7.7.5, any polynomial in (f) whose degree is the same as that of f can serve as a generator of (f). If we write $f = a_n t^n + a_{n-1} t^{n-1} + \cdots + a_0$, then $a_n \neq 0$ and we can create a new polynomial $m = a_n^{-1} f = t^n + (a_{n-1}/a_n) t^{n-1} + \cdots + a_0/a_n$ whose leading coefficient is one. Since the polynomial a_n^{-1} is in $\mathbb{Q}[t]$, it follows that $m \in (f)$. Also $\deg m = \deg f$, so $(m) = (f)$. A polynomial whose leading coefficient is the unity element of the ring of coefficients is called *monic*, and we see that every ideal in $\mathbb{Q}[t]$ has a monic generator.

Since $\mathbb{Q}[t]$ is an ED, it is also a UFD, so that every nonzero polynomial in $\mathbb{Q}[t]$ factors down into irreducible polynomials. What is more, this factorization is unique, at least up to a point. Any two factorizations of f that appear to be different can be seen upon inspection to have components that can be paired up as associates. Now associates are unit multiples of each other, and from Exercise 19 from Section 7.5, the units in $\mathbb{Q}[t]$ are the polynomials of degree zero. Thus, two polynomials in $\mathbb{Q}[t]$ are associates if and only if one is a nonzero constant multiple of the other.

Example 7.7.9. In $\mathbb{Q}[t]$, $f = t^3 + 3t^2 + 2t$ can be factored as $t(t+2)(t+1)$ or $(4t)(\frac{1}{3}t + \frac{2}{3})(\frac{3}{4}t + \frac{3}{4})$. Associate pairs are t and $4t$, $t+2$ and $\frac{1}{3}t + \frac{2}{3}$, and $t+1$ and $\frac{3}{4}t + \frac{3}{4}$.

We've said a lot about irreducible polynomials in $\mathbb{Q}[t]$, but we have not developed any criteria by which we can determine whether a polynomial is reducible or irreducible. By Theorem 7.5.18, if $\deg f = 1$ and $f = gh$, then one of g or h has degree zero, and is therefore a unit. Thus polynomials of degree one are irreducible in $\mathbb{Q}[t]$.

Another simple criterion for irreducibility involves evaluating $f \in \mathbb{Q}[t]$ at some $a \in \mathbb{Q}$. In Section 7.3, we said we weren't really interested in elements of $R[t]$ as functions where we would plug in values for t, but more as a string of symbols. Well, that was sort of a lie. As a ring in and of itself, $R[t]$ really is exactly as we described it in Section 7.3, and t really is just a formal symbol whose presence is used to define addition and multiplication of two elements in $R[t]$. However, for a given $f = a_n t^n + a_{n-1} t^{n-1} + \cdots + a_0 \in R[t]$, there are some good reasons we might want to choose a specific $\alpha \in R$ and calculate the value in R of the expression $a_n \alpha^n + a_{n-1} \alpha^{n-1} + \cdots + a_0$. The point is that sometimes the element of R that is produced by calculating $f(\alpha)$ makes an important statement about the polynomial f and its role in $R[t]$. Perhaps you spent time in high school algebra trying to factor polynomials into irreducibles, and one way you might have stumbled onto a linear factor of f was by discovering some number a such that $f(a) = 0$. For example, if $f = t^3 - t^2 + 2t - 2$, you might have noticed that $f(1) = 0$, so you wrote $f = (t-1)g$, which by division became $f = (t-1)(t^2 + 2)$. This worked because of the following theorem (Exercise 5), where the division algorithm on $\mathbb{Q}[t]$ makes itself very useful:

Theorem 7.7.10. *Suppose $f \in \mathbb{Q}[t]$ and $a \in \mathbb{Q}$. Then $(t - a) \mid f$ if and only if $f(a) = 0$.*

Theorem 7.7.10 makes the following almost immediate (Exercise 6):

Theorem 7.7.11. *Suppose $f \in \mathbb{Q}[t]$ and $\deg f \in \{2, 3\}$. Then f is reducible if and only if there exists $a \in \mathbb{Q}$ such that $f(a) = 0$.*

Whether a polynomial in $\mathbb{Q}[t]$ is reducible is not an easy question to answer in general. There are several criteria that can help answer the question for certain special polynomials, and you'll probably see them in your upper level algebra class. When we look at $\mathbb{Z}[t]$ we'll see an important relationship between reducibility in $\mathbb{Q}[t]$ and in $\mathbb{Z}[t]$.

If there's one important idea in mathematics that you should find yourself cluing into at this point in the game, it is the fact that certain properties of mathematical structures are the logical basis for other properties, and that these properties can sometimes be isolated and translated over to other structures that might be different in some ways. For example, we proved in Section 2.3 that $a \cdot 0 = 0$ for all $a \in \mathbb{Z}$. Then we noted in Section 7.1 that $a \cdot 0 = 0 \cdot a = 0$ in any ring by an argument identical to that for $\mathbb{Z}$. The point is that a few basic ring properties came together to make $a \cdot 0 = 0 \cdot a = 0$ true (basic properties of addition, including additive cancellation, and the distributive property), even without commutativity of multiplication. Thus, any mathematical structure in which we have additive cancellation and the distributive property will be a structure where $a \cdot 0 = 0 \cdot a = 0$.

The fact that $\mathbb{Q}[t]$ is an ED calls on the fact that $\mathbb{Q}$ is a field, but it does not require that it be a particularly special kind of field. If we go back and replace $\mathbb{Q}$ with an arbitrary field K in Theorems 7.7.8–7.7.11, exactly the same proofs work. This can be particularly interesting if we use the field $\mathbb{Z}_p$ for p prime. Here are restatements of Theorems 7.7.8–7.7.11 for an arbitrary field, and some examples to illustrate its application beyond $\mathbb{Q}[t]$:

Theorem 7.7.12. *Let K be a field. Then K with valuation $\deg f$ is a Euclidean domain.*

To illustrate Theorem 7.7.12, you'll verify the following in Exercise 7:

Example 7.7.13. Let $f_1 = 3t^2 + t + 2$ and $g_1 = 3t^4 + 3t^3 + t^2 + 2t + 4$ in $\mathbb{Z}_5[t]$. Then by polynomial division (like in high school algebra), there exist $q, r \in \mathbb{Z}_5[t]$ such that $g_1 = f_1 q + r$ and $\deg r < \deg f_1$. However, for $f_2 = 2t + 2$ and $g_2 = t^2$ in $\mathbb{Z}_6[t]$, it is impossible to write $g_2 = f_2 q + r$ for any $q, r \in \mathbb{Z}_6[t]$ where either $r = 0$ or $\deg r < \deg f_2$.

Theorem 7.7.14. *Let K be a field, $f \in K[t]$, and $a \in K$. Then $(t - a) \mid f$ if and only if $f(a) = 0$.*

Theorem 7.7.15. *Let K be a field, $f \in K[t]$, and suppose $\deg f \in \{2, 3\}$. Then f is reducible if and only if there exists $a \in K$ such that $f(a) = 0$.*

In Exercise 8 you will apply Theorem 7.7.15 to several polynomials in $\mathbb{Z}_3[t]$. Since $\mathbb{Z}_3$ is such a small field, determining whether there exists $a \in \mathbb{Z}_3$ such that $f(a) = 0$ is very quick.

7.7.3 $\mathbb{Z}[t]$ is a UFD

We waited to show that $\mathbb{Z}[t]$ is a UFD, and now is the time to tackle the question. We could have shown the existence of a factorization of a polynomial in $\mathbb{Z}[t]$ into irreducibles earlier, but uniqueness of this factorization up to order and association of the factors requires us to view elements of $\mathbb{Z}[t]$ as elements of $\mathbb{Q}[t]$, where factorizations are unique up to association. The reason we have some work to do to show uniqueness is that reducibility and association in $\mathbb{Q}[t]$ are different from reducibility and association in $\mathbb{Z}[t]$. Polynomials such as $2t + 6$ and $10t + 15$ are irreducible and associates in $\mathbb{Q}[t]$, but not in $\mathbb{Z}[t]$. First the easy part, which you'll show in Exercise 9:

Theorem 7.7.16. *Every nonzero, nonunit polynomial in $\mathbb{Z}[t]$ has a factorization into irreducible polynomials in $\mathbb{Z}[t]$.*

To make our way to uniqueness up to order and association, we need the following. You'll provide the climactic detail in Exercise 10. Remember that the term *primitive* applies only to polynomials of degree at least one.

Theorem 7.7.17 (Gauss' Lemma). *The product of two primitive polynomials is primitive.*

Proof: Suppose $f, g \in \mathbb{Z}[t]$ are both primitive polynomials. We show that fg is primitive by supposing $p \in \mathbb{N}$ is any prime, then showing there is some coefficient in fg that is not divisible by p.

Suppose $p \in \mathbb{N}$ is prime, and write

$$f = a_m t^m + \cdots + a_0 \quad \text{and} \quad g = b_n t^n + \cdots + b_0. \tag{7.63}$$

Since f and g are both primitive, there are coefficients in both f and g that are not divisible by p. Let a_k and b_l be the coefficients of the lowest powers of t in f and g, respectively, that are not divisible by p; that is, p divides all of $a_0, a_1, \ldots, a_{k-1}, b_0, b_1, \ldots, b_{l-1}$, but not a_k or b_l. Then by Exercise 10, p does not divide the coefficient of t^{k+l} in fg. Since this is true for all primes, fg is primitive. ∎

Theorem 7.7.18. *Suppose $f \in \mathbb{Z}[t]$. If f is reducible in $\mathbb{Q}[t]$, then it is reducible in $\mathbb{Z}[t]$.*

Theorem 7.7.18 says simply that if a polynomial has integer coefficients, and it can be factored into polynomials of degree at least one by viewing it as an element of $\mathbb{Q}[t]$ and resorting to rational coefficients in the factors, then you can adjust these factor coefficients and make them integers. We'll prove Theorem 7.7.18 here in a somewhat conversational way to keep the notation from getting too sloppy. Notice the point at which we apply Theorem 7.7.17.

Proof: Suppose $f \in \mathbb{Z}[t]$, and suppose $f = f_1 f_2$ for some $f_1, f_2 \in \mathbb{Q}[t]$, where $\deg f_k \geq 1$ for $1 \leq k \leq 2$. Let $d_1, d_2 \in \mathbb{Z}$ be the product of all the denominators of all the coefficients of f_1 and f_2, respectively, then let $g_1 = d_1 f_1$ and $g_2 = d_2 f_2$, so that $g_1, g_2 \in \mathbb{Z}[t]$. Next, let $c_1, c_2 \in \mathbb{Z}$ be the content of g_1 and g_2, respectively, and factor these out to write $g_1 = c_1 h_1$ and $g_2 = c_2 h_2$, where $h_1, h_2 \in \mathbb{Z}[t]$ are primitive. If we let c be the content of f, we may write $f = cg$, where $g \in \mathbb{Z}[t]$ is primitive. Thus, we have

$$(cd_1 d_2)g = (d_1 d_2)f = d_1 f_1 d_2 f_2 = g_1 g_2 = (c_1 c_2) h_1 h_2, \tag{7.64}$$

where $g, h_1, h_2 \in \mathbb{Z}$ are all primitive. By Theorem 7.7.17, $h_1 h_2$ is also primitive so the content of the left-hand side of Eq. (7.64) is $cd_1 d_2$ and the content of the right-hand side is $c_1 c_2$. Since $cd_1 d_2$ and $c_1 c_2$ are both positive integers, $cd_1 d_2 = c_1 c_2$, and $g = h_1 h_2$. Thus, $f = cg = ch_1 h_2$ and we have a proper factorization of f in $\mathbb{Z}[t]$. ∎

Now the result we've been waiting for:

Theorem 7.7.19. *The factorization of $f \in \mathbb{Z}[t]$ into irreducible polynomials in $\mathbb{Z}[t]$ is unique up to order and association of the factors.*

Proof: Suppose $f \in \mathbb{Z}[t]$ can be written as

$$f = p_1 p_2 \cdots p_m = q_1 q_2 \cdots q_n, \tag{7.65}$$

where all p_k and q_k are irreducible polynomials in $\mathbb{Z}[t]$. Note that all p_k and q_k are either prime constant polynomials, or primitive. Since $p_m \mid q_1 \cdots q_n$, then viewing p_m and all q_k as elements of $\mathbb{Q}[t]$, irreducible in $\mathbb{Q}[t]$ by Theorem 7.7.18, we have that $p_m \mid q_k$ (in $\mathbb{Q}[t]$) for some k by Theorem 7.6.13. Reordering the q_k, we may assume

$p_m \mid q_n$, and since q_n is irreducible in $\mathbb{Q}[t]$, we may write $(a/b)p_m = q_n$ for some $a/b \in \mathbb{Q}$. Thus, $ap_m = bq_n$, and since p_m and q_n are primitive, $a = \pm b$. Therefore, $q_n = \pm p_m$, so that p_m and q_n are associates in $\mathbb{Z}[t]$. Substituting $\pm p_m$ for q_n in Eq. (7.65) and canceling p_m, we have

$$p_1 \cdots p_{m-1} = \pm q_1 \cdots q_{n-1}. \tag{7.66}$$

By the inductive assumption, the remaining p_k and q_k may be reordered and paired as associates, so that the factorization of f into irreducibles in $\mathbb{Z}[t]$ is unique up to order and association of the factors. ∎

EXERCISES

1. Prove Theorem 7.7.5: An ED is a PID.

2. Prove Theorem 7.7.6: If D is an ED, then $d(u) = d(e)$ if and only if u is a unit.

3. Prove Theorem 7.7.7: Suppose D is a domain with a valuation d, and with the property that, for all nonzero $a, b \in D$, either $b \in (a)$, or there exist $m, n \in D$ such that $d(ma + nb) < d(a)$. Then D is a PID.[10]

4. Finish the proof of Theorem 7.7.8 by showing the following:

    (a) The polynomial $r_1 = g - fq - (b_n/a_m)t^{n-m}f$ is an element of S such that $\deg r_1 < \deg r$.

    (b) If $g = fq_1 + r_1$ and $g = fq_2 + r_2$ where $r_1 = 0$ or $\deg r_1 < \deg f$ and where $r_2 = 0$ or $\deg r_2 < \deg f$, then $q_1 = q_2$ and $r_1 = r_2$.[11]

5. Prove Theorem 7.7.10: Suppose $f \in \mathbb{Q}[t]$ and $a \in \mathbb{Q}$. Then $(t - a) \mid f$ if and only if $f(a) = 0$.

6. Prove Theorem 7.7.11: Suppose $f \in \mathbb{Q}[t]$ and $\deg f \in \{2, 3\}$. Then f is reducible if and only if there exists $a \in \mathbb{Q}$ such that $f(a) = 0$.[12]

7. Verify the claims in Example 7.7.13: First write $g_1 = f_1 q + r$ for $q, r \in \mathbb{Z}_5[t]$. Then show that if q and r are any polynomials in $\mathbb{Z}_6[t]$ such that either $r = 0$ or $\deg r < \deg f_2$, then $g_2 = f_2 q + r$ is impossible.[13]

8. Apply Theorem 7.7.15 to the following polynomials in $\mathbb{Z}_3[t]$ by either finding a proper factorization or explaining why they are irreducible:

    (a) $f_1 = t^2 + t + 1$
    (b) $f_2 = t^2 + t + 2$
    (c) $f_3 = t^3 + t^2 + 2$
    (d) $f_4 = t^3 + t + 2$

[10] Show that an arbitrary ideal I is principal by letting $a \in I$ be such that $d(a)$ is minimum. You can then show that any $b \in I$ must be a multiple of a.
[11] What can you say about $\deg(r_2 - r_1)$? What does Theorem 7.5.18 allow you to conclude?
[12] If f is reducible, what can you say about the degree of one of its factors?
[13] For the $\mathbb{Z}_6[t]$ claim, the coefficient of t^2 in $f_2 q$ must be one. Show that this is impossible.

9. Prove Theorem 7.7.16: Every nonzero, nonunit polynomial in $\mathbb{Z}[t]$ has a factorization into irreducible polynomials in $\mathbb{Z}[t]$.[14]

10. Finish the proof of Theorem 7.7.17 by showing p does not divide the coefficient of t^{k+l} in fg.[15]

7.8 Ring Morphisms

The theory of ring morphisms should seem like a pretty breezy topic after having studied group morphisms in Section 6.5. The only difference in the definition in the context of rings is that there are two binary operations to preserve. As we're doing increasingly often, we'll be fairly relaxed about notation for elements and operations in the two rings, unless we need it to avoid confusion.

Definition 7.8.1. Suppose R and S are rings and $\phi : R \to S$ is a function with the properties that $\phi(x+y) = \phi(x) + \phi(y)$ and $\phi(xy) = \phi(x)\phi(y)$ for all $x, y \in R$. Then ϕ is called a *morphism* from R to S. The terms *monomorphism, epimorphism, isomorphism,* and *automorphism* are defined in a way analogous to that for groups, and if there exists an isomorphism from R to S, we write $R \cong S$.

Here are some examples. Let's start with a trivial one:

Example 7.8.2. If R and S are any rings, the mapping defined by $\phi(x) = 0$ for all $x \in R$ is called the *trivial* morphism.

Example 7.8.3. If R is any ring, the identity mapping $i(x) = x$ is an automorphism.

Example 7.8.4. Define $\phi : \mathbb{Z} \to \mathbb{Z}_n$ by $\phi(x) = (n) + x$, the equivalence class of $x \bmod n$. Thus, ϕ maps x to its remainder upon division by n, and ϕ is an epimorphism:

$$\phi(x+y) = (n) + [x+y] = [(n)+x] + [(n)+y] = \phi(x) + \phi(y) \quad (7.67)$$

$$\phi(xy) = (n) + xy = [(n)+x][(n)+y] = \phi(x)\phi(y). \quad (7.68)$$

Furthermore, ϕ is onto. For if $k \in \mathbb{Z}_n$, then $\phi(k) = k$.

Example 7.8.5. For $\mathbb{Z}[\sqrt[n]{x}]$ in the form of Eq. 7.35, define $\phi : \mathbb{Z} \to \mathbb{Z}[\sqrt[n]{x}]$ by

$$\phi(k) = k + 0\sqrt[n]{x} + \cdots + 0\sqrt[n]{x^{n-1}}. \quad (7.69)$$

Then ϕ is a monomorphism. This example illustrates a subtle distinction between two ideas that deserves a comment. After we constructed $\mathbb{Z}[\sqrt[3]{2}]$ in Example 7.3.1 (see page 256), we said that $\mathbb{Z}$ is a subring of $\mathbb{Z}[\sqrt[3]{2}]$. If we had been a bit more rigorous and abstract in our construction of $\mathbb{Z}[\sqrt[3]{2}]$, we would have built it up, not as a set of

---
[14]Use strong induction on deg f and mimic the proof of Theorem 2.4.8. Theorem 2.4.8 itself takes care of the case deg $f = 0$.
[15]The coefficient of t^{k+l} is $\sum_{i=0}^{k+l} a_i b_{k+l-i}$. Look separately at the terms $0 \le i \le k-1$, $i = k$, and $k+1 \le i \le k+l$.

expressions of the form $a + b\sqrt[3]{2} + c\sqrt[3]{4}$, but as a set of ordered triples

$$S = \{(a, b, c) : a, b, c \in \mathbb{Z}\}, \tag{7.70}$$

where addition and multiplication are defined to coincide with the definitions in Example 7.3.1. Specifically, defining addition and multiplication by

$$\begin{aligned}(a, b, c) + (d, e, f) &= (a + d, b + e, c + f) \\ (a, b, c) \cdot (d, e, f) &= (ad + 2bf + 2ce, ae + bd + 2cf, af + be + cd)\end{aligned} \tag{7.71}$$

incorporates the behavior of $\sqrt[3]{2}$ into the operations, even though they are mere manipulations of ordered triples with no apparent presence of $\sqrt[3]{2}$. With these definitions, to say $\mathbb{Z}$ is a subring of $\mathbb{Z}[\sqrt[3]{2}]$ is technically not true, for it is not true that $\mathbb{Z} \subseteq \mathbb{Z}[\sqrt[3]{2}]$, as subrings must be. Whereas $\mathbb{Z}$ is a set of numbers, $\mathbb{Z}[\sqrt[3]{2}]$ is a set of ordered triples of numbers. However, it does not mean that the link between them is somehow illusory. Defining $\psi : \mathbb{Z} \to S$ by $\psi(n) = (n, 0, 0)$, we have an exact parallel to ϕ in Eq. 7.69. Whether you think of $\mathbb{Z}[\sqrt[3]{2}]$ as we originally defined it in Example 7.3.1 or as in Eqs. (7.70) and (7.71), we say that we *embed* $\mathbb{Z}$ monomorphically in $\mathbb{Z}[\sqrt[3]{2}]$. Imagine $\mathbb{Z}$ and $\mathbb{Z}[\sqrt[3]{2}]$ as separate, where $\mathbb{Z}$ is a set of elements of the form n, and $\mathbb{Z}[\sqrt[3]{2}]$ consists of elements of the form (a, b, c). Then $\phi(\mathbb{Z})$ is the set of all elements of $\mathbb{Z}[\sqrt[3]{2}]$ of the form $(a, 0, 0)$, and is isomorphic to $\mathbb{Z}$, hence structurally the same.

Example 7.8.5 illustrates the slight breach of rigor we committed in defining ring extensions in Section 7.3, so let's clear that up here. It illustrates a technicality we need to be aware of when we say that $\mathbb{Z}[\sqrt[3]{2}]$ is an extension of $\mathbb{Z}$. If R and S are rings with $R \subseteq S$, then saying S is an extension of R has the same meaning as saying R is a subring of S. However, if you start with R and want to extend it to some S, the standard, more rigorous way is to build S from scratch and then monomorphically embed R in S. Then when we say that S is an extension of R, we mean that S is the range of a monomorphism whose domain is R.

Example 7.8.6. Let R be a commutative ring, and fix some $\alpha \in R$. Define $\phi_\alpha : R[t] \to R$ by $\phi(f) = f(\alpha)$. Clearly ϕ_α is a function (quick mental exercise), but also it's an epimorphism (Exercise 1). This particularly important morphism is called the *evaluation at α* morphism. It maps every polynomial in $R[t]$ to its value at α.

Example 7.8.7. Define $\phi : \mathbb{Z}[i] \to \mathbb{Z}[i]$ by $\phi(a + bi) = a - bi$. Then ϕ is an automorphism (Exercise 3). This automorphism is called the *conjugation* morphism, for it sends a Gaussian integer $a + bi$ to its *complex conjugate* $a - bi$.

Example 7.8.8. Let R be a ring with unity element e, and define $\phi : \mathbb{Z} \to R$ by $\phi(n) = ne$ (Definition 7.2.8, page 251). Then ϕ is a morphism (Exercise 4).

7.8.1 Properties of ring morphisms

Since a ring morphism $\phi : R \to S$ preserves addition between R and S as abelian additive groups, our results from Section 6.5 apply to addition. Thus we gain the following for free from Theorems 6.5.6 and 6.5.7.

Theorem 7.8.9. *Suppose $\phi : R \to S$ is a ring morphism. Then*

1. $\phi(0) = 0$.
2. $\phi(-r) = -\phi(r)$ for all $r \in R$.
3. $\phi(nr) = n\phi(r)$ for all $r \in R$ and $n \in \mathbb{Z}$.

To apply something like Theorems 6.5.6 and 6.5.7 to ring multiplication, we have to make some modifications. First here is an observation you'll prove in Exercise 6.

Theorem 7.8.10. *If R is a ring with unity e_R, and if $\phi : R \to S$ is a ring morphism such that $\phi(e_R) = 0$, then ϕ is the trivial morphism.*

The contrapositive of Theorem 7.8.10 reveals that if $\phi : R \to S$ is a nontrivial morphism, then $\phi(e_R) \neq 0$. If S has unity e_S, then comparable to Theorem 6.5.6, you might be tempted to think that a nontrivial morphism $\phi : R \to S$ would have to satisfy $\phi(e_R) = e_S$. To prove such a property and the exponent rules related to it, multiplicative cancellation is necessary. Thus a proof of the next theorem becomes possible. Since the proof of the exponent rules in parts 2–4 would be identical to the proof of Theorem 6.5.7, you'll prove only parts 1 and 3 in Exercise 7.

Theorem 7.8.11. *Suppose R is a ring with unity e_R, S is a domain with unity e_S, and $\phi : R \to S$ is a nontrivial ring morphism. Then the following hold:*

1. $\phi(e_R) = e_S$.
2. For all $r \in R$ and $n \in \mathbb{W}$, $\phi(r^n) = [\phi(r)]^n$.
3. If $u \in R$ is a unit, then so is $\phi(u)$, and $\phi(u^{-1}) = [\phi(u)]^{-1}$.
4. If $u \in R$ is a unit, then $\phi(u^{-n}) = [\phi(u)]^{-n}$ for all $n \in \mathbb{N}$.

Theorems 7.8.9 and 7.8.11 reveal that $\phi : \mathbb{Z} \to D$ in Example 7.8.8 is the only nontrivial morphism from $\mathbb{Z}$ to any other domain. For suppose $\psi : \mathbb{Z} \to D$ is a nontrivial morphism. Since 1 is the unity in $\mathbb{Z}$, it must be that $\psi(1) = e$. Furthermore, for any $n \in \mathbb{Z}$, $\psi(n) = \psi(n \cdot 1) = n\psi(1) = ne$. Applying this result to morphisms $\phi : \mathbb{Z} \to \mathbb{Z}$, we see that the only nontrivial morphism from $\mathbb{Z}$ to $\mathbb{Z}$ is the identity. In Exercise 8 you'll use other parts of Theorems 7.8.9 and 7.8.11 to show that the only automorphism of $\mathbb{Q}$ is the identity.

If S is not a domain, Theorem 7.8.11 might not apply. For example, if $\phi : \mathbb{Z} \to \mathbb{Z}_{2 \times 2}$ is defined by $\phi(n) = \begin{bmatrix} n & 0 \\ 0 & 0 \end{bmatrix}$, then ϕ is a morphism, but $\phi(1) = \begin{bmatrix} 1 & 0 \\ 0 & 0 \end{bmatrix} \neq I_{2 \times 2}$. Furthermore, though 1 is a unit in $\mathbb{Z}$, $\phi(1)$ is not a unit in $\mathbb{Z}_{2 \times 2}$.

Much of our work with group morphisms involved their relationship to normal subgroups. The interesting relationships between ring morphisms and substructures involves ideals. Here's a parallel to Theorem 6.5.10. When you prove the left-sided case of Theorem 7.8.12 in Exercise 9, there will be parts of your proof that will follow immediately from your work in group theory and not require new arguments. For example, in your proof of part 1, you may simply point out that since R is an additive group with respect to addition, $\phi(R)$ is an additive subgroup of S by Theorem 6.5.10. That takes care of three of the requirements in showing that $\phi(R)$ is a subring of S.

Theorem 7.8.12. *Suppose* $\phi : R \to S$ *is a ring morphism. Then*

1. $\phi(R)$ *is a subring of* S.
2. *If* I *is a left (right) ideal of* R, *then* $\phi(I)$ *is a left (right) ideal of* $\phi(R)$.
3. *If* I *is a left (right) ideal of* S, *then* $\phi^{-1}(I)$ *is a left (right) ideal of* R.

Similar to groups, the kernel of a ring morphism is defined as

$$\mathrm{Ker}(\phi) = \{r \in R : \phi(r) = 0\}. \tag{7.72}$$

Right away, we then have the following.

Theorem 7.8.13. *If* $\phi : R \to S$ *is a ring morphism, then* $\mathrm{Ker}(\phi)$ *is a two-sided ideal of* R.

You'll prove Theorem 7.8.13 in Exercise 10. As with Theorem 7.8.12, you can save yourself some work by applying Theorem 6.5.12. The following needs no additional proof, for it follows from Theorem 6.5.13 applied to R as an additive group.

Theorem 7.8.14. *If* $\phi : R \to S$ *is a ring morphism, then* $\mathrm{Ker}(\phi) = \{0\}$ *if and only if* ϕ *is one-to-one.*

If R is any ring with unity element e, the morphism $\phi : \mathbb{Z} \to R$ defined in Example 7.8.8 can look either of two ways, depending on $\mathrm{char}\, R$. If $\mathrm{char}\, R = 0$, then by definition ne is never zero for nonzero $n \in \mathbb{Z}$. The following should be immediate (Exercise 11).

Theorem 7.8.15. *If* R *is a ring with unity and* $\mathrm{char}\, R = 0$, *then* $\phi : \mathbb{Z} \to R$ *defined by* $\phi(n) = ne$ *is one-to-one.*

Example 7.8.8 and Theorem 7.8.15 say that the integers can be monomorphically embedded in a ring R with unity and characteristic zero. We loosely say that R *contains* $\mathbb{Z}$, meaning it contains a subring generated by its unity that is isomorphic to $\mathbb{Z}$. For example, in $\mathbb{R}_{2 \times 2}$, the image of $\mathbb{Z}$ under ϕ is

$$\left\{ \ldots, \begin{bmatrix} -2 & 0 \\ 0 & -2 \end{bmatrix}, \begin{bmatrix} -1 & 0 \\ 0 & -1 \end{bmatrix}, \begin{bmatrix} 0 & 0 \\ 0 & 0 \end{bmatrix}, \begin{bmatrix} 1 & 0 \\ 0 & 1 \end{bmatrix}, \begin{bmatrix} 2 & 0 \\ 0 & 2 \end{bmatrix}, \ldots \right\}. \tag{7.73}$$

If $\mathrm{char}\, R = n \neq 0$, then there is a smallest $n \in \mathbb{Z}^+$ for which $ne = 0$. It might seem clear that R in this case will contain a subring that looks like $\mathbb{Z}_n$ instead of $\mathbb{Z}$. Before we can make an argument for this, however, we need to create the notion of a *quotient ring*, which we will do in Section 7.9.

EXERCISES

1. Let R be a commutative ring and fix $\alpha \in R$. Show that the function $\phi_\alpha : R[t] \to R$ defined by $\phi_\alpha(f) = f(\alpha)$ is an epimorphism.

2. Let R be a commutative ring with unity. Describe the evaluation morphisms ϕ_0 and ϕ_e.

3. Show that complex conjugation as defined in Example 7.8.7 is an automorphism of $\mathbb{Z}[i]$.

4. Let R be a ring with unity element e. Show that $\phi : \mathbb{Z} \to R$ in Example 7.8.8 defined by $\phi(n) = ne$ is a morphism.

5. Write $A \in \mathbb{Z}_{2\times 2}$ as $\begin{bmatrix} a_{11} & a_{12} \\ a_{21} & a_{22} \end{bmatrix}$. Define $\phi : \mathbb{Z}_{2\times 2} \to \mathbb{Z}$ by $\phi(A) = a_{11}$. Is ϕ a morphism?

6. Prove Theorem 7.8.10: If R is a ring with unity e_R, and if $\phi : R \to S$ is a ring morphism such that $\phi(e_R) = 0$, then ϕ is the trivial morphism.

7. Prove parts 1 and 3 of Theorem 7.8.11: Suppose R is a ring with unity e_R, S is a domain with unity e_S, and $\phi : R \to S$ is a *nontrivial* ring morphism. Then $\phi(e_R) = e_S$. Furthermore, if $u \in R$ is a unit, so is $\phi(u)$, and $\phi(u^{-1}) = [\phi(u)]^{-1}$.

8. Show that the only automorphism of $\mathbb{Q}$ is the identity.

9. Prove the left-sided case of Theorem 7.8.12: Suppose $\phi : R \to S$ is a ring morphism. Then

   (a) $\phi(R)$ is a subring of S.
   (b) If I is a left ideal of R, then $\phi(I)$ is a left ideal of $\phi(R)$.
   (c) If I is a left ideal of S, then $\phi^{-1}(I)$ is a left ideal of R.

10. Prove Theorem 7.8.13: If $\phi : R \to S$ is a ring morphism, then $\text{Ker}(\phi)$ is a two-sided ideal of R.

11. Prove Theorem 7.8.15: If R is a ring with unity and char $R = 0$, then $\phi : \mathbb{Z} \to R$ defined by $\phi(n) = ne$ is one-to-one.

7.9 Quotient Rings

Given a group G and $H \triangleleft G$, we built the quotient group G/H. In an analogous way, given that I is an ideal of R, we can build the quotient ring R/I. Part of the work in building R/I is exactly like that in building G/H, so the fact that R/I has ring properties R1–R10 has already been done in part. As with groups, building the quotient structure begins by defining a form of equivalence.

Theorem 7.9.1. *Let R be a ring and I an ideal of R. For $a, b \in R$ define $a \equiv_I b$ if $a - b \in I$. Then $\equiv_I$ is an equivalence relation on R.*

Since R is an abelian group with respect to its addition operation, I is a normal additive subgroup of R. Therefore, Theorem 7.9.1 is merely a restatement of Exercise 2a from Section 6.3 in its additive form and needs no additional proof. Thus we're ready to define the set R/I with its addition and multiplication operations, and show that it has all properties R1–R10. There are no surprises in the definitions of the binary operations on R/I. However, considering the need for H to be a normal subgroup of G in showing that the binary operation on G/H is well defined, little bells should be going off in your head as you consider the burden of showing that addition and multiplication on R/I are well defined. It should be no surprise that you'll exploit the fact that I is an ideal of R in at least part of this demonstration. What you might not see at first is what kind of ideal I needs to be and where you'll call on the fact that I is this kind of ideal. Since I is normal as an additive subgroup of R, addition is well defined by our work in Chapter 6. It's a different story for multiplication, though. If multiplication in R is not commutative, then a left ideal might not be a right ideal, and vice versa. Thus, we might wonder whether I needs merely to be either a left ideal or a right ideal, or whether it needs to be both. It turns out that I needs to be a two-sided ideal, and you'll see how you need that fact when you prove the following in Exercise 1.

Theorem 7.9.2. *Let R be a ring and I a two-sided ideal of R. For $a \in R$ write $I + a = [a]$, where $[a]$ is the equivalence class of a modulo I. Define addition $\oplus$ and multiplication $\otimes$ on $R/I = \{I + a : a \in R\}$ by*

$$(I + a) \oplus (I + b) = I + (a + b) \quad \text{and} \quad (I + a) \otimes (I + b) = I + (ab). \quad (7.74)$$

Then R/I is a ring under the operations $\oplus$ and $\otimes$.

We can talk our way through almost all properties R1–R10, so that your work in proving Theorem 7.9.2 will be minimal. Since R is an abelian additive group and I is a normal additive subgroup, R/I has properties R1–R6. Properties R8–R10 are immediate from the definitions of $\oplus$ and $\otimes$. The only property that takes any real work is R7, showing that $\otimes$ is well defined, and this is what you'll show in Exercise 1. True to form, we suppose $I + a = I + b$ and $I + c = I + d$ and use this to show that $I + ac = I + bd$. That is, supposing $a - b, c - d \in I$ should somehow allow you to show that $ac - bd \in I$. Multiplication of $a - b$ and $c - d$ by strategically chosen elements, together with the fact that I is the certain kind of ideal that you specified, should get you over the hump.

With R/I defined and shown to be a ring, we should be able to bypass all the chatty exposition we provided in Chapter 6 and jump right to theorems analogous to Theorems 6.5.15 and 6.5.16. Go back and take a look at these theorems, then try to state analogous theorems for rings before you read the theorems below.

Theorem 7.9.3. *Suppose R is a ring and I is a two-sided ideal in R. Then the mapping $\phi : R \to R/I$ defined by $\phi(r) = I + r$ is an epimorphism whose kernel is I.*

Almost everything in Theorem 7.9.3 follows directly from Theorem 6.5.15. The fact that R is an abelian additive group means I is a normal subgroup, so that ϕ behaves morphically as far as addition is concerned, is onto, and satisfies $\text{Ker}(\phi) = I$. The only remaining claim is that ϕ behaves morphically with respect to multiplication. But this is a one-liner (Exercise 2). Having thought your way through this, writing a complete proof of the following theorem should come naturally (Exercise 3).

Theorem 7.9.4. *Suppose R and S are rings and $\phi : R \to S$ is an epimorphism. Then $S \cong R/\text{Ker}(\phi)$.*

Let's return to the morphism in Example 7.8.8 defined by $\phi(n) = ne$. If R has nonzero characteristic, then there is a smallest $n \in \mathbb{N}$ for which $ne = 0$. By Theorem 7.8.13, $\text{Ker}(\phi)$ is an ideal of $\mathbb{Z}$. However, $\mathbb{Z}$ is a PID, so $\text{Ker}(\phi) = (k)$ for some $k \in \mathbb{N}$. Since $\phi(k) = 0$, and since n is the smallest positive integer for which $ne = 0$, it must be that $k \geq n$. However, because $n \in \text{Ker}(\phi)$, it must be that n is a multiple of k, so that $k \leq n$. Thus $k = n$. By Theorem 7.9.4, the range of ϕ is isomorphic to $\mathbb{Z}/(n)$. That is, R contains a subring isomorphic to $\mathbb{Z}_n$.

If we think of a ring with unity element e merely as an abelian additive group, and let S be the subgroup generated by e, the identity element of the *other* operation, then the form of S can be determined by looking at the additive form of Eq. (6.25).

$$S = \{ne : n \in \mathbb{Z}\}, \tag{7.75}$$

which is precisely the range of ϕ in Example 7.8.8. Thus, as far as addition is concerned, S is the smallest additive subgroup of R that contains e. If we can then show that S is closed under multiplication, we'll have that S is the smallest subring of R that contains e. From Exercise 9, Section 7.1, closure is immediate. Thus, in bits and pieces, we have proved the following:

Theorem 7.9.5. *Suppose R is a ring with unity. If char $R = 0$, then R contains a subring isomorphic to $\mathbb{Z}$. If char $R = n \neq 0$, then R contains a subring isomorphic to $\mathbb{Z}_n$. In either case, such is the smallest subring of R that contains e.*

Because polynomial rings over fields are EDs, they make for some particularly important quotient rings. Let's spend some time studying the quotient ring created by modding out the ideal generated by $f = 2t^3 - t + 5$ from $\mathbb{Q}[t]$. The ties back to $\mathbb{Z}$ and $\mathbb{Z}_n$ are uncanny, so we'll hold the quotient ring $\mathbb{Q}[t]/(f)$ up against $\mathbb{Z}_n$ as we dissect it. To be concrete, we let $n = 6$ and draw parallels between the relationship of $\mathbb{Q}[t]$ to $\mathbb{Q}[t]/(f)$ and the relationship of $\mathbb{Z}$ to $\mathbb{Z}_6$.

Since $\mathbb{Z}$ is an ED, every $a \in \mathbb{Z}$ can be written as $a = 6q + r$ for some $q, r \in \mathbb{Z}$ and $0 \leq r \leq 5$. Consequently, every $a \in \mathbb{Z}$ is equivalent mod 6 to some element of $\{0, 1, 2, 3, 4, 5\}$. Thus every element of $\mathbb{Z}_6$ is a coset that can be addressed by a unique representative element in $\{0, 1, 2, 3, 4, 5\}$. To perform addition and multiplication in $\mathbb{Z}_6$, a purist would write something like

$$[(6) + 4] + [(6) + 3] = (6) + 7 = (6) + 1, \tag{7.76}$$

or

$$[(6) + 5] \times [(6) + 4] = (6) + 20 = (6) + 2, \tag{7.77}$$

where the first step in these two calculations is application of the definitions of addition and multiplication in $\mathbb{Z}_6$ from Theorem 7.9.2, and the second step is an application of equivalence mod 6 to simplify the calculation to a standard form with a representative element from $\{0, 1, 2, 3, 4, 5\}$. As long as we realize that this is what we are doing, we can write the calculations in Eqs. (7.76) and (7.77) as

$$4 + 3 =_6 7 =_6 1 \quad \text{and} \quad 5 \times 4 =_6 20 =_6 2. \tag{7.78}$$

This form has the imagery of performing addition and multiplication in $\mathbb{Z}$, with the stipulation that any sum or product that gets kicked out of bounds (i.e., out of $\{0, 1, 2, 3, 4, 5\}$) is translated back into $\{0, 1, 2, 3, 4, 5\}$ by subtracting a multiple of 6. This will help us to see how we may view elements of $\mathbb{Z}_6$, and how they combine by addition and multiplication to produce other elements of $\mathbb{Z}_6$.

The way to visualize elements of $\mathbb{Q}[t]/(f)$ in terms of polynomials in $\mathbb{Q}[t]$ is strikingly similar. Since $\mathbb{Q}[t]$ is an ED, any $g \in \mathbb{Q}[t]$ can be written uniquely as $g = fq + r$ for some $q, r \in \mathbb{Q}[t]$ where either $r = 0$ or $\deg r < \deg f$. Thus, if we consider any $(f) + g \in \mathbb{Q}[t]/(f)$, there is a unique $r \in \mathbb{Q}[t]$ such that $(f) + r = (f) + g$ and either $r = 0$ or $\deg r < \deg f$. Therefore, elements of $\mathbb{Q}[t]/(f)$ may always be addressed by a representative element r where either $r = 0$ or $\deg r < \deg f$. Since we're using $f = 2t^3 - t + 5$, then we may write

$$\mathbb{Q}[t]/(f) = \{(f) + at^2 + bt + c : a, b, c \in \mathbb{Q}\}, \tag{7.79}$$

and know that every element of $\mathbb{Q}[t]/(f)$ can be written as some $(f) + at^2 + bt + c$. Furthermore, if $(f) + r_1, (f) + r_2 \in \mathbb{Q}[t]/(f)$ and $(f) + r_1 = (f) + r_2$ are written in the form of polynomials in Eq. (7.79), then $r_2 \equiv_{(f)} r_1$, so that $r_2 - r_1$ is a multiple of f. Now nonzero multiples of f cannot have degree less than $\deg f$, but $\deg(r_2 - r_1) \geq \deg f$ is impossible. Thus $r_1 = r_2$, and we have that different polynomials of the form $at^2 + bt + c$ will always generate different cosets of (f).

In the same way we view elements of $\mathbb{Z}_6$ simply as $\{0, 1, 2, 3, 4, 5\}$, we can view elements of $\mathbb{Q}[t]/(f)$ simply as polynomials of the form $at^2 + bt + c$. What we must look at is how they add and multiply. Instead of using coset notation and writing $[(f) + a_1t^2 + b_1t + c_1] + [(f) + a_2t^2 + b_2t + c_2]$, we can just write

$$[a_1t^2 + b_1t + c_1] + [a_2t^2 + b_2t + c_2] =_{(f)} (a_1 + a_2)t^2 + (b_1 + b_2)t + (c_1 + c_2). \tag{7.80}$$

Adding two such polynomials cannot produce a sum of any larger degree, so Eq. (7.80) is all that needs to be said about addition in $\mathbb{Q}[t]/(f)$. However, for multiplication, let's illustrate with a concrete example.

$$\begin{aligned}(4t^2 + 2)(3t^2 - 2t + 8) &=_{(f)} 12t^4 - 8t^3 + 38t^2 - 4t + 16 \\ &=_{(f)} (6t - 4)f + 44t^2 - 38t + 36 \\ &=_{(f)} 44t^2 - 38t + 36.\end{aligned} \tag{7.81}$$

If we multiply two elements of $\mathbb{Q}[t]/(f)$ as if they were polynomials in $\mathbb{Q}[t]$, and we produce a product of degree at least three, we can apply the division algorithm to subtract off an appropriate multiple of f from the product to produce an equivalent polynomial of the form $at^2 + bt + c$. You'll work through similar details for another example in Exercise 4.

Now for another very interesting example. Since $\mathbb{Z}_3$ is a field, $\mathbb{Z}_3[t]$ is an ED, and we can construct the quotient ring $\mathbb{Z}_3[t]/(f)$ for $f \in \mathbb{Z}_3[t]$ in a similar way. Let's use $f = t^3 + t + 2$, construct the quotient ring, and look at addition and multiplication. By exactly the same reasoning as in $\mathbb{Q}[t]$, $\mathbb{Z}_3[t]/(f) = \{at^2 + bt + c : a, b, c \in \mathbb{Z}_3\}$. Notice that this is a finite set. Each of a, b, and c can take on values from $\{0, 1, 2\}$, so $\mathbb{Z}_3[t]/(f)$ has 27 elements. Adding elements of $\mathbb{Z}_3[t]/(f)$ is easy:

$$(2t^2 + t + 2) + (t^2 + 2t + 2) =_{(f)} 3t^2 + 3t + 4 =_{(f)} 1. \tag{7.82}$$

Doing multiplication would look like the following if we simplify the product by way of the division algorithm:

$$\begin{aligned}(2t^2 + t + 2)(t^2 + 2t + 2) &=_{(f)} 2t^4 + 5t^3 + 8t^2 + 6t + 4 \\ &=_{(f)} 2t^4 + 2t^3 + 2t^2 + 1 \\ &=_{(f)} (2t + 2)f \\ &=_{(f)} 0.\end{aligned} \quad (7.83)$$

However, there is a slick way to simplify multiplication by making substitutions. In $\mathbb{Z}_3[t]/(f)$, $f \equiv_{(f)} 0$, or $t^3 + t + 2 \equiv_{(f)} 0$. This can also be written as $t^3 \equiv_{(f)} -t - 2 \equiv_{(f)} 2t + 1$. The upshot is that any t^3 produced in the process of multiplication can be replaced with $2t + 1$, thus bringing the degree of a product back down:

$$\begin{aligned}(2t^2 + t + 2)(t^2 + 2t + 2) &=_{(f)} 2t^4 + 5t^3 + 8t^2 + 6t + 4 \\ &=_{(f)} 2t^4 + 2t^3 + 2t^2 + 1 \\ &=_{(f)} (2t)t^3 + 2t^3 + 2t^2 + 1 \\ &=_{(f)} (2t)(2t + 1) + 2(2t + 1) + 2t^2 + 1 \\ &=_{(f)} 4t^2 + 2t + 4t + 2 + 2t^2 + 1 \\ &=_{(f)} 6t^2 + 6t + 3 \\ &=_{(f)} 0.\end{aligned} \quad (7.84)$$

You'll practice this technique in Exercises 5 and 6.

The last two theorems we want to present in this section and the examples following each illustrate some very interesting implications of the results we have worked so hard to develop. The two theorems are not results that would likely jump out at you as obvious, but they are elegant and not difficult to prove. You'll prove the first one in Exercise 8.

Theorem 7.9.6. *Suppose R is a commutative ring, and I is an ideal of R. Then R/I is an integral domain if and only if I is a prime ideal.*

We state our last theorem as an if-and-only-if theorem, but you will be required to prove only one direction.

Theorem 7.9.7. *Suppose R is a commutative ring with unity e, and I is an ideal of R. Then R/I is a field if and only if I is a maximal ideal.*

When you prove the $\Leftarrow$ direction of Theorem 7.9.7 in Exercise 9, you're going to suppose I is a maximal ideal of a commutative ring R with unity e, then show that R/I is a field. Now R/I is also a commutative ring with unity $I + e$. Also, since I is a proper ideal of R, R/I has more than one element, so you can choose some $I + a \in R/I$ such that $I + a \neq I + 0$, for which you must find an inverse in R/I. Since $a \notin I$, Theorem 7.4.14 and the maximality of I are just the things to help you do that.

Now let's put all this together in a very elegant construction. If K is a field, then $K[t]$ is an ED, hence a PID. If $f \in K[t]$ is irreducible (prime), then (f) is a prime ideal by

Theorem 7.6.16. By Theorem 7.6.8, (f) is also maximal. Therefore, $K[t]/(f)$ is a field. We can use these facts to do the following:

Example 7.9.8. Construct a field with nine elements.

Solution: Since t^2+1 is irreducible in $\mathbb{Z}_3[t]$, $\mathbb{Z}_3[t]/(t^2+1) = \{at+b : a, b \in \mathbb{Z}_3\}$ is a field with nine elements. For notational simplicity, we write $at + b = (a, b)$ and illustrate multiplication in Table 7.1. Notice $t^2 =_{(t^2+1)} 2$ and the manifestation of this in the table. Also, notice how the table reveals that every element has a multiplicative inverse.

×	(0, 0)	(0, 1)	(0, 2)	(1, 0)	(1, 1)	(1, 2)	(2, 0)	(2, 1)	(2, 2)
(0, 0)	(0, 0)	(0, 0)	(0, 0)	(0, 0)	(0, 0)	(0, 0)	(0, 0)	(0, 0)	(0, 0)
(0, 1)	(0, 0)	(0, 1)	(0, 2)	(1, 0)	(1, 1)	(1, 2)	(2, 0)	(2, 1)	(2, 2)
(0, 2)	(0, 0)	(0, 2)	(0, 1)	(2, 0)	(2, 2)	(2, 1)	(1, 0)	(1, 2)	(1, 1)
(1, 0)	(0, 0)	(1, 0)	(2, 0)	(0, 2)	(1, 2)	(2, 2)	(0, 1)	(1, 1)	(2, 1)
(1, 1)	(0, 0)	(1, 1)	(2, 2)	(1, 2)	(2, 0)	(0, 1)	(2, 1)	(0, 2)	(1, 0)
(1, 2)	(0, 0)	(1, 2)	(2, 1)	(2, 2)	(0, 1)	(1, 0)	(1, 1)	(2, 0)	(0, 2)
(2, 0)	(0, 0)	(2, 0)	(1, 0)	(0, 1)	(2, 1)	(1, 1)	(0, 2)	(2, 2)	(1, 2)
(2, 1)	(0, 0)	(2, 1)	(1, 2)	(1, 1)	(0, 2)	(2, 0)	(2, 2)	(1, 0)	(0, 1)
(2, 2)	(0, 0)	(2, 2)	(1, 1)	(2, 1)	(1, 0)	(0, 2)	(1, 2)	(0, 1)	(2, 0)

Table 7.1 Cayley table for multiplication in $\mathbb{Q}[t]/(t^2 + 1)$

∎

In Exercise 10, you'll construct a field with eight elements.

EXERCISES

1. Finish the proof of Theorem 7.9.2 by showing that R/I has property R7.

2. Finish the proof of Theorem 7.9.3 by showing that ϕ behaves morphically with respect to multiplication.

3. Prove Theorem 7.9.4: Suppose R and S are rings and $\phi: R \to S$ is an epimorphism. Then $S \cong R/\text{Ker}(\phi)$.

4. In $\mathbb{Q}[t]$, let $f = t^4+2t+1$. Construct the form of elements of $\mathbb{Q}[t]/(f)$, and illustrate addition and multiplication.

5. In $\mathbb{Z}_3[t]$, let $f = t^4 + 2t + 1$. Construct the form of elements of $\mathbb{Z}_3[t]/(f)$, and illustrate addition and multiplication using the fact that $t^4 =_{(f)} t + 2$.

6. From our work in this section, $\mathbb{Q}[t]/(t^2-2) = \{at+b : a, b \in \mathbb{Q}\}$. Use a trick similar to that in Exercise 5 to simplify $(at + b)(ct + d)$.

7. In Exercise 4 from Section 7.3, you showed $\mathbb{Q}[\sqrt{2}]$ is a field. Calculate and simplify $(b + a\sqrt{2})(d + c\sqrt{2})$, and compare to Exercise 6.

8. Prove Theorem 7.9.6: Suppose R is a commutative ring and I an ideal of R. Then R/I is an integral domain if and only if I is a prime ideal.

9. Prove one direction of Theorem 7.9.7: Suppose R is a commutative ring with unity e, and I is a maximal ideal of R. Then R/I is a field.

10. Construct a field with eight elements, providing complete Cayley tables for addition and multiplication.

Index

$>$, 8, 56
$\dot{\vee}$, 17
ϵ-neighborhood, 140
   deleted, 144
$\mathbb{N}_n$, 100
$\sqrt{2}$, 88, 168
$\subseteq$, 2
$\supseteq$, 2
$\vee$, 17
$\wedge$, 16

A1–A22 (Assumptions), 5–9
Abelian, 210
Absolute value, 57
Accumulation point, 144
Addition
   associativity, 6
   closure, 6
   commutativity, 6
   well-defined, 6
Additive identity, 6
Additive inverse, 6
   uniqueness, 54
Algebraic function, 100
Alternating group, 229, 230
AND, 15–18
Antisymmetric property, 92, 94, 95
Archimedean property, 136
Associate, 269, 276
Associativity
   $\wedge$, $\vee$, 19
   addition, 6
   multiplication, 7

Asymmetric property, 94
Automorphism, 237, 287
Axiom of choice, 95

Bijection, 103
Binomial theorem, 126–128
Bolzano-Weierstrass
   theorem, 165
Bound
   greatest lower, 137
   least upper, 134
Boundary, 142, 145
Bounded
   function, 170
   sequence, 151
   sets, 134

C1–C3 (Closure), 144
Cancellation
   addition, 53
   in a group, 214
   in an integral domain, 269
   multiplication, 55
Cardinality, 110–118
Cartesian product, 90, 118, 120
Cauchy sequence, 165–169
Cayley table, 209
Characteristic, 254, 269
Closed interval, 135
Closed set, 141, 144
Closure
   addition, 6

Closure (*Continued*)
  multiplication, 6
  of sets, 144
Cluster point, 144
Codomain, 97
Combination, 122
Commensurable, 88
Commutativity
  addition, 6
  multiplication, 7
Compact, 146–149, 198, 202
Complement, 4
Completeness, 134, 167, 168
Complex numbers, 211
Composite, 66
Composition, 107, 189
Conclusion, 21
Content, 271
Continuity, 187–198
  and open sets, 197
  at a point, 188
  on a set, 190
  one-sided, 194
  uniform, 200–203
Contradiction, 19, 45
Contrapositive, 22, 44
Convergence
  of Cauchy sequences, 166
  of sequences, 153–159
Converse, 22
Coset, 225
Countable sets, 113
Counterexample, 45
Counting techniques, 118–126
Cover, 146
Cycle notation, 228
Cyclic subgroup, 217

D1–D3 (Discontinuity), 189
D1–D3 (Greatest common divisor), 82, 274
Dedekind, Richard, 48
Degree, 259, 271
Deleted neighborhood, 144, 174
DeMorgan's law, 19, 20, 32, 43, 46, 51, 52
Difference of sets, 40
Dihedral group, 230
Direct proof, 41–44
Directed graph, 93
Discontinuity, 189
Disjoint sets, 39, 112
Disjoint union, 46
Disproving a statement, 45
Distributive property, 7
Divergence, 155
Divisibility, 81, 94, 251
Division algorithm, 80, 279

Divisor
  greatest common, 82, 274, 276
  in $\mathbb{Z}$, 81
  in a ring, 251
  of zero, 251
  proper, 268, 270

E1–E3 (Equivalence relation), 70, 92
ED, 279–286
Empty set, 111
Epimorphism, 237, 287
Equality
  of functions, 101
  of sets, 39
  properties, 5, 70
Equivalence class, 72
Equivalence mod n, 71
Equivalence relation, 70, 92, 291
Euclid, 87
Euclidean domain, 279–286
Even
  integer, 79
  permutation, 230
Exclusive OR, 17
Existential quantifier, 29–30
Extended real numbers, 184
Extension, 256–260
Exterior, 142, 145
Extreme value theorem, 198

F1–F2 (Function), 98
Field, 295
  an ED, 280
  defined, 254
  finite, 269
  ideals in, 263
Finite field, 269
Finite set, 111
  greatest lower bound, 137
  least upper bound, 137
Fixed point theorem, 199
Function
  algebraic, 100
  and finite sets, 121
  bounded, 170
  composition, 107, 189
  continuous, 187–198
  definition, 97
  equality, 101
  identity, 103, 188
  inverse, 108, 172
  limit, 173–186
  limit at infinity, 183
  limit of infinity, 185
  maximum, 198
  minimum, 198

monotone, 171, 172, 196
of a real variable, 170
one-sided limit, 180
one-to-one, 103, 108, 122, 172
onto, 102
range, 101
salt and pepper, 179
sequential limit, 181
well defined, 98
Fundamental theorem
of algebra, 199
of arithmetic, 66

G1–G2 (Greatest lower bound), 137
G1–G5 (Group), 210
Gauss' lemma, 284
Gauss, Carl Friedrich, 61
Gaussian integers, 257, 269
gcd, 82, 274, 276
Generated ideal, 262
Generated subgroup, 216
Greatest common divisor, 82, 274, 276
Greatest lower bound, 137
Group
alternating, 229, 230
defined, 210
dihedral, 230
isomorphism theorem, 240
morphism, 236–241
order, 210
permutation, 227
quotient, 223–225
symmetric, 227

H1–H3 (Subgroup), 213
Half-closed interval, 141
Half-open interval, 141
Heine-Borel theorem, 147
Hypothesis, 21

I1–I2 (Induction), 62
Ideals, 260–265
and equivalence relations, 291
generated, 262
in a field, 263
maximal, 264, 276, 295
prime, 264, 270, 276, 278, 295
principal, 263
Identity
additive, 6
function, 103, 188
group, 208
matrix, 247
multiplicative, 7
uniqueness, 214
If-then statement, 20–21

Iff, 22
Image, 97, 101
Index set, 47
Induction, 61–66
Inductive assumption, 63
Inductive step, 63
Infinite set, 113
Infinity, 184
Injection, 103
Integers
a PID, 275
an ED, 280
countability, 114
even, 79
Gaussian, 257, 269
mod n, 71, 84, 220–223, 245, 287
odd, 79
Integral domain, 267–271, 295
Interior, 142, 145
Intermediate value theorem, 195
Intersection, 4, 39, 141
over a family, 50
Interval(s)
closed, 135
half closed, 141
half open, 141
nested, 160
open, 135
Inverse
additive, 6
function, 108
in a group, 209
multiplicative, 7
statement, 22
uniqueness, 214
Irrational numbers, 87
Irreducible polynomial, 271
Irreflexive property, 94
Isomorphism, 287
Isomorphism theorem, 240, 292

J1–J2 (Induction), 64
Jump discontinuity, 189

K1–K2 (Induction), 66
Kernel, 239, 290, 292

L1–L2 (Least upper bound), 9, 135
Lagrange's theorem, 225
Least upper bound property, 89, 134–138, 161, 164, 167
Limit
at infinity, 183
of a function, 173–186
of a sequence, 155
of infinity, 185

Limit point, 143
Linear combination, 81
Logical equivalence, 18, 23

M1–M2 (Least upper bound), 135
Matrix ring, 245
Maximal ideal, 264, 276, 295
Maximum
  of a function, 198
  of a set, 138
Metric, 133, 167, 197
Minimum
  of a function, 198
  of a set, 138
Monomorphism, 237, 287
Monotone
  function, 171
  sequence, 150
Morphism
  group, 236–241
  ring, 287–290
Multiplication
  associativity, 7
  closure, 6
  commutativity, 7
  well defined, 6
Multiplicative identity, 7
Multiplicative inverse, 7

N1–N3 (Norm), 57–58
Negation, 15
  $\forall, \exists$, 33–36
  $\rightarrow$, 33
  $\land, \lor$, 32
Neighborhood, 140
  deleted, 144, 174
  of infinity, 182
Nested interval property, 134, 160–165, 167, 168
Norm, 57, 133
Normal subgroup, 224, 232–236, 238

O1–O3 (Order relation), 92
Odd
  integer, 79
  permutation, 230
One-sided continuity, 194
One-sided limit, 180
One-to-one, 103, 108, 122
One-to-one correspondence, 103
Onto, 102
Open cover, 146
Open interval, 135
Open set, 140, 197
OR, 15–18

Order
  of a group, 210
  of a group element, 218
Order relation, 92
  strict, 94
  total ordering, 94
  well ordering, 95

P1–P3 (Partition), 72
Paradox, 14
Partial ordering, 92
Partition, 72, 124
Peano's postulates, 76
Permutation, 122, 227
  even, 230
  odd, 230
Permutation group, 227
PID, 274–278, 280
Point at infinity, 184
Polynomial
  content, 271
  degree, 259, 271
  irreducible, 271
  primitive, 271
  ring, 258–260, 271, 282–284, 293–296
Pre-image, 97, 109
Prime
  element, 251, 270, 277, 278
  factorization, 66, 83, 284
  ideal, 264, 270, 276, 278, 295
  number, 66, 83
Primitive, 271
Principal ideal, 263
Principal ideal domain, 274–278, 280
Principal of mathematical induction, 61–66
Principle of zero products, 59
Product rule, 118
Proof
  by induction, 61–66
  contradiction, 45
  contrapositive, 44
  counterexample, 45
  direct, 41–44
Proper divisor, 81, 268, 270
Proper subset, 2
Pythagoras, 88

Q1–Q6 (Function question), 107
Quantifier
  existential, 29–30
  universal, 27–29
Quaternions, 211, 216, 260
Quotient group, 223–225
Quotient ring, 291–296

R1–R10 (Ring), 244
Range, 97, 101
Rationals, 114
Real numbers
    algebraic properties, 5–7
    assumptions, 5–9
    completeness, 167, 168
    extended, 184
    ordering properties, 7–9, 56–57
Reflexive property, 5, 70, 92, 94, 95
Relation, 90–95, 97
Removable discontinuity, 189
Ring
    characteristic, 254, 269
    defined, 244
    extensions, 256–260
    morphism, 287–290
    of matrices, 245
    of polynomials, 258–260, 271, 282–284, 293–296
    prime element, 251
    quotient, 291–296
    unit, 250
    unity, 244
Roots, 9, 86

S1–S3 (Strict order relation), 94
Salt and pepper function, 179
Sandwich theorem, 177
    for sequences, 160
Sequence(s), 149–159
    bounded, 151, 156, 163, 166
    Cauchy, 165–169
    convergence, 153–159
    monotone, 150, 161
Sequential limit, 181
Set(s), 1–5
    boundary, 142, 145
    bounded, 134, 147
    cardinality, 110–118
    Cartesian product, 90, 118, 120
    closed, 141, 144, 147
    closure, 144
    compact, 146–149, 198, 202
    complement, 4
    countable, 113
    cover, 146
    difference, 40
    disjoint, 39, 112
    element, 2
    empty, 111
    equality, 39
    exterior, 142, 145
    finite, 111
    image, 101
    index, 47
    infinite, 113
    interior, 142, 145
    intersection, 4, 39, 50
    maximum, 138
    minimum, 138
    open, 140, 197
    partition, 72, 74
    pre-image, 109
    symmetric difference, 40
    uncountable, 116
    union, 4, 39, 50, 116
    universal, 2
Statement(s), 13–14
    if-then, 20–21
    logical equivalence, 18, 23
    negation, 15
    stronger, 24
Strict order relation, 94
Strong induction, 65
Subcover, 146
Subgroup, 213
    cyclic, 217
    generated, 216
    normal, 224, 232–236, 238
Subring, 248
Subsequence, 162, 163
Subset, 2, 39, 112, 123
    proper, 2
Superset, 2
Surjection, 102
Symmetric difference, 40
Symmetric group, 227
Symmetric property, 6, 70, 92

T1–T3 (Topology), 198
T1–T4 (Total ordering), 94
Tautology, 19, 23
Topology, 197
Total ordering, 94
Transitive property
    in a relation, 92, 94, 95
    of equality, 6, 70, 92
    of integer divisibility, 81
    of subset inclusion, 46
Transposition, 229
Triangle inequality, 58
Trichotomy law, 8, 56

U1–U3 (Generated ideal), 262
U1–U3 (Generated subgroup), 216
UFD, 273–274, 284–286
Uncountability, 116
Uniform continuity, 200–203
Union, 4, 39, 116, 141
    Disjoint, 46
    over a family, 50

Unique existence, 30
Unique factorization domain, 273–274, 284–286
Uniqueness
    additive inverse, 54
    group identity, 214
    inverse in a group, 214
    least upper bound, 135
    multiplicative inverse, 55
    of prime factorization, 278
    ring unity, 250
Unit, 250, 280
    in $\mathbb{Z}_p$, 257
Unity, 244
Universal quantifier, 27–29
Universal set, 2

Valuation, 279
Venn diagram, 2

W1–W4 (Well ordering), 95
Well defined
    addition, 6, 78
    cardinality, 112
    function, 98
    multiplication, 6, 78
Well ordering, 95
Well-ordering principle, 8, 61

X1–X3 (Roots), 86

Y1–Y4 (Ideal), 261

Z1–Z4 (Integer assumptions), 76
Zermelo, Ernst, 95
Zero divisor, 251